THE THREE-BODY PROBLEM

How do three celestial bodies move under their mutual gravitational attraction? This problem was studied by Isaac Newton and other leading mathematicians over the last two centuries. Poincaré's conclusion, that the problem represents an example of chaos in nature, opens the new possibility of using a statistical approach. For the first time this book presents these methods in a systematic way, surveying statistical as well as more traditional methods.

The book begins by providing an introduction to celestial mechanics, including Lagrangian and Hamiltonian methods, and both the two- and restricted three-body problems. It then surveys statistical and perturbation methods for the solution of the general three-body problem, providing solutions based on combining orbit calculations with semi-analytic methods for the first time. This book should be essential reading for students in this rapidly expanding field and is suitable for students of celestial mechanics at advanced undergraduate and graduate level.

THE THREE-BODY PROBLEM

MAURI VALTONEN AND HANNU KARTTUNEN
Väisälä Institute for Space Physics and Astronomy,
University of Turku

CAMBRIDGE
UNIVERSITY PRESS

CAMBRIDGE
UNIVERSITY PRESS

University Printing House, Cambridge CB2 8BS, United Kingdom

One Liberty Plaza, 20th Floor, New York, NY 10006, USA

477 Williamstown Road, Port Melbourne, VIC 3207, Australia

314-321, 3rd Floor, Plot 3, Splendor Forum, Jasola District Centre, New Delhi - 110025, India

79 Anson Road, #06-04/06, Singapore 079906

Cambridge University Press is part of the University of Cambridge.

It furthers the University's mission by disseminating knowledge in the pursuit of education, learning and research at the highest international levels of excellence.

www.cambridge.org
Information on this title: www.cambridge.org/9780521852241

First published 2006

A catalogue record for this publication is available from the British Library

ISBN 978-0-521-85224-1 Hardback

Contents

Preface *page* ix
1 Astrophysics and the three-body problem 1
1.1 About the three-body problem 1
1.2 The three-body problem in astrophysics 5
1.3 Short period comets 8
1.4 Binary stars 12
1.5 Groups of galaxies 15
1.6 Binary black holes 17
2 Newtonian mechanics 20
2.1 Newton's laws 20
2.2 Inertial coordinate system 21
2.3 Equations of motion for N bodies 22
2.4 Gravitational potential 24
2.5 Constants of motion 25
2.6 The virial theorem 27
2.7 The Lagrange and Jacobi forms of the equations of motion 29
2.8 Constants of motion in the three-body problem 31
2.9 Moment of inertia 32
2.10 Scaling of the three-body problem 34
2.11 Integration of orbits 34
2.12 Dimensions and units of the three-body problem 38
2.13 Chaos in the three-body problem 39
2.14 Rotating coordinate system 43
Problems 45
3 The two-body problem 47
3.1 Equations of motion 47
3.2 Centre of mass coordinate system 48
3.3 Integrals of the equation of motion 49

3.4 Equation of the orbit and Kepler's first law 52
3.5 Kepler's second law 53
3.6 Orbital elements 54
3.7 Orbital velocity 57
3.8 True and eccentric anomalies 58
3.9 Mean anomaly and Kepler's equation 60
3.10 Solution of Kepler's equation 61
3.11 Kepler's third law 63
3.12 Position and speed as functions of eccentric anomaly 64
3.13 Hyperbolic orbit 66
3.14 Dynamical friction 68
3.15 Series expansions 70
Problems 78
4 Hamiltonian mechanics 80
4.1 Generalised coordinates 80
4.2 Hamiltonian principle 81
4.3 Variational calculus 82
4.4 Lagrangian equations of motion 85
4.5 Hamiltonian equations of motion 87
4.6 Properties of the Hamiltonian 89
4.7 Canonical transformations 92
4.8 Examples of canonical transformations 95
4.9 The Hamilton–Jacobi equation 95
4.10 Two-body problem in Hamiltonian mechanics: two dimensions 97
4.11 Two-body problem in Hamiltonian mechanics: three dimensions 103
4.12 Delaunay's elements 108
4.13 Hamiltonian formulation of the three-body problem 109
4.14 Elimination of nodes 111
4.15 Elimination of mean anomalies 113
Problems 113
5 The planar restricted circular three-body problem and other special cases 115
5.1 Coordinate frames 115
5.2 Equations of motion 116
5.3 Jacobian integral 119
5.4 Lagrangian points 123
5.5 Stability of the Lagrangian points 125
5.6 Satellite orbits 130
5.7 The Lagrangian equilateral triangle 133
5.8 One-dimensional three-body problem 136
Problems 139

6 Three-body scattering 141
- 6.1 Scattering of small fast bodies from a binary 141
- 6.2 Evolution of the semi-major axis and eccentricity 148
- 6.3 Capture of small bodies by a circular binary 152
- 6.4 Orbital changes in encounters with planets 154
- 6.5 Inclination and perihelion distance 157
- 6.6 Large angle scattering 162
- 6.7 Changes in the orbital elements 165
- 6.8 Changes in the relative orbital energy of the binary 169
- Problems 170

7 Escape in the general three-body problem 171
- 7.1 Escapes in a bound three-body system 171
- 7.2 A planar case 179
- 7.3 Escape velocity 180
- 7.4 Escaper mass 183
- 7.5 Angular momentum 184
- 7.6 Escape angle 188
- Problems 195

8 Scattering and capture in the general problem 197
- 8.1 Three-body scattering 197
- 8.2 Capture 203
- 8.3 Ejections and lifetime 207
- 8.4 Exchange and flyby 211
- 8.5 Rates of change of the binding energy 214
- 8.6 Collisions 216
- Problems 219

9 Perturbations in hierarchical systems 221
- 9.1 Osculating elements 221
- 9.2 Lagrangian planetary equations 222
- 9.3 Three-body perturbing function 225
- 9.4 Doubly orbit-averaged perturbing function 227
- 9.5 Motions in the hierarchical three-body problem 231
- Problems 239

10 Perturbations in strong three-body encounters 240
- 10.1 Perturbations of the integrals $\boldsymbol{k}$ and $\boldsymbol{e}$ 240
- 10.2 Binary evolution with a constant perturbing force 243
- 10.3 Slow encounters 246
- 10.4 Inclination dependence 260
- 10.5 Change in eccentricity 264
- 10.6 Stability of triple systems 268

10.7 Fast encounters 274
10.8 Average energy exchange 281
Problems 285
11 Some astrophysical problems 288
11.1 Binary black holes in centres of galaxies 288
11.2 The problem of three black holes 296
11.3 Satellite black hole systems 310
11.4 Three galaxies 310
11.5 Binary stars in the Galaxy 313
11.6 Evolution of comet orbits 320
Problems 327
References 329
Author index 341
Subject index 343

Preface

Classical orbit calculation in Newtonian mechanics has experienced a renaissance in recent decades. With the beginning of space flights there was suddenly a great practical need to calculate orbits with high accuracy. At the same time, advances in computer technology have improved the speed of orbit calculations enormously.

These advances have also made it possible to study the gravitational three-body problem with new rigour. The solutions of this problem go beyond the practicalities of space flight into the area of modern astrophysics. They include problems in the Solar System, in the stellar systems of our Galaxy as well as in other galaxies. The present book has been written with the astrophysical applications in mind.

The book is based on two courses which have been taught by us: *Celestial Mechanics* and *Astrodynamics*. The former course includes approximately Chapters 2–5 of the book, with some material from later chapters. It is a rather standard introduction to the subject which forms the necessary background to modern topics. The celestial mechanics course has been developed in the University of Helsinki by one of us (H. K.) over about two decades. The remainder of the book is based on the astrodynamics course which arose subsequently in the University of Turku. Much of the material in the course is new in the sense that it has not been presented at a textbook level previously.

In our experience there has been a continuous need for specialists in classical orbit dynamics while at the same time this area of study has received less attention than it used to in the standard astronomy curriculum. By writing this book we hope to help the situation and to attract new students to the research area, which is still modern after more than 300 years of studies.

We have been privileged to receive a great deal of help and encouragement from many colleagues. Especially we would like to thank Douglas Heggie, Kimmo Innanen and Bill Saslaw who have between them read nearly the whole manuscript and suggested numerous improvements. We also appreciate the comments by Victor Orlov, Harry Lehto and Tian-Yi Huang which have been most useful. Seppo Mikkola has generously provided research tools for the calculations in this book. Other members of the Turku research group have also helped us with illustrations. Most of all, we would like to thank Sirpa Reinikainen for typing much of the final text, including the great many mathematical formulae.

Financial support for this project has been provided by Finland's Society for Sciences and Letters and the Academy of Finland (project 'Calculation of Orbits'), which gave the opportunity for one of us (M. V.) to concentrate on writing the book for a period of two years. The generous support by the Department of Computer Science, Mathematics and Physics (in Barbados) and the Department of Physics (in Trinidad) of the University of the West Indies made it possible to carry out the writing in optimal surroundings. Parts of the text were originally published in Finnish: H. Karttunen, *Johdatus Taivaanmekaniikkaan*, Helsinki: Ursa, 2001.

Finally, M. V. would like to express his appreciation to Sverre Aarseth who taught him how to calculate orbits (and much else), and to his wife Kathleen, the Caribbean link, whose encouragement was vital for the accomplishment of the book.

1

Astrophysics and the three-body problem

1.1 About the three-body problem

The three-body problem arises in many different contexts in nature. This book deals with the classical three-body problem, the problem of motion of three celestial bodies under their mutual gravitational attraction. It is an old problem and logically follows from the two-body problem which was solved by Newton in his *Principia* in 1687. Newton also considered the three-body problem in connection with the motion of the Moon under the influences of the Sun and the Earth, the consequences of which included a headache.

There are good reasons to study the three-body gravitational problem. The motion of the Earth and other planets around the Sun is not strictly a two-body problem. The gravitational pull by another planet constitutes an extra force which tries to steer the planet off its elliptical path. One may even worry, as scientists did in the eighteenth century, whether the extra force might change the orbital course of the Earth entirely and make it fall into the Sun or escape to cold outer space. This was a legitimate worry at the time when the Earth was thought to be only a few thousand years old, and all possible combinations of planetary influences on the orbit of the Earth had not yet had time to occur.

Another serious question was the influence of the Moon on the motion of the Earth. Would it have long term major effects? Is the Moon in a stable orbit about the Earth or might it one day crash on us? The motion of the Moon was also a question of major practical significance, since the Moon was used as a universal time keeping device in the absence of clocks which were accurate over long periods of time. After Newton, the lunar theory was studied in the eighteenth century using the restricted problem of three bodies (Euler 1772). In the restricted problem, one of the bodies is regarded as massless in comparison with the other two which are in a circular orbit relative to each other. At about the same time, the first special solution of the general three-body problem was discovered, the Lagrangian equilateral triangle solution (Lagrange 1778). The theory of the restricted three-body problem was further developed by Jacobi (1836), and it was used for the purpose of identifying comets by Tisserand (1889, 1896) and reached its peak in the later nineteenth century with the work of Hill (1878) and Delaunay (1860). The 'classical' period reached its final phase with Poincaré (1892–1899).

In spite of these successes in special cases, the solution of the general three-body problem remained elusive even after two hundred years following the publication of *Principia*. In the general three-body problem all three masses are non-zero and their initial positions and velocities are not arranged in any particular way. The difficulty of the general three-body problem derives from the fact that there are no coordinate transformations which would simplify the problem greatly. This is in contrast to the two-body problem where the solutions are found most easily in the centre of mass coordinate system. The mutual force between the two bodies points towards the centre of mass, a stationary point in this coordinate system. Thus the solution is derived from the motion in the inverse square force field. Similarly, in the restricted three-body problem one may transfer to a coordinate system which rotates at the same rate around the centre of mass as the two primary bodies. Then the problem is reduced to the study of motions in two stationary inverse square force fields. In the general problem, the lines of mutual forces do not pass through the centre of mass of the system. The motion of each body has to be considered in conjunction with the motions of the other two bodies, which made the problem rather intractable analytically before the age of powerful computers.

At the suggestion of leading scientists, the King of Sweden Oscar II established a prize for the solution of the general three-body problem. The solution was to be in the form of a series expansion which describes the positions of the three bodies at all future moments of time following an arbitrary starting configuration. Nobody was able to claim the prize for many years and finally it was awarded in 1889 to Poincaré who was thought to have made the most progress in the subject even though he had not solved the specific problem. It took more than twenty years before Sundman completed the given task (Sundman 1912). Unfortunately, the extremely poor convergence of the series expansion discovered by Sundman makes this method useless for the purpose of calculating the orbits of the three bodies. Now that the orbits can be calculated quickly by computer, it is quite obvious why this line of research could not lead to a real solution of the three-body problem: the orbits are good examples of chaos in nature, and deterministic series expansions are utterly unsuitable for their description. Poincaré was on the right track in this regard and with the current knowledge was thus a most reasonable recipient of the prize.

At about the same time, a new approach began which has been so successful in recent years: the integration of orbits step by step. In orbit integration, each body,

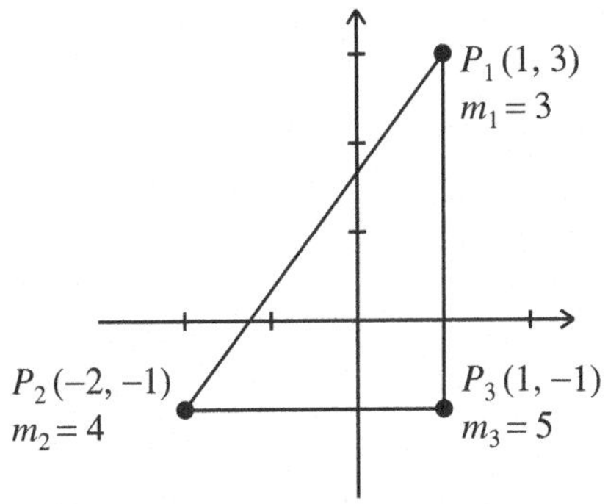

Figure 1.1 Initial configuration of the Pythagorean problem.

in turn, is moved forward by small steps. In the most basic scheme, the step is calculated on the basis of the accelerations caused by the two other bodies during that step while they are considered to remain fixed. There is an error involved when only one body is moved at a time, and others move later, but this can be minimised by taking short steps and by other less obvious means.

Burrau (1913) considered a well defined, but in no way special, initial configuration of three bodies which has since become known as the Pythagorean problem since the three bodies are initially at the corners of a Pythagorean right triangle. The masses of the three bodies are 3, 4 and 5 units, and they are placed at the corners which face the sides of the triangle of the corresponding length (Fig. 1.1). In the beginning the bodies are at rest. Burrau's calculation revealed the typical behaviour of a three-body system: two bodies approach each other, have a close encounter, and then recede again. Subsequently, other two-body encounters were calculated by Burrau until he came to the end of his calculating capacity. Only after the introduction of modern computers and new orbit integration methods was the celestial dance in the Pythagorean problem followed to its conclusion.

Later work has shown that the solution of the Pythagorean problem is quite typical of initially bound three-body systems. After many close two-body approaches, a configuration arises which leads to an escape of one body and the formation of a binary by the other two bodies (see Fig. 1.2, Szebehely and Peters 1967). A theoretical treatment of a three-body system of this kind is given in Chapter 7. In the following chapter, situations are discussed where a third body comes from a large distance, meets a binary, and perhaps takes the place of one of the binary members which escapes. Such orbits were calculated already in 1920 (Becker 1920). Sometimes the third body is always well separated from the binary; then the situation is best described by perturbations on the binary caused by the third body. Some examples of these systems are discussed in Chapter 10.

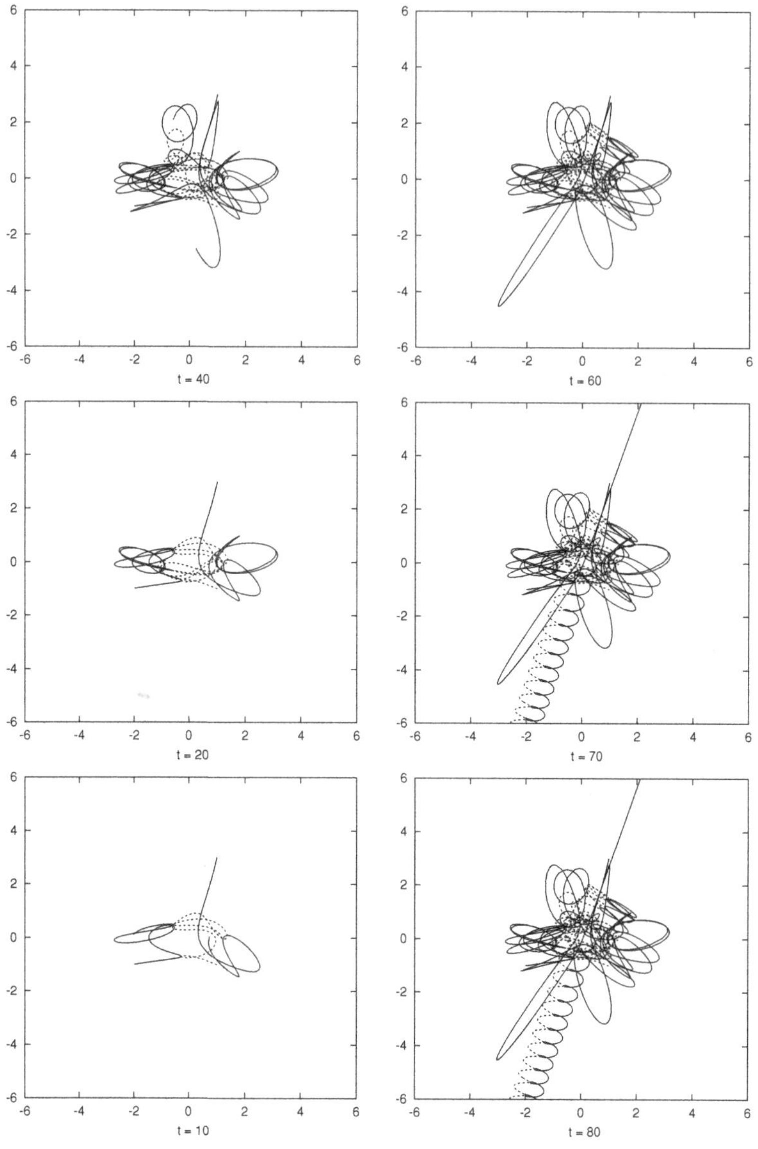

Figure 1.2 Trajectories of the Pythagorean problem. The orbits up to time 10, 20, 40, 60, 70 and 80 units are shown. The last two panels are identical since the escape has already happened and the bodies are outside the frame.

In recent years there has been increasing demand for solutions of the general three-body problem in various astrophysical situations. For example, binaries and their interactions with single stars play a major role in the evolution of star clusters (Aarseth 1973). Triple stellar systems are another obvious astrophysical three-body problem. Many other astrophysical bodies, ranging from compact bodies to galaxies, occur in triple systems. These will be discussed in Chapter 11.

Before then, we have to do some preparatory work which is partly found in standard textbooks on mechanics and celestial mechanics. The proper discussion of the three-body problem is thus delayed until Chapter 5 where special cases are first introduced. This leads to consideration of the general three-body problem in the later chapters, the main objective of this book.

1.2 The three-body problem in astrophysics

For more than 300 years there have been many different motivations to solve the three-body problem and many different techniques have been applied to it. In this book we have the rather limited purpose of looking at solutions of astrophysical significance. At the present time we can solve any given three-body problem, starting from the known positions, velocities and masses of the three bodies, by using a computer. There is of course the limitation of the accuracy of calculation which may be quite significant in some cases. But notwithstanding the accuracy, the solution of an astrophysical problem usually involves much more than a calculation of a single orbit. Typically we have to sample three-body orbits in a phase space of up to eleven dimensions. Then the calculation of orbits is only one tool; one has to have a deeper understanding of the three-body process to make sense of the limited amount of information that is derivable from orbit calculations.

Therefore we do not deal with the mathematical three-body problem. Fortunately, there are excellent books by Marchal (1990) and Hénon (1997, 2001) which deal with the mathematical aspects very thoroughly. As an example, periodic orbits of the general three-body problem are of great mathematical interest, but there are very few examples where they are important in astrophysics.

In problems of astrophysical importance, one may almost always identify a binary and a third body. A binary can be treated as a single entity with certain 'internal' properties (like a molecule). It is described by its component masses, by its energy and angular momentum, as well as by its orientation in space. This binary entity interacts with a third body once, or more frequently, which changes the internal properties of the binary. At the same time, the third body absorbs whatever energy is given out from the binary, in order to conserve the total energy. Similarly, conservation of angular momentum between the binary and the third body has to be satisfied.

Before we can take up the discussion of the three-body problem, we have to be familiar with binaries, i.e. the two-body problem. The two-body problem is treated in basic courses of mechanics and celestial mechanics. Therefore the discussion of Chapter 3 may appear as unnecessary repetition to some readers, and they may want to skip much of the first chapters. Also the Hamiltonian techniques of Chapter 4 are commonly studied in courses of advanced mechanics. They form such an essential part of the treatment of weak perturbations of binaries that it is necessary to introduce the Hamiltonian concepts also in this book.

A planet going around the Sun is an example of a binary. The third body could then be another planet, a moon, an asteroid or a comet. Because of the large differences between the masses of the bodies, from the dominant Sun down to asteroids of kilometre size or even smaller, Solar System dynamics is special in many ways. A very up to date treatment exists in this area (Murray and Dermott 1999). We discuss only a small class of Solar System problems which are related to stellar dynamics and therefore form a suitable introduction to later studies of three-body systems with more equal masses. Thus Chapter 5 contains many topics which readers may have encountered earlier.

At the present time, three-body astrophysics is primarily motivated by the need to understand the role of binaries in the evolution of stellar systems. For most of the time, a binary acts just like a single star in a stellar system. The distances between stars are large compared with the sizes of the stars and even compared with the sizes of close binary orbits. For a relatively brief moment a binary and a third star interact strongly, a 'new' binary forms, and a 'new' third star leaves the scene. The importance of the process lies in its ability to redistribute energy and angular momentum efficiently; the population of binaries may become more and more tightly bound as time and three-body scattering go on, while the population of single stars may gain speed and become 'heated'. This will have profound consequences on the structure and evolution of a star cluster; for example single stars and sometimes also binaries escape from the cluster; binary orbits may shrink to form contact binaries, and also triple stars may form where the third body remains bound to the binary. The end products of the three-body process may appear as sources of radio jets, X-rays, gamma rays or as other kinds of 'exotic' objects (Hut *et al.* 2003).

These various scenarios can be reproduced by numerical orbit calculations. The orbits of thousands of stars can be calculated in a simulation of a star cluster. Even though these simulations have now reached a great level of complication and trustworthiness (e.g. Heggie and Hut 2003, Aarseth 2003a), it is still useful to examine the three-body process to see how much is understood from elementary principles. Together with the simulations of large numbers of bodies one may attain a deeper understanding of the evolutionary process.

Chapter 7 starts with the discussion of initially strongly interacting three-body systems. We will learn that such systems have a limited lifetime, and we do not expect to find very many of them in nature. But they are important in the description of the intermediate state between the impact of a third body on a binary and the departure of a 'new' third body from a 'new' binary. The techniques used in Chapter 7, assuming a complete reshuffle of the positions and velocities of the three bodies in the phase space, work surprisingly well.

By 'work well' we mean that a large body of numerical orbit calculations can be described in a statistical sense by simple physical principles. That this should be so is not at all obvious at the outset. Therefore we devote considerable space to comparing numerical calculations with the theory. The beauty of it is that once the theory is established for certain parts of the parameter space, we have good reasons to expect that it applies more generally. Also we do not need to go back to calculating millions of orbits when a slightly different astrophysical problem arises but we can use the theory directly with a fair amount of confidence.

In addition to these practical considerations, it gives the reader a certain satisfaction to learn that simple analytical solutions of the three-body problem exist, even though only in a statistical sense and for a limited part of the parameter space. It will also become clear that these are the only solutions of any significance in large parts of the phase space due to the chaotic nature of the problem.

Chapters 9 and 10 try to cover the remaining parameter space, i.e. when the binary and the third body are so well separated throughout the interaction that the principle of complete chaos is not productive. At one extreme there is the very slow and gentle perturbation of the binary known as the Kozai mechanism. Then only the binary eccentricity and inclination change periodically while the orbital sizes are unaffected. At the other extreme we have a high speed intruder which gives the binary a 'shock treatment' during its brief encounter with the binary. In between, a binary is strongly perturbed at close encounters but is able to maintain its identity and not break up or exchange members. A stability boundary is derived which tells us where the perturbation treatment ends and the chaos theory begins.

In this way we can give a rather complete coverage of the astrophysical three-body problem. As in the case of the chaos theory of Chapters 7 and 8, also in the perturbation theory of Chapters 9 and 10, a great deal of space is dedicated to comparing numerical results with the theory. This is necessary since it is not always obvious, in the absence of exact theory, what approximations should lead to the best understanding of the experiments. Often we even find that in the final steps we just have to accept the guidance of the numerical experiments without clear justification of the theory. This is not because the theories could not be pushed any further but more because we like to keep the theory at a rather simple level (and it may appear quite complicated to some readers already as it is). But also we have to remember that in the general three-body problem with strong interactions no exact theory exists, and we should not spend too much effort towards this elusive goal.

Throughout the development of the theory we will look at some small astrophysical problems which are easily solved at this stage. In the final chapter a couple of larger issues are discussed which require solutions from different parts of the text as well as other astrophysical information. To give the reader an idea of what sorts of problems we are dealing with we outline a couple of astrophysical examples in the next sections.

1.3 Short period comets

The origin of short period comets is one of the oldest three-body problems. Lexell (1778, 1779) studied the motion of the comet found in 1770 by Messier, and suggested that the orbit had become elliptical with a period of 5.6 years, when the comet passed close to Jupiter. Later, Laplace (1799–1825, 1805) and Leverrier worked on the capture hypothesis. Tisserand (1889) and H. A. Newton (1891), among others, discussed details of the capture process. The idea, already put forward by Lexell, was that short period comets are created from long period comets which pass near Jupiter and lose energy during the encounter. Everhart (1969) carried out a major survey of close encounter orbits between comets and planets using computers. In spite of all these and later efforts, the origin of the short period comets is still an open question to a large extent.

There are more than two hundred known short period comets, even though it is well documented that comets fade away after 10^2–10^3 revolutions around the Sun and that they escape the solar system after 10^4–10^5 revolutions (Fig. 1.3). There must be a source of comets which constantly (or from time to time) replenishes the population. Two such sources have been suggested: the Oort Cloud of comets and the Kuiper Belt of asteroids and comets. Sometimes also other source regions, such as interstellar comets, have been mentioned. The processes which may keep the short period comet population intact are mostly related to the three-body problem.

The Oort Cloud (Oort 1950) is a collection of as many as 10^{12} comets loosely bound to the Solar System. The orbits of the comets are such that they generally do not enter the planetary region, and their orbits are mainly affected by the Sun, passing stars, gas clouds and the tidal field of the Galaxy as a whole. As a result of these influences, there is a more or less steady flux of 'new' comets which enter the planetary system for the first time. The flux may be rather uniform per pericentre interval (closest approach distance to the Sun, in AU) up to the distance of Jupiter; beyond that the flux is expected to rise but the difficulty of observing distant comet passages prevents observational confirmation of this expected trend.

Since we have some idea of the Oort Cloud flux of comets we may ask how much of this flux is captured to short period orbits and what are the orbital properties of the captured comets. An Oort Cloud comet comes in a highly eccentric orbit, and in the absence of planets it would return back to the Oort Cloud in the same elliptical orbit. But the orbit is influenced by one or more planets, and this influence is likely

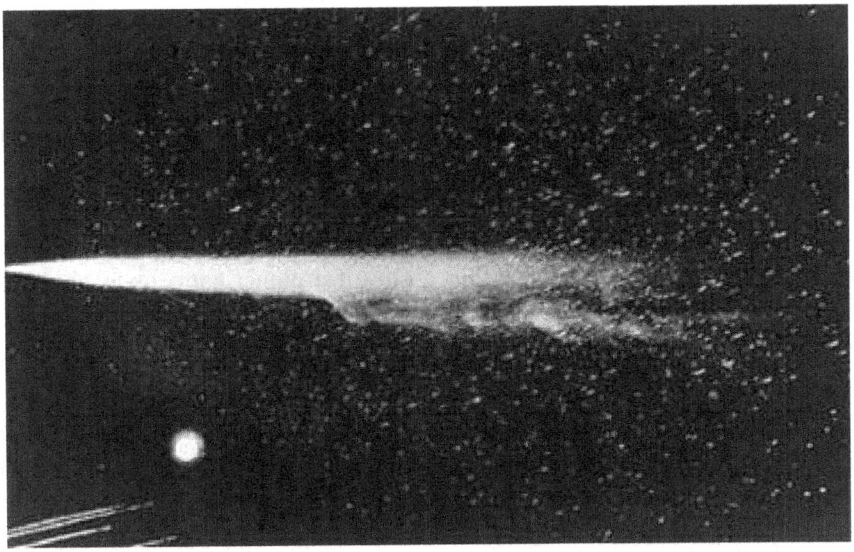

Figure 1.3 Halley's comet, the most famous periodic comet, photographed in May 13, 1910. The big round object is Venus and the stripes are city lights of Flagstaff. Image Lowell Observatory.

to change the orbit either to a hyperbolic escape orbit or to a more strongly bound short period elliptical orbit. Basically we need to solve the three-body problem consisting of the Sun, a planet and a comet. Since the comet has much smaller mass than the other two bodies, the problem is restricted only to the question of the motion of the comet. Generally we may assume that the planet goes around the Sun in a circular orbit, and remains in this orbit independent of what happens to the comet. This is an example of the restricted circular three-body problem which will be discussed in Chapter 5.

It is a straightforward procedure to make use of an orbit integrator, a computer code which calculates orbits, and to calculate the orbit of an Oort Cloud comet through the Solar System, past various planets, perhaps through millions of orbital cycles, until the comet escapes from the Solar System, or until it collides with the Sun or one of the planets, or until the comet disintegrates. The calculation can be long but it is possible with modern computers. However, questions remain: how representative is this orbit and how accurate is the solution? Indeed, we do not know the exact starting conditions for any Oort Cloud comets and neither can we carry out the calculation over the orbital time of millions of years without significant loss of accuracy.

To some extent these problems can be avoided by using the method of sampling. We take a large sample of orbits with different initial conditions and calculate them through the necessary period of time. We may then look at observational samples and compare them with samples obtained by orbit calculation. In practice we need millions of orbits in order to get satisfactory statistics of the captured short period comets. It becomes a major computational challenge.

We can learn quite a lot by studying the interaction of a comet with only one of the planets. A typical Oort Cloud comet has a high inclination relative to the plane of the Solar System. It comes from one side of the plane (say, above), dives through the plane, turns around below the plane, and crosses the plane again. A strong three-body interaction takes place only if a planet happens to be close to one of the crossing points at the right time. Most likely no planet is there at all, but when a close encounter happens, almost always it is only with one planet only at one of the two crossing points. Therefore we have a three-body problem.

But even then the exact nature of the three-body encounter is unclear. The best we can do is to develop a statistical theory of how the comet is likely to react to the presence of the planet close to the crossing point. Öpik (1951) developed such a theory where the comet is assumed to follow an exact two-body orbit around the Sun until it comes to the sphere of gravitational influence of the planet. At this point the comet starts to follow an exact two-body orbit relative to the planet. After leaving the sphere of influence of the planet, the orbit is again a (different) two-body orbit around the Sun. A theory along these lines will be discussed in Chapter 6, together with some more recent work. It is the most basic form of a solution to the three-body problem. Notice that we will be discussing probabilities; this is the recurring theme of the solutions of astrophysical three-body problems.

The statistical properties of comets which we will have to confront are primarily the distribution of their orbital sizes (semi-major axes), their perihelion distances and the distribution of orbital inclinations relative to the Solar System plane. In terms of orbital sizes, the comets can be classified as being either Jupiter family comets (orbital period below 20 years), Halley type comets (period above 20 but below 200 years) or long period comets (longer periods than 200 years). Oort Cloud comets have orbital periods in excess of a million years. Within these groups, the inclination distributions vary (Fig. 1.4). One of the aims of a successful theory of short period comets is to explain their observed numbers (28 Halley type, 183 Jupiter family comets; Marsden and Williams 1999) in relation to the rate of comets coming from the Oort Cloud. Also it should explain how the differences in the inclination distributions originate.

If it turns out that a decrease in the semi-major axis of a comet is associated with a decrease in its inclination (and this will be shown in Chapter 6) then we might propose that comets evolve from Oort Cloud comets to Halley type comets to Jupiter family comets by successively decreasing the semi-major axis and the inclination of the orbit. How this works out in detail will be discussed in Chapter 11.

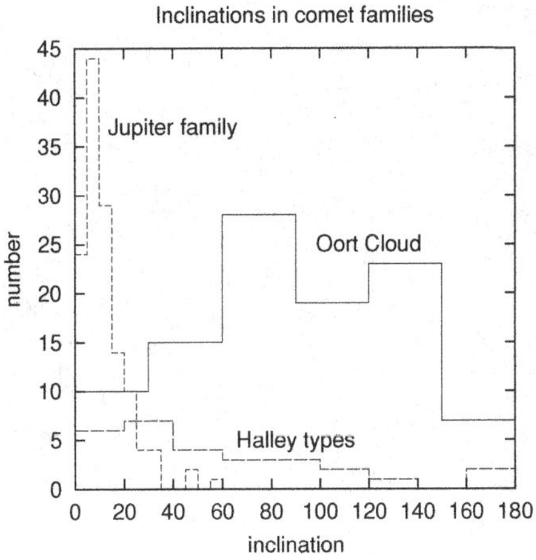

Figure 1.4 The inclination distributions in three classes of comets: Oort Cloud comets, the orbits of which agree with random orientations (number proportional to sin i), Halley type comets which are more often in direct orbits than in retrograde orbits, and Jupiter family comets which are in direct orbits close to the Solar System plane.

Another idea which is as old as the Oort Cloud theory is the concept of a disk of comets left over from the formation of the Solar System. This is usually called the Kuiper Belt or Edgeworth–Kuiper Belt (Edgeworth 1949, Kuiper 1951). These comets reside mostly by the orbit of Neptune or beyond (Transneptuneans). Comets do not develop a tail at this distance from the Sun and they can be observed only if the nucleus of the comet (the solid body from which the tail originates) is quite large, greater than about 100 km in diameter. Therefore our knowledge of the comets at the Kuiper Belt is limited to the very largest bodies among them. By extrapolation it has been deduced that there may exist as many as 10^7 'ordinary' comets in the Kuiper Belt and in its vicinity (Levison and Duncan 1997), even though only somewhat over 300 very bright ones are known. Some of the Transneptunean comets certainly approach the Sun at some point in their orbital evolution and become visible by their bright tails; however, it is difficult to know exactly which comets have this origin.

The Oort Cloud comets are very loosely bound to the Solar System and therefore the Oort Cloud may require replenishment. A more tightly bound and dynamically more robust cloud of comets has been suggested to lie inside the Oort Cloud, the so-called Inner Oort Cloud (Hills 1981). Numerical simulations of the origin of the Oort Cloud by ejection of cometary bodies from the region of the outer planets also suggest the existence of the Inner Oort Cloud (Duncan *et al.* 1987). Passing stars and molecular clouds perturb the orbits of the Inner Oort Cloud comets and

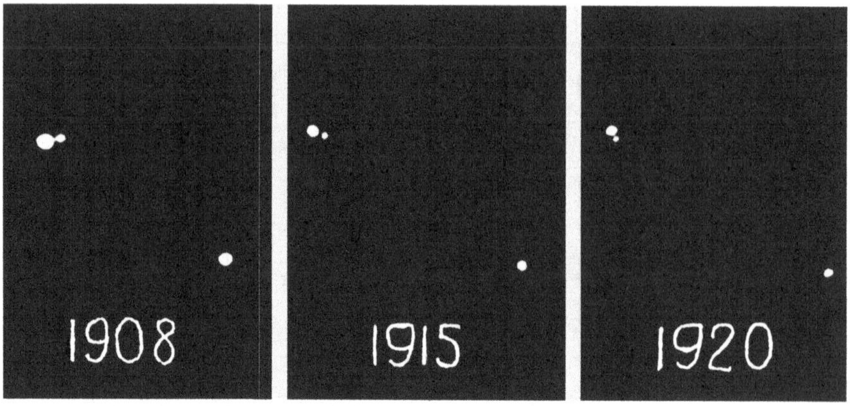

Figure 1.5 In the course of years the components of the binary star Krüger 60 are seen to orbit their common centre of mass. Image Yerkes Observatory

some of them end up in the Oort Cloud proper. The Inner Oort Cloud is yet another potential source region for short period comets.

1.4 Binary stars

About half of the stars in our Galactic neighbourhood are binary stars (Fig. 1.5). Binary stars have been studied since the eighteenth century when William Herschel discovered that stars may appear in physical pairs. Combined with Newton's laws of gravity and dynamics, binary stars have been used in the studies of stellar masses, radii and other properties. Yet even today, we do not have a full understanding of the origin of even the most basic properties of the binaries.

The study of binary stars is an interesting field in its own right but it also has wider implications. Binaries appear as mass points when seen from far away, but in close encounters with single stars or other binaries their internal motions may release or absorb energy. Therefore they behave somewhat similarly to molecules in gas dynamics. The role of binaries is very crucial in understanding the dynamics of stellar systems.

Binaries may have very different semi-major axes and orbital eccentricities, ranging from two stars in contact with each other and in a circular orbit, to a wide and typically eccentric binary with a semi-major axis of the order of 0.1 pc. Corresponding orbital periods range from a fraction of a day to millions of years. One of the basic questions in binary star studies is to understand these distributions: the distributions of binary periods, of binary orbit eccentricities and of mass ratios of the binary components.

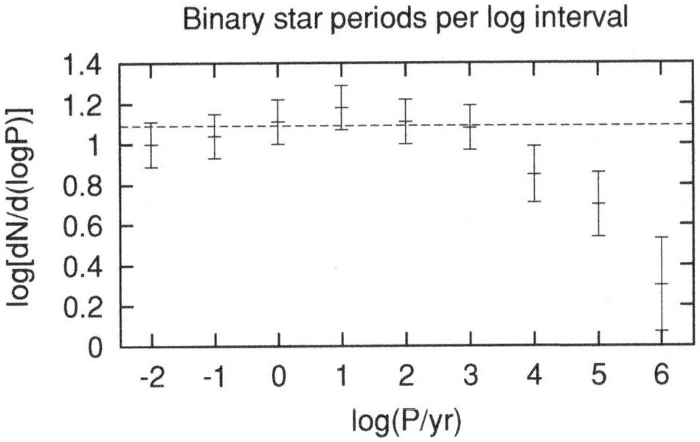

Figure 1.6 The distribution of binary star periods P, plotted per logarithmic interval d log P (Abt 1977). The horizontal dashed line represents the so-called Öpik law.

Because of the wide range of physical sizes of binary systems, studies of their distributions require different methods on different scales. Various selection effects also complicate these studies. When all the studies on different scales and with corrections for selection effects have been combined, the following results have been obtained. The period distribution is flat over five orders of magnitude when plotted per logarithmic interval of the period P (Fig. 1.6; Abt 1977). A distribution of this type is sometimes called 'scale free'. Written in the usual way the distribution is

$$f(P)\,\mathrm{d}P \propto P^{-1}\,\mathrm{d}P \tag{1.1}$$

which corresponds to

$$f(|E_B|)\,\mathrm{d}|E_B| \propto |E_B|^{-1}\,\mathrm{d}|E_B| \tag{1.2}$$

in terms of the binary orbital energy E_B. This form of binary energy distribution is known as Öpik's law (Öpik 1924). One of our aims is to explain the origin of this law; we will come back to it in Chapter 11.

The distribution of orbital eccentricities e depends on the period: short period orbits have low eccentricities while long period orbits are more eccentric with the weight of the distribution at the high end of the range of e (Fig. 1.7; Duquennoy and Mayor 1991). In the latter case the data are consistent with

$$f(e)\,de = 2e\,de, \tag{1.3}$$

an equilibrium distribution in stellar dynamics (Jeans 1919, Ambartsumian 1937). Short period orbits are of low eccentricity (Campbell 1910). For contact binaries it is easy to understand why the orbit is circular: tidal friction between the two stars

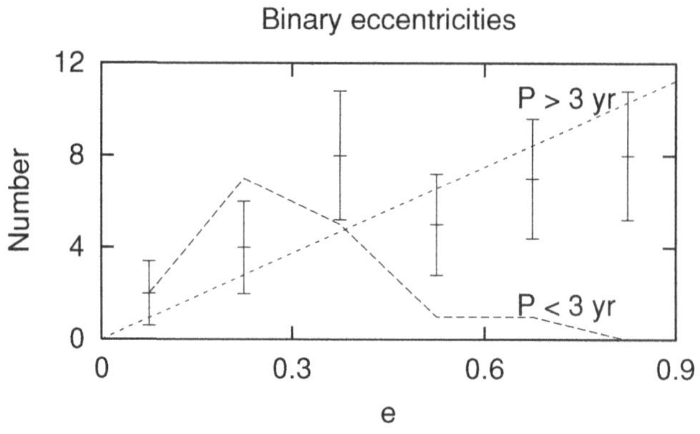

Figure 1.7 The observed distribution of eccentricities in a sample of binary stars. The data for binaries of longer orbital periods (over 3 years) are shown by points with error bars while the data for binaries of short orbital periods (less than 3 years) are indicated by the broken line (Duquennoy and Mayor 1991). The straight dotted line represents a possible interpretation of the long orbital period data.

operates most strongly during the pericentre of the orbit, when the two stars are closest to each other. The resulting loss of orbital speed at the pericentre leads to a rather circular orbit. But the low eccentricity orbits are dominant even among binaries whose orbital size is as large as one astronomical unit (AU), the distance between the Earth and the Sun. Tidal friction does not play any role at these separations since the radius of the Sun (and a typical main sequence star) is only about 0.005 AU. We have to look for other explanations for the different eccentricity distributions.

The mass ratio distribution depends on the type of the primary star (Abt 1983). The most massive stars (primaries of spectral type O) tend to be in rather equal mass pairs, $m_2/m_1 \gtrsim 0.3$ while the reverse is true for less massive upper main sequence stars. For the low mass, solar type primaries the mass ratio peaks at about $m_2/m_1 = 0.25$ (Duquennoy and Mayor 1991).

To some extent the differences can be understood by the form of the mass distribution function for single stars:

$$f(m) \propto m^{-\alpha} \tag{1.4}$$

where $\alpha = 2.35$ over most of the range (Salpeter 1955). If the more massive component is taken to be $m_1 = 1$, then the distribution for m_2 should follow Eq. (1.4), and the same should also apply to the mass ratio $m = m_2/m_1$. However, observations give a less steep function. For example, Kuiper (1935a, b) found that $f(m) = 2(1 + m)^{-2}$ gives a good fit to data on close binaries while Trimble (1990) prefers $f(m) \propto m^{-1}$. Thus the masses of the binary components are more equal

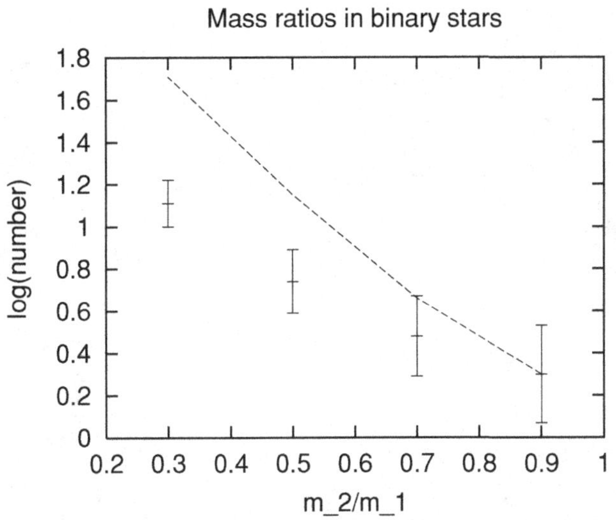

Figure 1.8 The distribution of the mass ratio m_2/m_1 ($m_2 < m_1$) in a B spectral type binary star sample (points with error bars; Evans 1995). In comparison, the dashed line gives the expected distribution if the companion of the B-type star has been picked at random among ordinary stars lighter than the primary.

than we would predict by picking a companion for the primary at random from the single star mass function (Fig. 1.8; Evans 1995).

On the other hand, the mass function for low mass stars (below the mass of the Sun) has a lower power-law index $\alpha \simeq 1.25$. Therefore the mass ratio distribution is also expected to be less steep, $f(m) \propto m^{-1.25}$. This agrees with the observed flatter distribution of m for solar type primaries. Quantitative comparisons are difficult because the single star mass function below about 0.5 solar mass is poorly known.

At the other end, the mass function for O-type stars is steeper than Eq. (1.4) with $\alpha = 2.35$. A power-law index $\alpha = 3.2$ may be used at the high mass end, above $m \simeq 10M_\odot$ (Mihalas and Binney 1981). This leads to an even steeper mass ratio distribution, in contrast with observations which go in the opposite direction. Thus there is much more to the mass ratio distribution than simply picking a pair at random. Even if the pairs might have formed like this initially, there has obviously been strong evolution which has modified the mass ratios. We will discuss the three-body process in Chapter 11 as a possible cause of this evolution.

1.5 Groups of galaxies

Just like stars, galaxies also appear in binaries, triples, quadruples and other small groups, as well as in large groups and clusters (Fig. 1.9). Binaries and their interactions with third galaxies may have consequences for the evolution of these systems. There is also important information about galaxies themselves which

Figure 1.9 Coma Cluster of galaxies. Picture taken with the KPNO 4-m Mayall Telescope. Image NOAO/AURA/NSF.

requires an understanding of the multiple systems. First and foremost, the total masses of galaxies, including their dark halos, are best derived from the study of multiple systems.

The method which is commonly used is to assume that the multiple system is in a bound state, i.e. that none of the members is currently escaping from the grouping. In addition, some assumptions are made about the nature of the orbits (e.g. circular, radial or intermediate) as well as about projections onto the plane of the sky. With these assumptions it is possible to connect observed properties, i.e. projected separations and radial velocities with inferred quantities such as the total mass of the system.

The total amount of light emitted by all the member galaxies is also measurable. Then the calculated mass (in solar mass) may be divided by the measured light output (in solar luminosity), and the resulting number is called the mass-to-light ratio. It is usually assumed that it is rather narrowly confined to a value around 30–50, at least for spiral galaxies. It is still an unsolved problem in astronomy as to how frequently galaxies in a complete volume sample fall into this range of mass-to-light ratio, and to what extent there is variation from one class of galaxies to another.

The existence of binary galaxies complicates matters. Their interaction with third galaxies may lead to escape orbits. It is very difficult to infer observationally whether a particular galaxy is in an escape orbit. The natural assumption is that any galaxy associated with a group is bound to it; this assumption overestimates the true mass by a factor of two or so in cases where there are escapers among the galaxies. This will be discussed in the final chapter. There we will also derive the maximum speed of escape based on the three-body theory.

1.6 Binary black holes

Black holes represent a form of matter which has collapsed on itself, in principle into a single point. In practice, the event horizon at the Schwarzschild radius R_g, defined as

$$R_g = \frac{2GM}{c^2}, \tag{1.5}$$

where M is the mass of the body and c the speed of light, represents the radius of the body. In a non-rotating black hole, this radius separates the internal space of the black hole from the outside world; no communication is possible from the inside of the Schwarzschild radius to the outside.

For many purposes we may treat black holes as mass points, and calculate their motions relative to each other as we do for ordinary bodies in Newtonian physics. Close to the Schwarzschild radius the dynamics differs very much from Newtonian dynamics, but far away, say at $100R_g$, only small corrections to Newtonian theory are required. In this chapter and in the following we will only consider the latter situations.

There are black holes of stellar masses, and supermassive black holes of millions of solar masses. These we are fairly certain of. There may be a whole mass range of black holes, but for various reasons we tend to observe only these two extremes. In this book we will only discuss the supermassive variety.

Supermassive black holes have been identified in the centres of many galaxies. Possibly every galaxy of sufficient size has or has had a supermassive black hole in its centre. Black holes are surrounded by disks of gas, and associated with them are phenomena which sometimes produce extremely bright radiation. For example, jets of relativistic particles are thought to flow out in two opposite directions, along the rotation axis of the disk. These jets may appear very bright, especially when seen end-on.

When galaxies merge, their central supermassive black holes also approach each other, but generally do not merge. They form a binary system of supermassive black holes. Eventually there may also be a merger of the binary, but typically the time

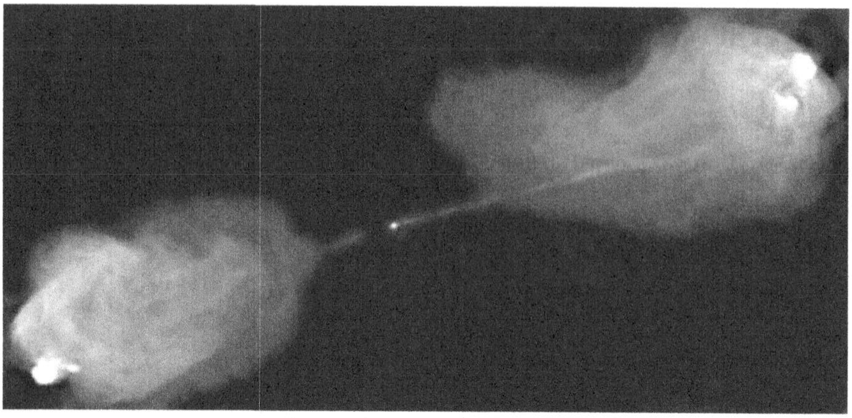

Figure 1.10 Radio galaxy Cygnus A, as seen in radio waves. The two bright emission regions straddle the galaxy which is centred on the spot of radio emission at the centre of the picture. Image courtesy NRAO/AUI.

scale is very long, of the order of the current age of the Universe (Hubble time). Occasionally the binary merger may happen in a shorter time scale through an increase of eccentricity in the binary, and a collision of black holes at the pericentre of the binary orbit (Aarseth 2003b). The problem of calculating this time scale will be discussed in the final chapter using three-body theory.

Multiple mergers of galaxies also take place, and these may produce triple or quadruple supermassive black hole systems. The evolution of the triple systems is also discussed in the last chapter. Then we will find that supermassive black holes should occasionally fly out of their parent galaxies, or sometimes remain as oscillators in the galaxy. These are processes not well understood at present since no unique identification of a supermassive black hole outside a galaxy has yet been made. However, the currently popular ΛCDM cosmology leads us to expect that the black hole escape process was extremely common in the early universe when large galaxies were assembled from their small progenitors.

Supermassive black holes are most likely associated with strong radio sources, radio galaxies and quasars. Radio galaxies typically have a double lobe structure where the regions of strongest radio emission lie on diametrically opposite sides of the galaxy (Fig. 1.10). Two categories of theories are usually advanced to explain the double radio source phenomenon: beam theory (Blandford and Rees 1974) and slingshot theory (Saslaw *et al.* 1974). According to the former theory, supermassive black holes remain stationary in the centres of galaxies, or at most move around each other in binary orbits, and send out beams of relativistic plasma in two oppositely directed beams. These beams supposedly reach outside the galaxy and cause radio emission there. In the slingshot model supermassive black holes themselves are thrown out of the centre of the galaxy via the three-body process. Due to conservation of linear momentum, a single black hole escapes in one direction and the binary in the opposite direction from the galaxy, and each produces radio emission along its escape path.

As long as no supermassive black holes have been directly observed or their existence in the radio lobes has not been disproven, it is difficult to verify either theory. However, there are indirect ways to do so. The most clear-cut test is related to the escape direction of radio lobes. In the beam theory the direction is along the rotational axis of the disk of gas (the accretion disk) which surrounds the black hole. Even though the disk usually cannot be imaged directly, it is a fair assumption that it arises from a more extended disk of gas in the galaxy via accretion. Therefore the observed orientation of the extended gas in the galaxy should also give information about the accretion disk of the central black hole. The projection of a circular disk in the sky is an ellipse; the direction of the minor axis of the ellipse shows the projection of the axis of symmetry of the disk.

Another way to find the orientation of the accretion disk is to observe small scale jets which emanate at right angles to the disk. Again the jet line shows the projection of the axis of symmetry in the sky. In the beam theory the radio lobes should lie further along the jet line, and the projected angle in the sky between the jet line and the double lobe radio axis should be zero. Thus the radio axis should be in the direction of the minor axis of the image of the extended gas.

It is also possible that the distribution of stars in the galaxy is spheroidal and flattened along the same axis as the gas disk. This is certainly true of spiral galaxies, but may also pertain to elliptical galaxies which are hosts of double radio sources. The alignment of the gas disk and the stellar disk could have its origin in mergers of galaxies: the angular momentum of both components should be strongly influenced by the orbital angular momentum of the binary galaxy which has merged.

In contrast to the beam theory, the slingshot theory predicts the escape of radio lobes along or close to the central plane of the galaxy, and perpendicular to the minor axis of the gas distribution and perpendicular also to the jet directions. In Chapter 7 we will discuss the escape directions in the three-body problem in more detail, and also compare them with observations.

2
Newtonian mechanics

This chapter introduces the basic concepts of Newtonian mechanics. We will emphasize the areas which are most useful in the three-body problem, and also familiarise ourselves with a system of units and scaling laws. The calculation of orbits using Newton's laws is a central theme of this book, and therefore a brief introduction to the methods follows. It is not the purpose of this work to teach the latest orbit calculation techniques; therefore only a brief introduction is given. Finally, we discuss the connection of Newtonian mechanics to chaos. It may come as a surprise that the introduction of just one more body to the well behaved two-body system brings about a chaotic, unpredictable dynamical system. This was realised by Poincaré well before the concept of deterministic chaos became a popular topic.

2.1 Newton's laws

We begin with the fundamental laws of mechanics, as given by Newton in his *Principia* in 1687, although in a more modern form.

First law If there are no external forces, an object will maintain its state of motion, i.e. it will stay at rest or continue rectilinear motion at constant velocity.

Second law The rate of change of the momentum of an object is proportional to the applied force. If the force is $\boldsymbol{F}$, momentum $\boldsymbol{p}$, mass m and the radius vector $\boldsymbol{r}$, then

$$\boldsymbol{F} = \dot{\boldsymbol{p}} = \frac{\mathrm{d}}{\mathrm{d}t}(m\dot{\boldsymbol{r}}). \tag{2.1}$$

Third law If a body A exerts a force $\boldsymbol{F}$ on a body B, the body B will exert a force $-\boldsymbol{F}$ on body A.

Following these fundamental laws Newton presents a corollary, which essentially states that the forces acting on a body can be evaluated independently of each other.

In more modern language this means that in the presence of several forces $\boldsymbol{F}_1$, $\boldsymbol{F}_2$, ..., $\boldsymbol{F}_n$ the body will behave as if there were only a single force $\boldsymbol{F}$, which is the vector sum of the individual forces: $\boldsymbol{F} = \boldsymbol{F}_1 + \boldsymbol{F}_2 + \cdots + \boldsymbol{F}_n$.

These laws determine how objects will move in the presence of forces. Another law was discovered by Newton to provide the force:

Law of gravity If the bodies A and B have masses m_A and m_B, respectively, and if their mutual distance is r, A will act on B with a force that is directed towards A and has a magnitude

$$Gm_{\mathrm{A}}m_{\mathrm{B}}/r^2, \tag{2.2}$$

where G is a constant, the *constant of gravity*, the value of which depends on the units chosen.

When we calculate the gravitational force or acceleration due to an object, or the effect of gravity on a body, we assume that all the mass is concentrated in the centre of mass, i.e. the objects are pointlike. This is valid also for real bodies if they are spherically symmetric, which is true for many celestial objects.

This is almost all one needs to know about physics; from here on, mainly mathematical methods are required.

2.2 Inertial coordinate system

Vectors and rectilinear motions appearing in Newton's laws are geometric entities, independent of any coordinate frames. In practice, however, our calculations are carried out in some coordinate frame. If an object is at rest, then none of its coordinates changes with time. Since things look different in different frames, the definition of the coordinate frame is fundamental.

An *inertial frame* is a coordinate frame in which Newton's laws hold true. If we had an absolute frame of the whole universe, at least all the frames moving at constant velocity with respect to this absolute frame would be inertial. The concept of such an absolute frame has, however, turned out to be rather problematic. Yet we can define that a frame is inertial if experiments show that Newton's laws are valid in that frame.

Often the situation depends on the required accuracy. An observer on the surface of the Earth is in circular motion around the Earth's axis, the Earth orbits the Sun, and the Sun orbits the centre of the Milky Way galaxy. Accelerations are involved in all these motions, and thus the frames moving with the Earth or the Sun are not inertial. Yet we can use them as if they were. The deviation from the inertial frame has to be taken into account only if (1) we are interested in such a long period of time that the curvature of the trajectory becomes evident, or (2) we require such a high accuracy that the accelerations affect the results.

2.3 Equations of motion for N bodies

Assume we have n point masses with radius vectors $\boldsymbol{r}_i$ and masses m_i. The total mass of the system is

$$M = \sum_{i=1}^{n} m_i \tag{2.3}$$

and the radius vector $\boldsymbol{R}$ of the centre of mass

$$\boldsymbol{R} = \frac{1}{M}\left(\sum_{i=1}^{n} m_i \boldsymbol{r}_i\right). \tag{2.4}$$

The gravitational force affecting the object i is

$$\begin{aligned} \boldsymbol{F}_i &= -G \sum_{j=1, j\neq i}^{n} m_i m_j \frac{1}{(\boldsymbol{r}_i - \boldsymbol{r}_j)^2} \frac{\boldsymbol{r}_i - \boldsymbol{r}_j}{|\boldsymbol{r}_i - \boldsymbol{r}_j|} \\ &= -G \sum_{j=1, j\neq i}^{n} m_i m_j \frac{\boldsymbol{r}_i - \boldsymbol{r}_j}{|\boldsymbol{r}_i - \boldsymbol{r}_j|^3}. \end{aligned} \tag{2.5}$$

From Newton's second law

$$\boldsymbol{F}_i = \frac{\mathrm{d}}{\mathrm{d}t}(m_i \dot{\boldsymbol{r}}_i) = m_i \ddot{\boldsymbol{r}}_i. \tag{2.6}$$

Equating these two forces and dividing by m_i the *equation of motion* of the object i becomes:

$$\ddot{\boldsymbol{r}}_i = -G \sum_{j=1, j\neq i}^{n} m_j \frac{\boldsymbol{r}_{ij}}{r_{ij}^3}, \tag{2.7}$$

where

$$\begin{aligned} \boldsymbol{r}_{ij} &= \boldsymbol{r}_i - \boldsymbol{r}_j, \\ r_{ij} &= |\boldsymbol{r}_i - \boldsymbol{r}_j|. \end{aligned} \tag{2.8}$$

In terms of rectangular xyz coordinates and basis vectors $\hat{\boldsymbol{e}}_x$, $\hat{\boldsymbol{e}}_y$ and $\hat{\boldsymbol{e}}_z$, the position of m_i is

$$\boldsymbol{r}_i = x_i \hat{\boldsymbol{e}}_x + y_i \hat{\boldsymbol{e}}_y + z_i \hat{\boldsymbol{e}}_z. \tag{2.9}$$

Defining the gradient operator ∇_i as:

$$\nabla_i = \hat{\boldsymbol{e}}_x \frac{\partial}{\partial x_i} + \hat{\boldsymbol{e}}_y \frac{\partial}{\partial y_i} + \hat{\boldsymbol{e}}_z \frac{\partial}{\partial z_i}, \tag{2.10}$$

the potential energy V of our system is

$$\begin{aligned} V &= -G\sum_{j=1}^{n}\sum_{k=j+1}^{n}\frac{m_j m_k}{r_{jk}} \\ &= -\frac{1}{2}G\sum_{j=1}^{n}\sum_{k=1,k\neq j}^{n}\frac{m_j m_k}{r_{jk}}, \end{aligned} \tag{2.11}$$

and

$$\nabla_i V = -\frac{1}{2}G\sum_{j=1}^{n}\sum_{k=1,k\neq j}^{n} m_j m_k \nabla_i\left(\frac{1}{r_{jk}}\right).$$

Since r_{jk} depends only on the coordinates of the bodies j and k, only the terms with $j = i$ or $k = i$ are non-zero:

$$\begin{aligned} \nabla_i V &= -\frac{1}{2}G\sum_{k=1,k\neq i}^{n} m_i m_k \nabla_i\left(\frac{1}{r_{ik}}\right) - \frac{1}{2}G\sum_{j=1,j\neq i}^{n} m_j m_i \nabla_i\left(\frac{1}{r_{ji}}\right) \\ &= -G\sum_{j=1,j\neq i}^{n} m_j m_i \nabla_i\left(\frac{1}{r_{ij}}\right), \end{aligned}$$

where

$$\begin{aligned} \nabla_i\left(\frac{1}{r_{ij}}\right) &= \nabla_i((x_i - x_j)^2 + (y_i - y_j)^2 + (z_i - z_j)^2)^{-1/2} \\ &= -\frac{(x_i - x_j)\hat{\boldsymbol{e}}_x + (y_i - y_j)\hat{\boldsymbol{e}}_y + (z_i - z_j)\hat{\boldsymbol{e}}_z}{r_{ij}^3} \\ &= -\frac{\boldsymbol{r}_{ij}}{r_{ij}^3}. \end{aligned}$$

Then immediately

$$\nabla_i V = G\sum_{j=1,j\neq i}^{n} m_i m_j \frac{\boldsymbol{r}_{ij}}{r_{ij}^3}. \tag{2.12}$$

Comparing this with the equation of motion (2.7) one may write

$$m_i\ddot{\boldsymbol{r}}_i = -\nabla_i V, \quad i = 1, \ldots, n, \tag{2.13}$$

where

$$V = -\frac{1}{2}G\sum_{j=1}^{n}\sum_{k=1,k\neq j}^{n}\frac{m_j m_k}{r_{jk}}. \tag{2.14}$$

2.4 Gravitational potential

When studying the effect of a system on an external test body, the potential instead of the potential energy is often used. The potential

$$U(\boldsymbol{r}) = -G \sum_{j=1}^{n} \frac{m_j}{|\boldsymbol{r} - \boldsymbol{r}_j|} \tag{2.15}$$

has the dimension of energy per unit mass, and the potential energy of a mass m at $\boldsymbol{r}$ is $U(\boldsymbol{r})m$. The equation of motion of this object is

$$\ddot{\boldsymbol{r}} = -\nabla U. \tag{2.16}$$

A point mass can be replaced by a continuous mass distribution, the density of which at $\boldsymbol{r}$ is denoted by $\rho(\boldsymbol{r})$. The potential at $\boldsymbol{r}_0$ due to this mass distribution is

$$U(\boldsymbol{r}_0) = -G \int \frac{\rho(\boldsymbol{r})\,\mathrm{d}V}{|\boldsymbol{r} - \boldsymbol{r}_0|}, \tag{2.17}$$

where $\mathrm{d}V$ is the volume element and the integration is extended over the whole mass distribution. Evaluation of the potential of an arbitrary object often leads to complicated integrals.

Example 2.1 Find the potential of a homogeneous disk with radius R at a distance z along the axis of the disk.

Let the surface density of the disk be ρ (kg/m^2). Then its total mass is $\rho\pi R^2$. In cylindrical coordinates (r, ϕ) the potential at a distance z is

$$\begin{aligned}
U(z) &= -G \int_0^R \mathrm{d}r \int_0^{2\pi} \frac{\rho r\,\mathrm{d}\phi}{\sqrt{z^2 + r^2}} \\
&= -2\pi G\rho \int_0^R \frac{r\,\mathrm{d}r}{\sqrt{z^2 + r^2}} \\
&= -2\pi G\rho \left.\sqrt{z^2 + r^2}\right|_0^R \\
&= -2\pi G\rho \left(\sqrt{z^2 + R^2} - \sqrt{z^2}\right) \\
&= -2\pi G\rho \frac{R^2}{\sqrt{z^2 + R^2} + z} \\
&= -\frac{GM}{z} \frac{2}{\sqrt{1 + (R/z)^2} + 1}.
\end{aligned}$$

If the distance z is much greater than the diameter of the disk, we can replace the square root with the two first terms of its Taylor expansion:

$$\begin{aligned} U(z) &\approx -\frac{GM}{z} \frac{1}{1+\frac{1}{4}(R/z)^2} \\ &\approx -\frac{GM}{z}\left[1-\frac{1}{4}\left(\frac{R}{z}\right)^2\right]. \end{aligned}$$

The potential of a point mass would be $-GM/z$. It is the shape of the mass distribution that gives rise to the second term. Thus we cannot replace such an object with a point mass.

The gradient of the potential gives the force (per unit mass). In this case only the component parallel to the z axis remains. Using the previous series approximation,

$$F_z = -\frac{\mathrm{d}U}{\mathrm{d}z} \approx \frac{GM}{z^2}\left[1-\frac{3}{4}\left(\frac{R}{z}\right)^2\right].$$

Example 2.2 Potential of an infinite plane.

In the previous example we let the radius R of the disk grow without limit. The potential obviously approaches infinity. However, this does not necessarily mean that the force will also become infinite. When $R \to \infty$ in the expression of the potential $U(z)$, $U(z) \to -2\pi G\rho(R-z)$ and we get $F = -2\pi G\rho$. This does not depend on the distance at all. The gravitational attraction of an infinite plane is everywhere constant and proportional to the density of the disk. For example, when calculating stellar motions with respect to the galactic plane we can approximate the galaxy as an infinite plane to get very simple equations of motion.

2.5 Constants of motion

The equation of motion of a body is a second order, vector differential equation. Thus the total solution involves six integration constants, which in celestial mechanics are usually called the *integrals* of the equation of motion. The general solution of n bodies thus requires $6n$ integrals. Some of them can be easily evaluated.

Summing all of the equations of motion gives:

$$\sum_{i=1}^{n} m_i \ddot{\boldsymbol{r}}_i = -G \sum_{i=1}^{n} \sum_{j=1, j\neq i}^{n} m_i m_j \frac{\boldsymbol{r}_{ij}}{r_{ij}^3}.$$

The sum on the right hand side consists of pairs $(\boldsymbol{r}_i - \boldsymbol{r}_j)/r_{ij}^3$ and $(\boldsymbol{r}_j - \boldsymbol{r}_i)/r_{ji}^3$. These terms cancel each other, and hence

$$\sum_{i=1}^{n} m_i \ddot{\boldsymbol{r}}_i = 0.$$

We integrate this twice with respect to time and get

$$\sum_{i=1}^{n} m_i \boldsymbol{r}_i = \boldsymbol{a}t + \boldsymbol{b}, \tag{2.18}$$

where $\boldsymbol{a}$ and $\boldsymbol{b}$ are constant vectors. Comparison with the definition of the centre of mass (2.4) shows that the sum on the left hand side is $M\boldsymbol{R}$. Thus we have

$$\boldsymbol{R} = \frac{\boldsymbol{a}t + \boldsymbol{b}}{M}. \tag{2.19}$$

Thus the centre of mass or *barycentre* moves at constant velocity along a straight line. The obvious consequence of this is that the barycentric coordinate frame is an inertial frame.

The total angular momentum of the system is

$$\boldsymbol{L} = \sum_{i=1}^{n} m_i \boldsymbol{r}_i \times \dot{\boldsymbol{r}}_i. \tag{2.20}$$

The time derivative of this is

$$\begin{aligned}
\dot{\boldsymbol{L}} &= \frac{\mathrm{d}}{\mathrm{d}t}\left(\sum_{i=1}^{n} m_i \boldsymbol{r}_i \times \dot{\boldsymbol{r}}_i\right) \\
&= \sum_{i=1}^{n} (m_i \dot{\boldsymbol{r}}_i \times \dot{\boldsymbol{r}}_i) + \sum_{i=1}^{n} (m_i \boldsymbol{r}_i \times \ddot{\boldsymbol{r}}_i) \\
&= -G \sum_{i=1}^{n} \sum_{j=1, j\neq i}^{n} m_i m_j \frac{\boldsymbol{r}_i \times (\boldsymbol{r}_i - \boldsymbol{r}_j)}{r_{ij}^3} \\
&= G \sum_{i=1}^{n} \sum_{j=1, j\neq i}^{n} m_i m_j \frac{\boldsymbol{r}_i \times \boldsymbol{r}_j}{r_{ij}^3}.
\end{aligned}$$

Again the terms on the right hand side cancel each other pairwise, because the vector product is anticommutative. Thus

$$\dot{\boldsymbol{L}} = 0, \tag{2.21}$$

i.e. the total angular momentum is constant.

Finally the total energy of the system is derived. Taking the scalar product of the equation of motion of the ith body and $\dot{\boldsymbol{r}}_i$, and adding all equations, one obtains:

$$\begin{aligned}\sum_{i=1}^{n} m_i \ddot{\boldsymbol{r}}_i \cdot \dot{\boldsymbol{r}}_i &= -\sum_{i=1}^{n} \dot{\boldsymbol{r}}_i \cdot \nabla_i V \\ &= -\sum_{i=1}^{n}\left(\dot{x}_i \frac{\partial V}{\partial x_i} + \dot{y}_i \frac{\partial V}{\partial y_i} + \dot{z}_i \frac{\partial V}{\partial z_i}\right) \\ &= -\frac{\mathrm{d}V}{\mathrm{d}t}.\end{aligned}$$

Integration with respect to time gives

$$\frac{1}{2}\sum_{i=1}^{n} m_i \dot{\boldsymbol{r}}_i \cdot \dot{\boldsymbol{r}}_i = -V + E, \tag{2.22}$$

where E is a constant of integration. The left hand side of the equation gives the kinetic energy of the system

$$T = \frac{1}{2}\sum_{i=1}^{n} m_i v_i^2, \tag{2.23}$$

where v_i is the speed of object i. Thus Eq. (2.22) can be written as

$$E = T + V, \tag{2.24}$$

which is the familiar law of total energy conservation.

Thus far we have found ten integrals for the equations of motion: vectors $\boldsymbol{a}$ and $\boldsymbol{b}$ describing the trajectory of the centre of mass, the total angular momentum $\boldsymbol{L}$, and the total energy E. For the complete solution, $6n - 10$ additional constants are required. The missing constants are easily found when $n = 2$. When $n > 2$, no additional independent constants are known. In fact, it has been shown that there are no additional independent constants that are algebraic functions of time, coordinates and velocities.

2.6 The virial theorem

In addition to constants of motion, statistical properties can be derived. The 'virial' of the system is defined as

$$A = \sum_{i=1}^{n} m_i \boldsymbol{r}_i \cdot \dot{\boldsymbol{r}}_i. \tag{2.25}$$

The time derivative of this is

$$\dot{A} = \sum_{i=1}^{n} (m_i \dot{\boldsymbol{r}}_i \cdot \dot{\boldsymbol{r}}_i + m_i \ddot{\boldsymbol{r}}_i \cdot \boldsymbol{r}_i)$$
$$= 2T + \sum_{i=1}^{n} \boldsymbol{F}_i \cdot \boldsymbol{r}_i .$$

Consider the time average in the interval $t \in [0, \tau]$. Denoting this average value by $\langle\rangle$,

$$\langle \dot{A} \rangle = \frac{1}{\tau} \int_0^\tau \dot{A}\, \mathrm{d}t = \langle 2T \rangle + \left\langle \sum_{i=1}^{n} \boldsymbol{F}_i \cdot \boldsymbol{r}_i \right\rangle . \tag{2.26}$$

If the system remains bounded, its virial cannot increase without limit. Thus the integral in (2.26) remains finite. When the length of the interval increases, $\langle \dot{A} \rangle$ approaches zero, and thus

$$2 \langle T \rangle + \left\langle \sum_{i=1}^{n} \boldsymbol{F}_i \cdot \boldsymbol{r}_i \right\rangle = 0. \tag{2.27}$$

This is the *virial theorem* in its general form. If the forces $\boldsymbol{F}_i$ are due to mutual gravitational forces only, the sum becomes

$$\sum_{i=1}^{n} \boldsymbol{F}_i \cdot \boldsymbol{r}_i = -G \sum_{i=1}^{n} \sum_{j=1, j\neq i}^{n} m_i m_j \frac{\boldsymbol{r}_i - \boldsymbol{r}_j}{r_{ij}^3} \cdot \boldsymbol{r}_i .$$

It is possible to write the double sum using only the vector differences $\boldsymbol{r}_i - \boldsymbol{r}_j$. We start by writing the sum twice but by changing the order of the summation indices i and j. We add the two identical sums and divide by two:

$$\sum_{i=1}^{n} \boldsymbol{F}_i \cdot \boldsymbol{r}_i$$
$$= -\frac{1}{2} G \left[\sum_{i=1}^{n} \sum_{j=1, j\neq i}^{n} \left(m_i m_j \frac{\boldsymbol{r}_i - \boldsymbol{r}_j}{r_{ij}^3} \cdot \boldsymbol{r}_i + m_j m_i \frac{\boldsymbol{r}_j - \boldsymbol{r}_i}{r_{ji}^3} \cdot \boldsymbol{r}_j \right) \right]$$
$$= -\frac{1}{2} G \sum_{i=1}^{n} \sum_{j=1, j\neq i}^{n} m_i m_j \frac{\boldsymbol{r}_i - \boldsymbol{r}_j}{r_{ij}^3} \cdot (\boldsymbol{r}_i - \boldsymbol{r}_j)$$
$$= -\frac{1}{2} G \sum_{i=1}^{n} \sum_{j=1, j\neq i}^{n} \frac{m_i m_j}{r_{ij}} .$$

The double sum $\sum_{i=1}^{n} \sum_{j=1}^{n}$, where $i \neq j$, gives every term twice, due to the symmetry relative to interchange between i and j. In order to have every term only once, the latter sum should start from $j = i + 1$. We make this change and multiply the above expression by two:

$$\sum_{i=1}^{n} \boldsymbol{F}_i \cdot \boldsymbol{r}_i = -G \sum_{i=1}^{n} \sum_{j=i+1}^{n} \frac{m_i m_j}{r_{ij}} = V.$$

Thus

$$\langle T \rangle = -\frac{1}{2} \langle V \rangle . \tag{2.28}$$

The virial theorem (2.28) can be used to study, for example, the stability of clusters of stars. The kinetic energy can be determined from the velocity dispersion and the potential energy from the mass distribution derived from the brightness of the objects. If the kinetic energy is much higher than the average value given by the virial theorem, the system is unstable. The virial theorem indicates that clusters of galaxies should be unstable, which contradicts observational evidence about their stability. This is one reason to believe that such systems must contain large amounts of dark matter.

2.7 The Lagrange and Jacobi forms of the equations of motion

For three bodies, Eq. (2.7) gives us

$$\ddot{\boldsymbol{r}}_1 = -G \left(m_2 \frac{\boldsymbol{r}_{12}}{r_{12}^3} + m_3 \frac{\boldsymbol{r}_{13}}{r_{13}^3} \right) \tag{2.29}$$

$$\ddot{\boldsymbol{r}}_2 = -G \left(m_3 \frac{\boldsymbol{r}_{23}}{r_{23}^3} + m_1 \frac{\boldsymbol{r}_{21}}{r_{21}^3} \right) . \tag{2.30}$$

In the *Lagrangian formulation* of the equations of motion the coordinate differences are used instead of the coordinates themselves. The acceleration between bodies 1 and 2 is obtained by subtracting (2.30) from (2.29):

$$\begin{aligned} \ddot{\boldsymbol{r}}_{12} &= -G \left[(m_2 + m_1) \frac{\boldsymbol{r}_{12}}{r_{12}^3} + m_3 \left(\frac{\boldsymbol{r}_{13}}{r_{13}^3} + \frac{\boldsymbol{r}_{32}}{r_{32}^3} \right) \right] \\ &= -G \left[M \frac{\boldsymbol{r}_{12}}{r_{12}^3} + m_3 \left(\frac{\boldsymbol{r}_{13}}{r_{13}^3} + \frac{\boldsymbol{r}_{32}}{r_{32}^3} + \frac{\boldsymbol{r}_{21}}{r_{21}^3} \right) \right] \\ &= G \left(m_3 \boldsymbol{W} - M \frac{\boldsymbol{r}_{12}}{r_{12}^3} \right) , \end{aligned} \tag{2.31}$$

where

$$\boldsymbol{W} = \frac{\boldsymbol{r}_{12}}{r_{12}^3} + \frac{\boldsymbol{r}_{23}}{r_{23}^3} + \frac{\boldsymbol{r}_{31}}{r_{31}^3} . \tag{2.32}$$

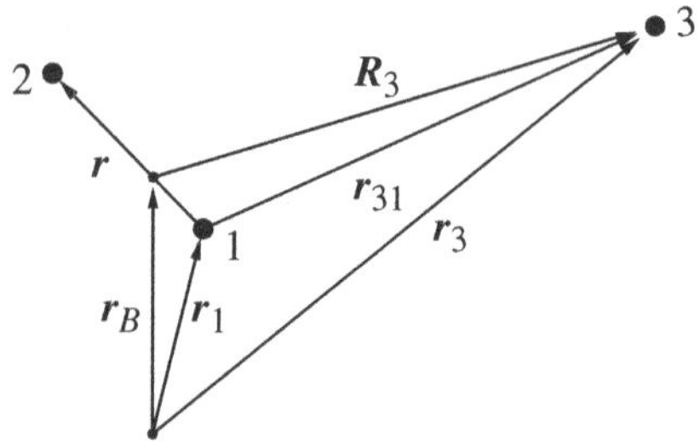

Figure 2.1 Coordinates of the Jacobi system: the bodies 1 and 2 form a binary while the third body (3) is more distant.

The corresponding equations for the acceleration between bodies 2 and 3 as well between bodies 3 and 1 are obtained from (2.31) by changing the subscripts accordingly.

In the *Jacobi system of coordinates*, the three-body system is considered to be composed of two parts: a clearly defined binary, and a somewhat distant third body (Fig. 2.1). The centre of mass of the binary is used as a reference point whose position is $\boldsymbol{r}_B$. The relative position vector of the binary is $\boldsymbol{r}_{21} = \boldsymbol{r}$. By definition of the centre of mass (Eq. (2.4))

$$\begin{aligned}\boldsymbol{r}_B &= \frac{1}{m_1+m_2}(m_1\boldsymbol{r}_1 + m_2\boldsymbol{r}_2)\\ &= \frac{m_1}{m_1+m_2}\boldsymbol{r}_1 + \frac{m_2}{m_1+m_2}\boldsymbol{r}_2 + \frac{m_2}{m_1+m_2}\boldsymbol{r}_1 - \frac{m_2}{m_1+m_2}\boldsymbol{r}_1\\ &= \boldsymbol{r}_1 + \frac{m_2}{m_1+m_2}\boldsymbol{r}\end{aligned} \tag{2.33}$$

or, alternatively,

$$\begin{aligned}\boldsymbol{r}_B &= \frac{m_1}{m_1+m_2}\boldsymbol{r}_1 + \frac{m_2}{m_1+m_2}\boldsymbol{r}_2 + \frac{m_1}{m_1+m_2}\boldsymbol{r}_2 - \frac{m_1}{m_1+m_2}\boldsymbol{r}_2\\ &= \boldsymbol{r}_2 - \frac{m_1}{m_1+m_2}\boldsymbol{r}.\end{aligned} \tag{2.34}$$

Let us call the position vector of body 3 relative to the centre of mass $\boldsymbol{R}_3 = \boldsymbol{r}_3 - \boldsymbol{r}_B$. Then

$$\begin{aligned}\boldsymbol{r}_{31} &= \boldsymbol{r}_3 - \boldsymbol{r}_1 = \boldsymbol{R}_3 + \boldsymbol{r}_B - \boldsymbol{r}_1 = \boldsymbol{R}_3 + \frac{m_2}{m_1+m_2}\boldsymbol{r},\\ \boldsymbol{r}_{32} &= \boldsymbol{r}_3 - \boldsymbol{r}_2 = \boldsymbol{R}_3 + \boldsymbol{r}_B - \boldsymbol{r}_2 = \boldsymbol{R}_3 - \frac{m_1}{m_1+m_2}\boldsymbol{r}.\end{aligned} \tag{2.35}$$

Also

$$\begin{aligned}\frac{m_1}{m_1+m_2}\boldsymbol{r}_{31} + \frac{m_2}{m_1+m_2}\boldsymbol{r}_{32} &= \left(\frac{m_1}{m_1+m_2} + \frac{m_2}{m_1+m_2}\right)\boldsymbol{R}_3 + \left(\frac{m_1 m_2}{(m_1+m_2)^2} - \frac{m_1 m_2}{(m_1+m_2)^2}\right)\boldsymbol{r} \\ &= \boldsymbol{R}_3. \end{aligned} \tag{2.36}$$

Using Eqs. (2.31) and (2.36), the equations of motion are:

$$\ddot{\boldsymbol{r}} = -G\left[(m_1+m_2)\frac{\boldsymbol{r}}{r^3} + m_3\left(\frac{\boldsymbol{r}_{31}}{r_{31}^3} - \frac{\boldsymbol{r}_{32}}{r_{32}^3}\right)\right] \tag{2.37}$$

and

$$\begin{aligned}\ddot{\boldsymbol{R}}_3 &= \frac{m_1}{m_1+m_2}\ddot{\boldsymbol{r}}_{31} + \frac{m_2}{m_1+m_2}\ddot{\boldsymbol{r}}_{32} \\ &= G\left[m_2\frac{m_1}{m_1+m_2}\boldsymbol{W} - M\frac{m_1}{m_1+m_2}\frac{\boldsymbol{r}_{31}}{r_{31}^3}\right. \\ &\quad \left. + m_1\frac{m_2}{m_1+m_2}(-\boldsymbol{W}) - M\frac{m_2}{m_1+m_2}\frac{\boldsymbol{r}_{32}}{r_{32}^3}\right] \\ &= -GM\left(\frac{m_1}{m_1+m_2}\frac{\boldsymbol{r}_{31}}{r_{31}^3} + \frac{m_2}{m_1+m_2}\frac{\boldsymbol{r}_{32}}{r_{32}^3}\right). \end{aligned} \tag{2.38}$$

If $r \ll R_3$, $\boldsymbol{r}_{31} \approx \boldsymbol{r}_{32} \approx \boldsymbol{R}_3$. Then the equations of motion (2.37) and (2.38) are simply

$$\ddot{\boldsymbol{r}} = -G(m_1+m_2)\frac{\boldsymbol{r}}{r^3}, \tag{2.39}$$

and

$$\ddot{\boldsymbol{R}}_3 = -GM\frac{\boldsymbol{R}_3}{R_3^3}. \tag{2.40}$$

These are two separate two-body equations of motion the solutions of which will be discussed in the next chapter. The solutions are Keplerian orbits, an inner orbit (Eq. (2.39)) and an outer orbit (Eq. (2.40)). As Eqs. (2.39) and (2.40) are not exact, one is really dealing with two perturbed Keplerian orbits.

2.8 Constants of motion in the three-body problem

The angular momentum for a three-body system is (Eq. (2.20))

$$\boldsymbol{L} = m_1\boldsymbol{r}_1 \times \dot{\boldsymbol{r}}_1 + m_2\boldsymbol{r}_2 \times \dot{\boldsymbol{r}}_2 + m_3\boldsymbol{r}_3 \times \dot{\boldsymbol{r}}_3. \tag{2.41}$$

In the Lagrangian form the angular momentum in the centre of mass coordinate system reads

$$L = \frac{m_1 m_2 m_3}{M} \left(\frac{\boldsymbol{r}_{12} \times \dot{\boldsymbol{r}}_{12}}{m_3} + \frac{\boldsymbol{r}_{23} \times \dot{\boldsymbol{r}}_{23}}{m_1} + \frac{\boldsymbol{r}_{31} \times \dot{\boldsymbol{r}}_{31}}{m_2} \right) \tag{2.42}$$

and in the Jacobi form

$$\boldsymbol{L} = \mathcal{M}(\boldsymbol{r} \times \dot{\boldsymbol{r}}) + m(\boldsymbol{R}_3 \times \dot{\boldsymbol{R}}_3) \tag{2.43}$$

where $\mathcal{M}$ and m are the reduced masses:

$$\begin{aligned} \mathcal{M} &= \frac{m_1 m_2}{m_1 + m_2}, \\ m &= \frac{m_3(m_1 + m_2)}{M} \end{aligned} \tag{2.44}$$

(Problem 2.5).

The plane through the centre of mass and perpendicular to $\boldsymbol{L}$ is called the *invariable plane*. When $\boldsymbol{L} = 0$, Eq. (2.43) tells us that the motion is confined to a plane which contains all four vectors $\boldsymbol{r}, \dot{\boldsymbol{r}}, \boldsymbol{R}_3$ and $\dot{\boldsymbol{R}}_3$.

The energy integral for the three-body problem is

$$\begin{aligned} E &= \frac{1}{2}\left(m_1 v_1^2 + m_2 v_2^2 + m_3 v_3^2\right) \\ &\quad - G\left(\frac{m_1 m_2}{r_{21}} + \frac{m_1 m_3}{r_{13}} + \frac{m_2 m_3}{r_{23}}\right). \end{aligned} \tag{2.45}$$

The corresponding equations in the Lagrangian and Jacobi forms are

$$\begin{aligned} E = \frac{m_1 m_2 m_3}{M} \Bigg[&\frac{1}{2}\left(\frac{v_{23}^2}{m_1} + \frac{v_{31}^2}{m_2} + \frac{v_{12}^2}{m_3}\right) \\ &- GM\left(\frac{1}{m_3 r_{21}} + \frac{1}{m_2 r_{13}} + \frac{1}{m_1 r_{23}}\right)\Bigg] \end{aligned} \tag{2.46}$$

and

$$E = \frac{1}{2}\left(\mathcal{M}v^2 + mV_3^2\right) - G\left(\frac{m_1 m_2}{r} + \frac{m_1 m_3}{r_{31}} + \frac{m_2 m_3}{r_{23}}\right), \tag{2.47}$$

respectively, where $v = |\dot{\boldsymbol{r}}|$ and $V_3 = |\dot{\boldsymbol{R}}_3|$ (Problem 2.6).

2.9 Moment of inertia

In the centre-of-mass coordinate system, the moment of inertia of a three-body system is

$$I = m_1 r_1^2 + m_2 r_2^2 + m_3 r_3^2. \tag{2.48}$$

In the Lagrangian form it becomes

$$I = \frac{m_1 m_2 m_3}{M}\left(\frac{r_{21}^2}{m_3} + \frac{r_{13}^2}{m_2} + \frac{r_{32}^2}{m_1}\right), \tag{2.49}$$

and in the Jacobi form

$$I = \mathcal{M}r^2 + mR_3^2 \tag{2.50}$$

(Problem 2.7).

Differentiating the moment of inertia twice with respect to time,

$$\dot{I} = 2m_1 \boldsymbol{r}_1 \cdot \dot{\boldsymbol{r}}_1 + 2m_2 \boldsymbol{r}_2 \cdot \dot{\boldsymbol{r}}_2 + 2m_3 \boldsymbol{r}_3 \cdot \dot{\boldsymbol{r}}_3,$$
$$\ddot{I} = 2m_1\left(v_1^2 + \boldsymbol{r}_1 \cdot \ddot{\boldsymbol{r}}_1\right) + 2m_2\left(v_2^2 + \boldsymbol{r}_2 \cdot \ddot{\boldsymbol{r}}_2\right) + 2m_3\left(v_3^2 + \boldsymbol{r}_3 \cdot \ddot{\boldsymbol{r}}_3\right)$$

from which it follows that

$$\frac{1}{2}\ddot{I} = 2T + V = 2E - V \tag{2.51}$$

(Problem 2.8). Equation (2.51) is known as the *Lagrange–Jacobi identity.*

The moment of inertia is a measure of compactness of the three-body system, in the sense that the more compact the system, the smaller is the moment of inertia. When the moment of inertia increases, a greater and greater sphere is required to surround the system. When in addition $\ddot{I} > 0$, the bounding sphere increases at an accelerated pace. In practice, $\ddot{I} > 0$ often means the escape of one body from the other two.

In the evolution of a three-body system there are periods when $\langle \ddot{I} \rangle \approx 0$. During those periods the system is bounded in a relatively small volume, and the virial theorem (Eq. (2.28)) is approximately satisfied. From numerical orbit calculations (see Fig. 1.2) the potential energy V is known to fluctuate a great deal, and since E is a constant in Eq. (2.51), $\ddot{I}$ must fluctuate on either side of zero. Sooner or later I goes through a deep minimum after which $\ddot{I}$ remains positive over a period which is sufficiently long to permit an escape of one of the bodies. Over that period, the time averages of T and V satisfy

$$\langle 2T \rangle > \langle -V \rangle . \tag{2.52}$$

Later, when the escape velocity of the third body $\dot{\boldsymbol{R}}_3$ asymptotically approaches a constant value, Eq. (2.28) is satisfied again. Therefore the general behaviour of the moment of inertia I in the three-body systems, as revealed by numerical orbit calculations, together with the Lagrange–Jacobi identity, clearly leaves open the possibility of an escape in the three-body problem. It does not prove that an escape must happen, and indeed not every minimum of I leads to an escape, but when the possibility of escape is offered often enough, it is easy to believe that the evolution of a three-body system ends in an escape of one of the bodies, just as in the Pythagorean problem (Fig. 1.2).

2.10 Scaling of the three-body problem

One of the nice features of the solutions of the three-body problem, as well as the N-body problem, is that the solutions can be freely scaled from one physical system to another. For example, the Pythagorean problem could represent the motions of three stars, of 3, 4 and 5 solar masses, starting from mutual separations of 3, 4 and 5 parsecs from each other, or three isolated planets of 3, 4 and 5 Earth masses, starting from 3, 4 and 5 AU (astronomical units) apart. It can be easily shown (Problem 2.9) that if $\boldsymbol{r}_1(t), \boldsymbol{r}_2(t)$ and $\boldsymbol{r}_3(t)$ represent the solution (orbits) of a three-body problem, and k is a real number, then also $k^2\boldsymbol{r}_1(t)$, $k^2\boldsymbol{r}_2(t)$ and $k^2\boldsymbol{r}_3(t)$ is a solution of the same problem, as long as the time t is scaled as k^3. In the new solution the velocity v scales as $k^2/k^3 = k^{-1}$, the energy E scales as v^2, i.e. as k^{-2}, and the angular momentum L scales as $\boldsymbol{r} \times \boldsymbol{v}$, i.e. as $k^2k^{-1} = k$. The product EL^2 is scale free, a fact which will be used in the following to normalise the angular momenta.

2.11 Integration of orbits

It is generally possible to calculate the orbits in N-body systems by using a computer. Here, a brief outline of the principles commonly used is given (Aarseth 1971). There are numerous methods in contemporary use, but the details are beyond the scope of this book. Good expositions of the latest methods are available for example in Aarseth (2003a). Usually the orbit calculation procedure is referred to as *integration of orbits*. To calculate the orbits, Eq. (2.7) is to be solved for all bodies in the system. As long as the current positions of the bodies are known, the accelerations at that moment are given by the solutions of Eq. (2.7). If the current velocities are also known, the motions at that moment are fully specified. The new position $\boldsymbol{r}_1$ is obtained from the current position $\boldsymbol{r}$ after a small time step Δt by the Taylor series

$$\boldsymbol{r}_1 = \boldsymbol{r} + \dot{\boldsymbol{r}}\Delta t + \frac{1}{2}\boldsymbol{F}\Delta t^2 + \frac{1}{6}\dot{\boldsymbol{F}}\Delta t^3 + \frac{1}{24}\ddot{\boldsymbol{F}}\Delta t^4 + \cdots. \tag{2.53}$$

Here we have written $\boldsymbol{F} = \ddot{\boldsymbol{r}}$. The derivatives $\dot{\boldsymbol{F}}$ and $\ddot{\boldsymbol{F}}$ are obtained from Eq. (2.7) by differentiation:

$$\begin{aligned}
\dot{\boldsymbol{F}} &= -G\sum_{j=1,j\neq i}^{n} m_j\left[\frac{\dot{\boldsymbol{r}}_{ij}}{r_{ij}^3} - 3\frac{\boldsymbol{r}_{ij}(\boldsymbol{r}_{ij}\cdot\dot{\boldsymbol{r}}_{ij})}{r_{ij}^5}\right],\\
\ddot{\boldsymbol{F}} &= -G\sum_{j=1,j\neq i}^{n} m_j\left(\frac{\ddot{\boldsymbol{r}}_{ij}}{r_{ij}^3} - 6\frac{\dot{\boldsymbol{r}}_{ij}(\boldsymbol{r}_{ij}\cdot\dot{\boldsymbol{r}}_{ij})}{r_{ij}^5}\right.\\
&\quad\left. + 3\frac{\boldsymbol{r}_{ij}}{r_{ij}^5}\left[5\frac{(\boldsymbol{r}_{ij}\cdot\dot{\boldsymbol{r}}_{ij})^2}{r_{ij}^2} - \dot{\boldsymbol{r}}_{ij}\cdot\dot{\boldsymbol{r}}_{ij} - \boldsymbol{r}_{ij}\cdot\ddot{\boldsymbol{r}}_{ij}\right]\right).
\end{aligned} \tag{2.54}$$

Similarly, the new value of the velocity $\dot{\boldsymbol{r}}_1$ is obtained from

$$\dot{\boldsymbol{r}}_1 = \dot{\boldsymbol{r}} + \boldsymbol{F}\,\Delta t + \frac{1}{2}\dot{\boldsymbol{F}}\Delta t^2 + \frac{1}{6}\ddot{\boldsymbol{F}}\Delta t^3 + \cdots. \tag{2.55}$$

By repeating the short time step for all of the bodies we may hope to calculate the time evolution of the whole system.

There are obvious problems with this scheme. Complicated calculations are carried out at every time step which consume computer time and which cause round-off errors. Thus, in order to finish the calculation in a reasonable amount of computer time, the time step should not be very small. But then the convergence of the Taylor series becomes poorer. We realise that during a relatively long time step the acceleration can change considerably from what it was at the beginning of the step, because in reality all bodies move simultaneously. Some form of *prediction* of motions of all bodies has to be introduced prior to taking the time step. The prediction can be based on a low order Taylor series which does not require complicated calculations:

$$\boldsymbol{r}_1 = \boldsymbol{r} + \dot{\boldsymbol{r}}\,\Delta t + \frac{1}{2}\boldsymbol{F}\,\Delta t^2 + \frac{1}{6}\hat{\boldsymbol{F}}\,\Delta t^3. \tag{2.56}$$

We define $\hat{\boldsymbol{F}}$ as the backward difference

$$\hat{\boldsymbol{F}} = \frac{\boldsymbol{F} - \boldsymbol{F}_{-1}}{\Delta t_{-1}} \tag{2.57}$$

where Δt_{-1} is the length of the preceding time step, and $\boldsymbol{F}_{-1}$ is the value of acceleration at the beginning of that step. In general, the computer algorithm becomes faster when the derivatives (2.54) are constructed from the corresponding backward differences. The values of $\boldsymbol{F}$ only need to be stored for the several previous steps, and the derivatives of $\boldsymbol{F}$ are then calculated by simple operations from these quantities. In particular, the prediction makes use of the values of the acceleration at two successive points where the acceleration has to be calculated in any case, and thus the prediction requires only modest extra calculation effort.

Other features which save computer time are the use of *variable time step* and *individual time step*. At some points in the orbit a body may be advanced by long time steps, at other more critical times a small step is necessary to preserve accuracy. Every body has a different situation in this regard; therefore it is useful to assign a different, individual, time step to each of them. The prediction of the orbits of all bodies up to a common moment of time then becomes crucial. When the acceleration and its derivatives change rapidly, it is necessary to go to a short time step. Therefore one may calculate the length of the step by requiring that the ratio of successive terms in the Taylor series is of the order of a small number $\eta \ll 1$.

For example,

$$\frac{\frac{1}{6}|\dot{\boldsymbol{F}}|\Delta t^3}{\frac{1}{2}|\boldsymbol{F}|\Delta t^2} = \eta,$$

from which it follows that

$$\Delta t = \eta \frac{3|\boldsymbol{F}|}{|\dot{\boldsymbol{F}}|}. \tag{2.58}$$

Usually a somewhat more complicated rule for Δt is used which takes account of the higher derivatives of $\boldsymbol{F}$ but the general principle is the same. One may also set other simple demands on Δt based on practical experience with computations.

A special situation arises when two bodies come very close to each other. Then a rule like Eq. (2.58) leads to such a small step that the computation may virtually terminate before the two bodies pass each other. The source of the trouble is the $1/r^3$ factor in Eq. (2.7). Let us write this equation again just for two bodies, with accelerations caused by other bodies lumped together as a perturbing acceleration $\boldsymbol{f}$:

$$\ddot{\boldsymbol{r}} = -\mu \frac{\boldsymbol{r}}{r^3} + \boldsymbol{f}. \tag{2.59}$$

Here $\mu = G(m_a + m_b)$; m_a and m_b are the masses of the two bodies. One attempts to transform this to another form where there are no singular terms which approach infinity when $r \to 0$. This kind of transformation is called *regularisation* of the equation of motion. The common first step is to introduce an auxiliary variable τ by

$$\frac{\mathrm{d}\tau}{\mathrm{d}t} = \frac{1}{r}. \tag{2.60}$$

If we represent differentiation relative to τ by primes Eq. (2.59) becomes

$$\boldsymbol{r}'' = \frac{r'}{r}\boldsymbol{r}' - \mu \frac{\boldsymbol{r}}{r} + r^2 \boldsymbol{f} \tag{2.61}$$

(Sundman 1912; Problem 2.10). This already helps somewhat since now the $1/r^3$ factor is replaced by $1/r$. In order to get rid of the remaining $1/r$ we introduce two auxiliary variables h and $\boldsymbol{e}$:

$$\begin{aligned} h &= \frac{1}{2}\frac{\boldsymbol{r}'^2}{r^2} - \frac{\mu}{r}, \\ \boldsymbol{e} &= -\frac{\boldsymbol{r}}{r} - \frac{1}{\mu r^2}\left[(\boldsymbol{r} \times \boldsymbol{r}') \times \boldsymbol{r}'\right]. \end{aligned} \tag{2.62}$$

It is then easy to see that Eq. (2.61) transforms to

$$\mathbf{r}'' = 2h\mathbf{r} - \mu\mathbf{e} + r^2\mathbf{f}. \tag{2.63}$$

When $\mathbf{f} = 0$, the quantities h and $\mathbf{e}$ are constants. We see this by evaluating

$$h' = \frac{\mathrm{d}}{\mathrm{d}\tau}\left(\frac{1}{2}\frac{\mathbf{r}'^2}{r^2} - \frac{\mu}{r}\right) = -\frac{r'\mathbf{r}'^2}{r^3} + \frac{\mathbf{r}'\cdot\mathbf{r}''}{r^2} + \frac{\mu}{r^2}r',$$

where

$$\begin{aligned}\frac{\mathbf{r}'\cdot\mathbf{r}''}{r^2} &= \frac{\mathbf{r}'}{r^2}\cdot\left(\frac{r'}{r}\mathbf{r}' - \mu\frac{\mathbf{r}}{r} + r^2\mathbf{f}\right)\\ &= \frac{r'\mathbf{r}'^2}{r^3} - \frac{\mu}{r^2}r' + \mathbf{r}'\cdot\mathbf{f}.\end{aligned}$$

Thus

$$h' = \mathbf{r}'\cdot\mathbf{f}. \tag{2.64}$$

After some calculations we find also

$$\mathbf{e}' = -\frac{1}{\mu}\left[\mathbf{r}'(\mathbf{r}\cdot\mathbf{f}) + \mathbf{f}(\mathbf{r}\cdot\mathbf{r}') - 2\mathbf{r}h'\right] \tag{2.65}$$

(Problem 2.11). Therefore a regularized equation of motion has been derived which has no singular terms when $\mathbf{r} \to 0$ as long as $\mathbf{f} \to 0$. Quantities h and $\mathbf{e}$ arise again in the next chapter where their physical significance is clarified (Eqs. (3.11) and (3.13)).

The above regularisation scheme (Burdet 1967, Heggie 1973) is not the only possibility. Even though the Burdet–Heggie regularisation is known to function perfectly well in numerical solutions of the N-body problem (Saslaw *et al.* 1974, Heggie 1975, Valtonen and Heggie 1979), actually another transformation called the K-S regularisation is more widely used (Kustaanheimo 1964, Kustaanheimo and Stiefel 1965). However, the important point to us is that there are methods to handle close two-body encounters without slowing down the orbit calculation and without loss of numerical accuracy.

How does one then measure the *accuracy* of the orbit calculation? One may test the method with a problem where the analytic solution is known, such as the two-body problem (Chapter 3, Problem 3.8). Alternatively, one may follow the conservation of quantities which are known to be strictly constant in the N-body problem, i.e. the centre of mass and momentum, the total energy and angular momentum. Perhaps the most stringent test is based on the time-reversability of the orbits. If at any point along the orbit the velocities of all bodies are reversed, the bodies should retrace their trajectories exactly. Computational errors make this unlikely to happen in numerical orbit integration. The ability to retrace the orbit backward is a very demanding test for the accuracy of the method.

2.12 Dimensions and units of the three-body problem

In the general three-body problem, 21 independent variables are to be specified: three position and three velocity components plus the mass value for each body. The position and the motion of the centre of mass, as well as the total mass and the linear scale (scale factor k above) of the system are not of much interest, so that the number of 'interesting' variables is reduced to 13. The same can be said about the spatial orientation of the system, described by two angles specifying the direction of the total angular momentum. Therefore the number of dimensions of the three-body problem is 11.

The magnitude of the total angular momentum L_0 and the value of the total energy E_0 are important parameters. The combination

$$L_0^2|E_0|/G^2m_0^5 \tag{2.66}$$

is a dimensionless quantity. Here m_0 signifies the average mass of the three bodies, to be defined more specifically in Chapter 7 (Eq. (7.28)). If this quantity is given the value 6.25,

$$L_0 \equiv L_{\max} = 2.5Gm_0^{2.5}/|E_0|^{0.5}. \tag{2.67}$$

$L_{\max}$ is an important standard of reference for the angular momenta of three-body systems (Chapter 7).

The reason why the total mass M and the linear scale factor r are quantities of little interest is that the three-body problem is scale free as mentioned above. If one three-body problem with a given mass scale (total mass) M, distance scale r and time scale t has been solved, then it is obvious (Eq. (2.7) and Section 2.10) that this solution also applies to any other mass scale M and distance scale r. The time scale $t \sim M^{-1/2}r^{3/2}$ and the velocity scale $v \sim M^{1/2}r^{1/2}$ transformations ensure this. Therefore M and r are 'uninteresting' parameters.

In numerical orbit calculations it is usual to fix the system of units by putting the gravitational constant $G = 1$. Writing this constant in the common astronomical units of solar mass ($M_\odot = 1.99 \times 10^{30}$ kg), parsec (1 pc $= 3.086 \times 10^{16}$ m) and km/s, we get

$$G = 4.3 \times 10^{-3}\,(\text{km/s})^2\,\text{pc}/M_\odot. \tag{2.68}$$

$G = 1$ ties the units such that corresponding to pc and km/s the mass unit is $1/(4.3 \times 10^{-3} M_\odot) = 233 M_\odot$. Since

$$1\ \mathrm{km/s} \approx 1\ \mathrm{pc/Myr} \tag{2.69}$$

the corresponding unit of time is very nearly one megayear (1 Myr $= 10^6$ yr). Obviously the calculation can be carried out in other units but this does not bring a new dimension to the three-body problem. In practice, the time is frequently given in the natural time unit of the three-body system which is the *crossing time* $T_{\rm cr} = GM^{5/2}/(2|E_0|)^{3/2}$ (see Eq. (8.26) and the discussion there). This is roughly the time that it takes a body to travel through the three-body system.

In principle, the three-body problem may be solved by going through the 11-dimensional phase space with a fine tooth comb, and by calculating the orbital evolution at each phase space point. Let us briefly consider how practical this proposal is. Construct a grid in the phase space composed of 100 values in each coordinate and calculate the orbits at each grid point. It is in total $100^{11} = 10^{22}$ orbits. If every orbit takes 1 second to calculate in a fast machine, the project would last 10^{22} seconds, i.e. about 3×10^{14} yr. Even so, the solution would be rather rough since a mesh of 100 grid points is far too coarse to cover the phase space adequately. Therefore it is not possible to find a complete solution of the three-body problem by 'brute force', even if we have perfect integration methods and powerful computers in use.

The alternative, an analytical solution for example in the form of a series has been proposed and searched for. It has been estimated that $10^{8\,000\,000}$ terms are required in the Sundman series to represent a three-body solution with the accuracy commonly used in the ephemeris calculations. The computer time requirement would be impossible, and certainly the representation could not be said to be simple. It is still possible that, in the future, computers that use mathematical manipulators (e.g. Maple, Mathematica) may be used to identify important terms in long series expansions.

In recent years it has become clear that solutions of the three-body problem can be found and represented in simple form by using a combination of analytical and computational methods. Analytical methods give the basic functional forms which can be tested and improved by computer experiments. They also provide useful physical insights into the problem. We will describe this avenue of research mainly in Chapters 7 to 10. Before that, some computer experiments will be described which are useful to develop the analytical theory.

2.13 Chaos in the three-body problem

A two-dimensional description of the state of the system is usually called a *map*. In the three-body problem the free dimensions require reduction in order to make

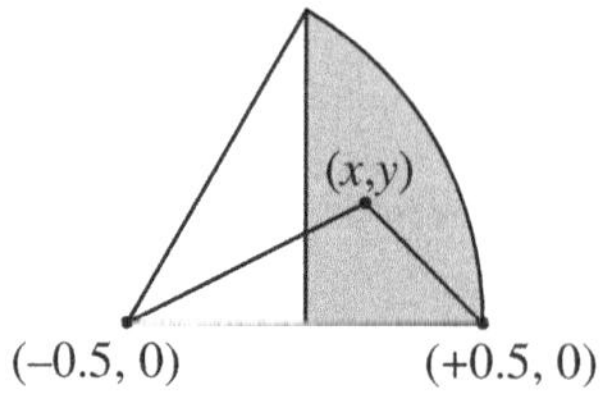

Figure 2.2 In the Agekyan–Anosova map two bodies are placed on the horizontal axis one unit apart. The third body is at (x, y), inside a 'curved triangle'. The map gives a unique description of the shape of the three-body configuration.

use of a map description. We may start by studying a planar problem where all the position and velocity vectors lie in the plane defined by the three bodies. The number of dimensions in the two-dimensional problem is $3 \times 4 + 3 = 15$, which is reduced to 9 when we neglect the centre of mass (4 coordinates) and the two scale-free parameters. There is now only one parameter related to orientation, the rotation around the angular momentum vector. This leaves us with 8 'interesting' dimensions. This can be further reduced by two if we limit ourselves to equal mass systems.

In the Agekyan–Anosova map (Agekyan and Anosova 1967) the largest of the three separations r_{12}, r_{23}, r_{31} is mapped on the horizontal axis and is assigned a unit length. The two bodies have coordinates $(-0.5, 0)$ and $(+0.5, 0)$, respectively. Then the spatial configuration ('shape') of the system is uniquely determined by a point (x, y), $x > 0$, $y > 0$, representing the third body. This point lies inside a 'curved triangle' (Fig. 2.2). The Agekyan–Anosova (AA for short) map therefore ignores the velocities of the bodies, but it may be used to describe the evolution of the 'shape' of one or many such systems. By choosing the velocities to be zero initially, the AA map gives a full description of the initial values of the three-body systems under study.

The end result of an orbit calculation, no matter where the initial position lies in the AA map, is the breakup of the system into two parts which escape from each other hyperbolically, i.e. the largest separation becomes infinite when time approaches infinity. In the end the system is always made up of a binary and an escaping third body. The only exceptions are orbits where two bodies collide; these are found in lines in the AA map and are of measure zero in the initial value space. The same is true for triple collisions which take place at certain points of the AA map (Tanikawa and Umehara 1998). Therefore the point (x, y) will always be attracted to the $(+0.5, 0)$ corner in the end. This point is referred to as an *attractor* of the map. When the evolution of a typical point (x, y) is followed, it repeatedly approaches the corner $(+0.5, 0)$ and spends an extraordinary amount of time in that neighbourhood, apparently for no reason, even before the final asymptotic plunge towards the attractor. Therefore this point is called a *strange attractor*. The term strange attractor was first introduced in connection with turbulent flows (Ruelle and Takens 1971), but it is now commonly used to describe unpredictable behaviour in deterministic systems.

Even though each point in the AA map gives rise to a unique orbit, the orbits in the neighbourhood of any one point are generally quite different from the original orbit, in an unpredictable way. This situation is described by the term *chaos* (Li and Yorke 1975). Already in 1890 Poincaré recognised the phenomenon and conjectured that three-body orbits are non-integrable. Indeed, the extreme sensitivity to initial conditions raises the question of whether it is actually always possible to integrate the correct orbit. In an effort to complete as many orbits as possible in the AA map, using 16-digit accuracy, and by requiring that the time reversed orbit does not deviate from the original (forward in time) orbit in relative amount by more than 10^{-3}, Anosova *et al.* found that 60% of the orbits are 'predictable' and 40% 'unpredictable', i.e. non-integrable to this accuracy (Anosova *et al.* 1994).

Notwithstanding this question of the absolute correctness of the integrated orbit, we proceed to integrate orbits in all parts of the AA map using the best methods currently available (Mikkola and Aarseth 1990, 1993, 1996, Lehto *et al.* 2000). Figure 2.3 shows one of the properties of these orbits, the lifetime of the three-body system. Lifetime is defined as the time interval from the beginning of the orbit calculation up to the time of departure of the third body in its hyperbolic escape orbit. In some parts of the AA map, the departure is immediate and the lifetime is zero. These areas of the AA map represent only a few per cent of the total area of the map. Because of symmetry with respect to time, these orbits are continuations of hyperbolic approach orbits by the same third body. They are called *flyby* orbits.

Generally, the distribution of lifetime gives an impression of great randomness but not of complete chaos. There are islands of stability where neighbouring orbits have lifetimes similar to each other and which form some obvious structures. This situation is usually referred to as *weak chaos*. When one takes a closer look at these structures they show substructures which again break down to more substructures with greater magnification. This is an example of *fractal geometry* which is encountered commonly in nature (Mandelbrot 1982). It is not really a surprise to find fractal geometry in the AA map because strange attractors are usually associated with fractals.

An important quantity which describes the properties of fractals is the *fractal dimension*. If, for example, a set of points forms some structure in a two-dimensional map (called imbedding space), and if the number of points inside a circle of radius r is $C(r)$, then a relation $C(r) \propto r$ tells us that the structure is in fact a line (i.e. one-dimensional). Similarly, if our imbedding space is three-dimensional and $C(r) \propto r^2$, then the structure is two-dimensional. In general, write $C(r) \propto r^D$, where now, by

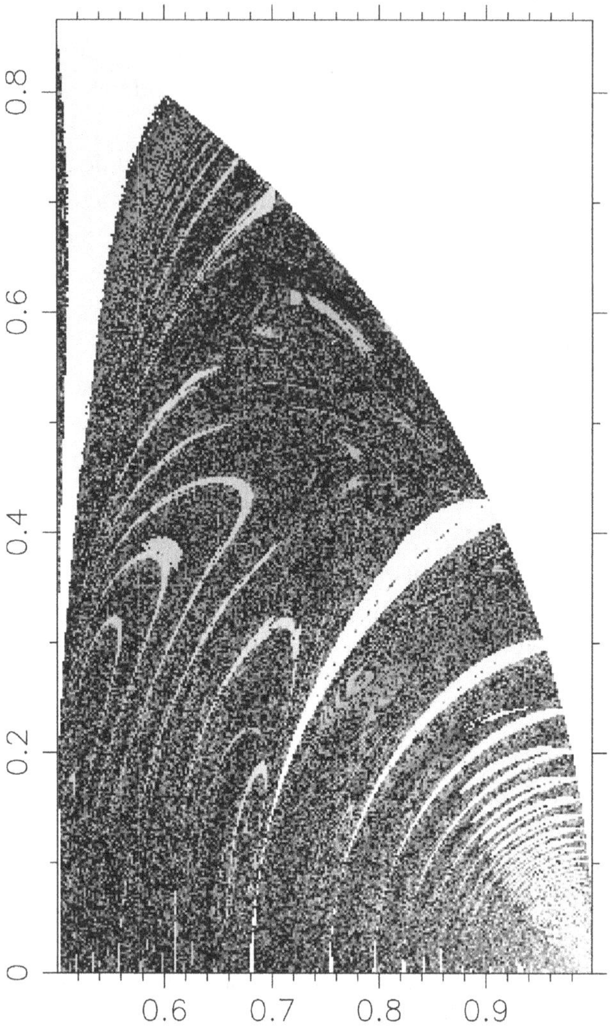

Figure 2.3 The lifetime of the triple system as a function of the initial position in the AA map. Lifetimes are short in light areas of the map, long in the dark areas.

definition, the number D is the fractal dimension. For fractals, this number is typically not a whole number. A good example is the Hénon map where points are added to a two-dimensional map by a simple recursive formula. Its fractal dimension is between one and two, indicating that the fractal structure is a little more than a line, having a bit of the second dimension (Hénon 1976a, 1982).

For the general three-body problem, the fractal dimension has been found to be a little above two, ≈ 2.1 (Heinämäki *et al.* 1998). This tells us that the fractal structures are a little more than two-dimensional, and that their embedding dimension is three. In other words, there are only three major physical parameters in the problem, instead of 11. Two of these have been clearly identified in numerical experiments and they are the energy E_0 and the angular momentum L_0 of the system (Valtonen 1974). For given values of the masses, the third obvious parameter is the binary binding energy E_B in configurations where a binary and a third body can be clearly identified. This result is extremely significant for the development of the theory for the three-body problem, to be examined in Chapter 7.

Another way to characterise a chaotic or weakly chaotic system is to calculate its *Lyapunov exponents*. Suppose $d(t)$ describes the separation of two orbits in the AA map as a function of time t. Then for typical chaotic systems, the separation increases exponentially from a small initial value:

$$d(t) \propto e^{\sigma t}, \tag{2.70}$$

where σ is the Lyapunov exponent. If the time is measured in crossing times the value of the Lyapunov exponent is typically $\sigma \simeq 1/2$ (Ivanov and Chernin 1991, Ivanov *et al.* 1995, Chernin and Valtonen 1998, Heinämäki *et al.* 1999). The inverse of the Lyapunov exponent, σ^{-1}, is frequently called the e-folding time.

We may also characterise the divergence of orbits in the AA map by looking at the so called '*phase drop*'. It is a tight bundle of points in phase space, in this case in the AA map. One follows the orbits initiated at each point inside the drop and finds out where they go in the AA map at subsequent times. At later times a convex contour is drawn which envelopes all the points and the relative area $\Delta\Gamma$ inside the envelope is measured. As the orbits diverge, $d(t) \simeq \exp(0.5t)$ (t in crossing times) and the area $\Delta\Gamma \simeq d(t)^2$. Defining the quantity

$$S = \ln(\Delta\Gamma), \tag{2.71}$$

it is obvious that $S \simeq t$. The quantity S is called the *Kolmogorov–Sinai entropy*. In the three-body systems it increases more or less like the time in units of the crossing time (Heinämäki *et al.* 1999). The Kolmogorov–Sinai entropy measures the loss of information about the initial conditions or the growth of disorder as time goes on, and in this sense it corresponds to the entropy concept used in thermodynamics. Here is a connection between thermodynamics and the three-body problem: a large number of three-body systems behaves like a large number of molecules in a single thermodynamic system. This analogy is helpful in describing the solutions of the three-body problem.

2.14 Rotating coordinate system

Chapters 5 and 6 will be concerned with dynamics in a rotating coordinate system. Then additional forces appear which are described below.

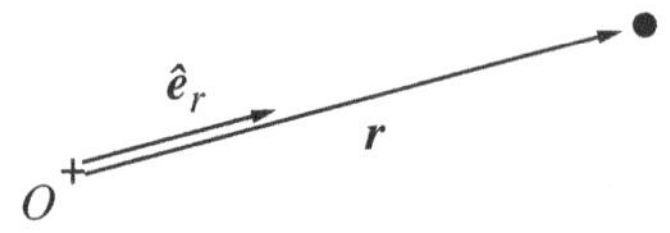

Figure 2.4 The unit vector $\hat{\boldsymbol{e}}_r$ points in the direction of the radius vector.

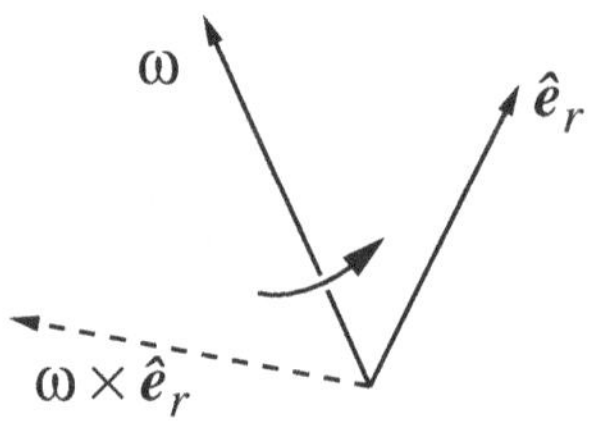

Figure 2.5 The vector $\boldsymbol{\omega}$ points along the rotation axis of a rotating coordinate frame.

Let a particle of mass m be in a coordinate system which rotates with a constant angular velocity $\boldsymbol{\omega}$ with respect to the inertial coordinate system. The position vector of the particle is $\boldsymbol{r} = r\hat{\boldsymbol{e}}_r$ where $\hat{\boldsymbol{e}}_r$ is the unit vector in the direction of $\boldsymbol{r}$ (Fig. 2.4). If $\mathrm{d}/\mathrm{d}t$ is written for the derivatives in the inertial system and $\partial/\partial t$ for the derivatives in the rotating system, then

$$\frac{\mathrm{d}\boldsymbol{r}}{\mathrm{d}t} = \frac{\partial \boldsymbol{r}}{\partial t} + r\frac{\mathrm{d}\hat{\boldsymbol{e}}_r}{\mathrm{d}t} \tag{2.72}$$

where $\partial \boldsymbol{r}/\partial t = (\mathrm{d}r/\mathrm{d}t)\hat{\boldsymbol{e}}_r$. The rate of rotation of the unit vector $\hat{\boldsymbol{e}}_r$ equals the magnitude of the angular speed. The direction of the change of $\hat{\boldsymbol{e}}_r$ is perpendicular to both $\boldsymbol{\omega}$ and $\hat{\boldsymbol{e}}_r$ (Fig. 2.5). Therefore

$$\frac{\mathrm{d}\hat{\boldsymbol{e}}_r}{\mathrm{d}t} = \boldsymbol{\omega} \times \hat{\boldsymbol{e}}_r \tag{2.73}$$

and

$$\frac{\mathrm{d}\boldsymbol{r}}{\mathrm{d}t} = \frac{\partial \boldsymbol{r}}{\partial t} + \boldsymbol{\omega} \times \boldsymbol{r}. \tag{2.74}$$

In terms of an operator we may write

$$\frac{\mathrm{d}}{\mathrm{d}t} = \frac{\partial}{\partial t} + \boldsymbol{\omega} \times . \tag{2.75}$$

The second derivative is obtained by applying the operator twice:

$$\frac{\mathrm{d}^2\boldsymbol{r}}{\mathrm{d}t^2} = \frac{\partial^2 \boldsymbol{r}}{\partial t^2} + 2\boldsymbol{\omega} \times \frac{\partial \boldsymbol{r}}{\partial t} + \boldsymbol{\omega} \times (\boldsymbol{\omega} \times \boldsymbol{r}). \tag{2.76}$$

After moving the last two terms to the left hand side of the equation we see that in a rotating coordinate system particles move as if they were subject to two additional forces: the *Coriolis force*

$$-2m\boldsymbol{\omega} \times \dot{\boldsymbol{r}} \tag{2.77}$$

and the *centrifugal force*

$$-m\boldsymbol{\omega} \times (\boldsymbol{\omega} \times \boldsymbol{r}). \tag{2.78}$$

These appear in addition to the $m\mathrm{d}^2\boldsymbol{r}/\mathrm{d}t^2$ force. A further force term appears if $\boldsymbol{\omega}$ is not constant.

Problems

Problem 2.1 A coordinate frame moving with the Earth is not quite inertial because of the rotation and orbital motion of the Earth around the Sun and the motion of the Sun around the galactic centre. Find the accelerations due to these motions.

Problem 2.2 A star is at a distance z above the galactic plane, and its velocity is $v = 0$. Due to the gravitational pull of the Milky Way it will begin to oscillate with respect to the galactic plane. Find the period of oscillation.

Problem 2.3 Show that the potential of a homogeneous sphere equals that of a point of the same mass. What happens inside a hollow spherical shell?

Problem 2.4 Replace the gravitational force by a general central force

$$f(r, \theta, \phi)\hat{\boldsymbol{e}}_r.$$

This is directed along the line joining the two bodies but the magnitude varies in an arbitrary way. What will happen to the trajectory of the centre of mass and the angular momentum?

Problem 2.5 Derive the expressions (2.42) and (2.43) of the angular momentum.

Problem 2.6 Derive the expressions (2.46) and (2.47) of the total energy.

Problem 2.7 Show that the moment of inertia is given by expressions (2.49) and (2.50).

Problem 2.8 Prove the Lagrange–Jacobi identity (2.51).

Problem 2.9 Assume that $\boldsymbol{r}_1(t)$, $\boldsymbol{r}_2(t)$ and $\boldsymbol{r}_3(t)$ is a solution of a three-body problem. Show that also $k^2\boldsymbol{r}_1(t)$, $k^2\boldsymbol{r}_2(t)$ and $k^2\boldsymbol{r}_3(t)$ is a solution of the same problem, where k is a real number and the time t is scaled as k^3.

Problem 2.10 Derive the equation (2.61).

Problem 2.11 Prove the equation (2.65). Hint: make use of the derivations of Eqs. (10.3) and (10.4).

Problem 2.12 If the gravitational constant $G = 1$, show that the unit of mass is $233 M_\odot$ if the unit of distance is one parsec and the unit of velocity is km/s. Show also that the corresponding unit of time is close to one million years. Confirm that the orbits can always be scaled so that the total energy $E = -1$ (if the energy is negative) and the total mass is a constant, e.g. $M = 3$ units.

3

The two-body problem

In this chapter we study a system consisting of two bodies. It is the most complex case that allows a complete analytical solution of the equations of motion. The solution is derived in many elementary textbooks. The treatment here in Sections 3.1–3.12 follows Karttunen *et al.* (2003).

For definiteness, we assume that the bodies are the Sun and a planet. Equally well they could be the components of a binary star or the Earth and an artificial satellite.

3.1 Equations of motion

Let the masses of the Sun and the planet be m_1 and m_2, respectively, and their radius vectors in some inertial frame $\boldsymbol{r}_1$ and $\boldsymbol{r}_2$ (Fig. 3.1). Let $\boldsymbol{r}$ be the position of the planet with respect to the Sun: $\boldsymbol{r} = \boldsymbol{r}_2 - \boldsymbol{r}_1$. The length of the vector $\boldsymbol{r}$ is r. The equations of motion of the Sun and the planet are then

$$\begin{aligned} \ddot{\boldsymbol{r}}_1 &= G\frac{m_2\boldsymbol{r}}{r^3}, \\ \ddot{\boldsymbol{r}}_2 &= -G\frac{m_1\boldsymbol{r}}{r^3}. \end{aligned} \tag{3.1}$$

Since we are mainly interested in the relative motion of the planet with respect to the Sun, we subtract the Sun's equation from the planet's equation of motion and get

$$\ddot{\boldsymbol{r}}_2 - \ddot{\boldsymbol{r}}_1 = -G\frac{m_1 + m_2}{r^3}\boldsymbol{r}.$$

Thus the equation for the relative motion is

$$\ddot{\boldsymbol{r}} = -\mu\frac{\boldsymbol{r}}{r^3}, \tag{3.2}$$

where

$$\mu = G(m_1 + m_2). \tag{3.3}$$

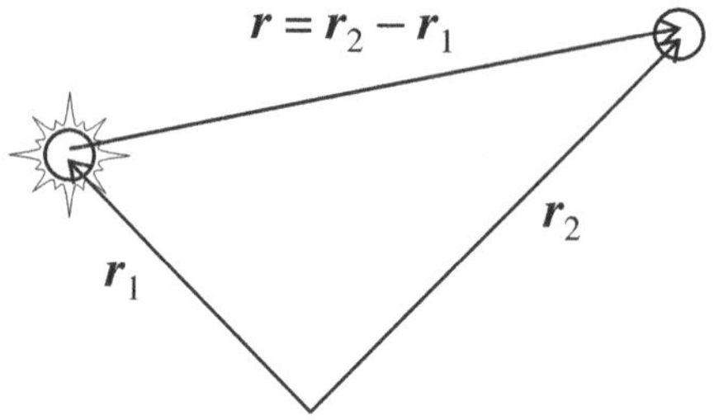

Figure 3.1 Radius vectors of the two-body problem.

If one of the masses is negligible (as in the case of an artificial satellite orbiting the Earth), the frame is inertial. Otherwise the coordinate frame of relative motion is accelerated. Even so the equation of motion (3.2) has the same form as in inertial coordinates except that the mass of the central body is replaced by the total mass of the system.

3.2 Centre of mass coordinate system

The position of the centre of mass, $\boldsymbol{R}$, is obtained from

$$M\boldsymbol{R} = m_1\boldsymbol{r}_1 + m_2\boldsymbol{r}_2, \tag{3.4}$$

where $M = m_1 + m_2$. If the radius vectors of the bodies with respect to the centre of mass are $\boldsymbol{R}_1$ and $\boldsymbol{R}_2$, we have

$$m_1\boldsymbol{R}_1 + m_2\boldsymbol{R}_2 = 0, \tag{3.5}$$

whence

$$\boldsymbol{R}_2 = -\frac{m_1}{m_2}\boldsymbol{R}_1. \tag{3.6}$$

The relative position of the planet is

$$\boldsymbol{r} = \boldsymbol{R}_2 - \boldsymbol{R}_1 = -\frac{m_1}{m_2}\boldsymbol{R}_1 - \boldsymbol{R}_1 = -\frac{m_1 + m_2}{m_2}\boldsymbol{R}_1. \tag{3.7}$$

We can now express the vectors $\boldsymbol{R}_1$ and $\boldsymbol{R}_2$ as functions of the relative radius vector $\boldsymbol{r}$:

$$\begin{aligned} \boldsymbol{R}_1 &= -\frac{m_2}{m_1 + m_2}\boldsymbol{r}, \\ \boldsymbol{R}_2 &= \frac{m_1}{m_1 + m_2}\boldsymbol{r}. \end{aligned} \tag{3.8}$$

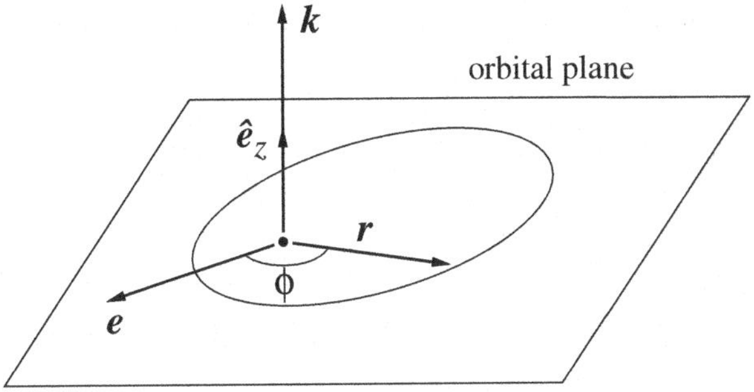

Figure 3.2 Vectors $\boldsymbol{e}$ and $\boldsymbol{k}$.

We see that the positions with respect to the centre of mass and the relative position differ only by a constant factor. Thus all these orbits are similar. Taking derivatives of Eq. (3.8) we see that the same proportions are valid also for velocities.

3.3 Integrals of the equation of motion

For the complete solution of the two-body equation of motion we need 12 integration constants. Six of them describe the trajectory of the centre of mass. From the remaining more interesting six constants three are needed for the angular momentum and one for the energy in the barycentric frame. Since we prefer to use the heliocentric frame, which is not inertial, we have to derive the six missing constants carefully.

Instead of angular momentum, it is customary to use the angular momentum per unit mass of the planet (in the rest of this chapter, this is called simply the 'angular momentum'):

$$\boldsymbol{k} = \boldsymbol{r} \times \dot{\boldsymbol{r}}. \tag{3.9}$$

The time derivative of this is

$$\dot{\boldsymbol{k}} = \dot{\boldsymbol{r}} \times \dot{\boldsymbol{r}} + \boldsymbol{r} \times \ddot{\boldsymbol{r}} = -\mu \frac{\boldsymbol{r} \times \boldsymbol{r}}{r^3} = 0. \tag{3.10}$$

Thus $\boldsymbol{k}$ is a constant vector. By definition, it is always perpendicular to the radius and velocity vectors. Thus the motion of the planet is always confined to the same plane perpendicular to $\boldsymbol{k}$ and containing the Sun (Fig. 3.2).

More integrals are found by taking the time derivative of the vector product $\boldsymbol{k} \times \dot{\boldsymbol{r}}$:

$$\begin{aligned}\frac{\mathrm{d}}{\mathrm{d}t}(\boldsymbol{k} \times \dot{\boldsymbol{r}}) &= \boldsymbol{k} \times \ddot{\boldsymbol{r}} \\ &= -(\boldsymbol{r} \times \dot{\boldsymbol{r}}) \times \frac{\mu \boldsymbol{r}}{r^3} \\ &= -\frac{\mu}{r^3}((\boldsymbol{r} \cdot \boldsymbol{r})\dot{\boldsymbol{r}} - (\boldsymbol{r} \cdot \dot{\boldsymbol{r}})\boldsymbol{r}) \\ &= -\mu \left(\frac{\dot{\boldsymbol{r}}}{r} - \frac{\dot{r}\boldsymbol{r}}{r^2} \right) \\ &= \frac{\mathrm{d}}{\mathrm{d}t}\left(-\frac{\mu \boldsymbol{r}}{r} \right),\end{aligned}$$

whence

$$\frac{\mathrm{d}}{\mathrm{d}t}\left(\boldsymbol{k} \times \dot{\boldsymbol{r}} + \frac{\mu \boldsymbol{r}}{r} \right) = 0.$$

Integration over time gives

$$\boldsymbol{k} \times \dot{\boldsymbol{r}} + \mu \boldsymbol{r}/r = -\mu \boldsymbol{e}, \tag{3.11}$$

where $\boldsymbol{e}$ is a constant vector (cf. Eqs. (2.62) and (2.65)). The constant factor $-\mu$ is needed to get a natural geometric interpretation for $\boldsymbol{e}$, as will be seen later.

We have found two constant vectors, i.e. six constants. However, $\boldsymbol{k}$ and $\boldsymbol{e}$ are not independent, as can be seen by evaluating the scalar product $\boldsymbol{k} \cdot \boldsymbol{e}$:

$$\begin{aligned}\boldsymbol{k} \cdot \boldsymbol{e} &= -\frac{1}{\mu}\left(\boldsymbol{k} \cdot (\boldsymbol{k} \times \dot{\boldsymbol{r}}) + \frac{\boldsymbol{k} \cdot \mu \boldsymbol{r}}{r} \right) \\ &= -\frac{1}{\mu}\left((\boldsymbol{k} \times \boldsymbol{k}) \cdot \dot{\boldsymbol{r}} + \frac{\mu(\boldsymbol{k} \cdot \boldsymbol{r})}{r} \right) \\ &= -\frac{(\boldsymbol{k} \cdot \boldsymbol{r})}{r}.\end{aligned}$$

Since $\boldsymbol{k}$ is perpendicular to the orbital plane, we have $\boldsymbol{k} \cdot \boldsymbol{r} = 0$, and thus

$$\boldsymbol{k} \cdot \boldsymbol{e} = 0. \tag{3.12}$$

This shows that the vector $\boldsymbol{e}$ is in the orbital plane, and gives only two independent constants of integration.

Yet another constant is found by taking the time derivative of the kinetic energy per unit mass:

$$\frac{\mathrm{d}}{\mathrm{d}t}\left(\frac{1}{2}v^2\right) = \frac{\mathrm{d}}{\mathrm{d}t}\left(\frac{1}{2}\dot{\boldsymbol{r}}\cdot\dot{\boldsymbol{r}}\right) = \dot{\boldsymbol{r}}\cdot\ddot{\boldsymbol{r}} = -\frac{\mu\dot{\boldsymbol{r}}\cdot\boldsymbol{r}}{r^3} = -\frac{\mu\dot{r}}{r^2}$$
$$= \frac{\mathrm{d}}{\mathrm{d}t}\left(\frac{\mu}{r}\right),$$

from which

$$\frac{\mathrm{d}}{\mathrm{d}t}\left(\frac{1}{2}v^2 - \frac{\mu}{r}\right) = 0$$

or

$$\frac{1}{2}v^2 - \frac{\mu}{r} = h, \tag{3.13}$$

where v is the velocity or the length of the vector $\dot{\boldsymbol{r}}$. The constant h is called the *energy integral* (cf. Eqs. (2.62) and (2.64)).

The energy integral can be expressed in terms of the previously found constants. To show this, we write the definition of $\boldsymbol{e}$ as

$$\mu\boldsymbol{r}/r + \mu\boldsymbol{e} = \dot{\boldsymbol{r}} \times \boldsymbol{k}$$

and take the scalar product of this equation with itself. Since $\dot{\boldsymbol{r}}$ and $\boldsymbol{k}$ are mutually perpendicular, the length of $\dot{\boldsymbol{r}} \times \boldsymbol{k}$ is the product of the lengths of $\dot{\boldsymbol{r}}$ and $\boldsymbol{k}$:

$$\mu^2\left(1 + 2\frac{\boldsymbol{r}\cdot\boldsymbol{e}}{r} + e^2\right) = v^2k^2. \tag{3.14}$$

The definition of the constant h gives

$$v^2 = 2h + 2\mu/r \tag{3.15}$$

and $\boldsymbol{r}\cdot\boldsymbol{e}$ can be evaluated using the definition (3.11) of $\boldsymbol{e}$:

$$\begin{aligned}\boldsymbol{r}\cdot\boldsymbol{e} &= -\frac{1}{\mu}\left(\boldsymbol{r}\cdot\boldsymbol{k}\times\dot{\boldsymbol{r}} + \mu\frac{\boldsymbol{r}\cdot\boldsymbol{r}}{r}\right)\\ &= -\frac{1}{\mu}(\boldsymbol{k}\cdot\dot{\boldsymbol{r}}\times\boldsymbol{r} + \mu r)\\ &= \frac{k^2}{\mu} - r.\end{aligned} \tag{3.16}$$

Substituting the expressions (3.15) and (3.16) for v^2 and $\boldsymbol{r}\cdot\boldsymbol{e}$ into (3.14) we get

$$\mu^2\left(1 + \frac{2k^2}{\mu r} - 2 + e^2\right) = 2hk^2 + 2\frac{\mu k^2}{r},$$

or

$$h = \frac{\mu^2}{2k^2}(e^2 - 1). \tag{3.17}$$

The constants we have found this far describe the orientation and shape of the orbit, but they do not tell where the planet is at a given moment. For this purpose we need one more integral. We use the energy integral

$$h = \frac{1}{2}|\dot{\boldsymbol{r}}|^2 - \frac{\mu}{r}$$

to find the length of the velocity vector

$$\left|\frac{\mathrm{d}\boldsymbol{r}}{\mathrm{d}t}\right| = \sqrt{2(h + \mu/r)}.$$

This is a separable differential equation, which can be integrated:

$$\int_{r_0}^{r} \frac{|\mathrm{d}\boldsymbol{r}|}{\sqrt{2(h + \mu/r)}} = \int_{\tau}^{t} \mathrm{d}t = t - \tau. \tag{3.18}$$

The constant τ here is the moment of time when the position of the planet is $\boldsymbol{r}_0$. Now we have the sixth integral that gives the zero point of time. If $\boldsymbol{r}_0$ is the position of the planet at perihelion, τ gives the time of the perihelion passage, and it is then called the *perihelion time*.

3.4 Equation of the orbit and Kepler's first law

We have shown that the orbit lies in the plane perpendicular to $\boldsymbol{k}$, but we still do not know the detailed geometry of the orbit. The equation for the orbit is obtained from the product $\boldsymbol{r} \cdot \boldsymbol{e}$, evaluated before. According to the definition of the scalar product we have

$$\boldsymbol{r} \cdot \boldsymbol{e} = re\cos\phi, \tag{3.19}$$

where ϕ is the angle between the vectors $\boldsymbol{r}$ and $\boldsymbol{e}$. Combining this with (3.16) we get

$$re\cos\phi = k^2/\mu - r$$

or

$$r = \frac{k^2/\mu}{1 + e\cos\phi}. \tag{3.20}$$

This is the equation of the orbit in polar coordinates. It gives the distance r of the planet from the Sun as a function of the angle ϕ measured from the direction of

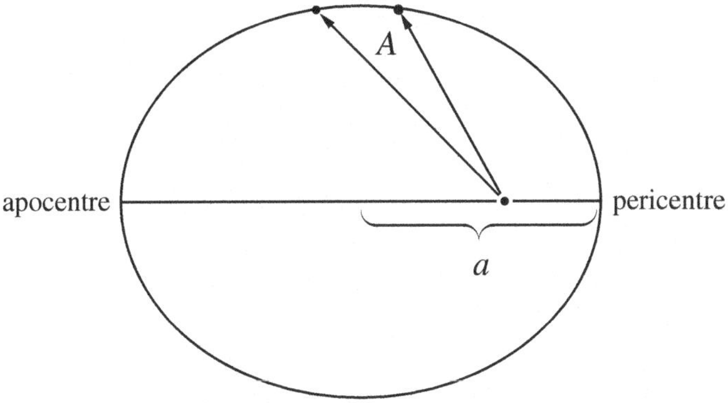

Figure 3.3 Surface velocity is the area A swept by the radius vector per unit time. In a two-body system the surface velocity is constant.

$\boldsymbol{e}$. Equation (3.20) is the general equation of a *conic section* when r is measured from a focus of the orbit. The eccentricity e of the orbit is the length of the vector $\boldsymbol{e}$, and the semi-latus rectum or parameter $p = k^2/\mu$. The angle ϕ is called the *true anomaly*.

Kepler's first law states that planets orbit the Sun along elliptic orbits, and the Sun is in one of the two foci of the ellipse. We have now shown that this follows from Newton's laws, even in a more general form: the orbit can be any conic section, an ellipse, parabola, or hyperbola.

3.5 Kepler's second law

In polar coordinates (r, ϕ) the velocity vector of the planet can be expressed as

$$\dot{\boldsymbol{r}} = \dot{r}\hat{\boldsymbol{e}}_r + r\dot{\phi}\hat{\boldsymbol{e}}_\phi, \tag{3.21}$$

where $\hat{\boldsymbol{e}}_r$ and $\hat{\boldsymbol{e}}_\phi$ are unit vectors parallel with and perpendicular to the radius vector, respectively. In polar coordinates the angular momentum $\boldsymbol{k}$ is

$$\boldsymbol{k} = r\hat{\boldsymbol{e}}_r \times (\dot{r}\hat{\boldsymbol{e}}_r + r\dot{\phi}\hat{\boldsymbol{e}}_\phi) = r^2\dot{\phi}\hat{\boldsymbol{e}}_z, \tag{3.22}$$

where $\hat{\boldsymbol{e}}_z$ is a unit vector perpendicular to the orbital plane. This shows that the length of $\boldsymbol{k}$ must be $r^2\dot{\phi}$. But the area swept by the radius vector in time $\mathrm{d}t$ is

$$\mathrm{d}A = \int_0^r r\,\mathrm{d}r\,\mathrm{d}\phi = \frac{1}{2}r^2\,\mathrm{d}\phi,$$

and thus the *surface velocity* $\dot{A}$ (Fig. 3.3) is

$$\dot{A} = \frac{1}{2}r^2\dot{\phi}.$$

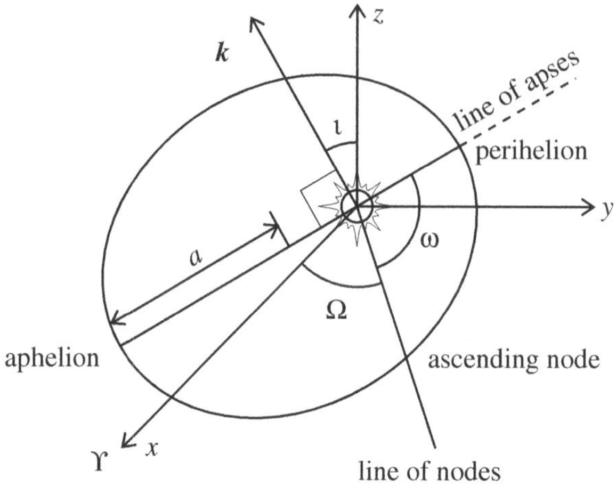

Figure 3.4 The orbital elements of an elliptic orbit. Note that we have two planes which intersect along the line of nodes. The normal of one of the planes is the z axis while the second plane (the orbital plane of the planet) is inclined to it by an angle ι.

Comparing this with the length of $\boldsymbol{k}$ we see that

$$\dot{A} = \frac{1}{2}k = \text{constant}. \tag{3.23}$$

Thus we have arrived at Kepler's second law which says that the surface velocity of the planet is constant.

Kepler's second law holds even in a much more general case. Since $\boldsymbol{r} \times \ddot{\boldsymbol{r}}$ vanishes for all central forces (for which $\ddot{\boldsymbol{r}}$ is parallel to $\boldsymbol{r}$), we can see from (3.10) that the angular momentum remains constant, and the motion obeys the second law.

3.6 Orbital elements

The integration constants derived above fully determine the orbit. They are convenient when studying the physics of the motion but they are not the best ones for example for computing the position of the planet. The constants may be chosen in many ways. For numerical calculations, the radius and velocity vectors at a given instant of time are often the most useful quantities. In the following we will introduce a third possibility, the conventional set of orbital elements, which describe the geometry of the orbit (Fig. 3.4).

Let us adopt a fixed rectangular heliocentric frame, and make the xy plane coincide with the orbital plane of the Earth, the ecliptic.

Since $\boldsymbol{k}$ is perpendicular to the orbital plane of the planet, the ratios of its components fully determine the orientation of the orbital plane. The same information is contained in two angles. We choose one of these angles to be the angle between $\boldsymbol{k}$ and the positive z axis and call it the *inclination* ι. (Inclination is often denoted by i. Here, however, ι is preferred in order to avoid the awkward notation $\dot{i}$ for the time derivative.) The second one is the angle between the positive x axis and the *line of nodes*. The line of nodes is defined as the intersection of the xy plane and orbital plane. The *nodes* are the two points where the planet crosses the xy plane. As a reference point we choose the node where the planet crosses the xy plane from below, i.e. where the z coordinate becomes positive. This angle is called the *longitude of the ascending node* Ω.

The vector $\boldsymbol{e}$ lies in the orbital plane of the planet. From the equation of the conic section we see that the distance r attains its minimum when $\phi = 0$, and hence ϕ gives the angular distance from the perihelion. Thus the vector $\boldsymbol{e}$ points to the direction of the perihelion. The direction of $\boldsymbol{e}$ may also be determined by specifying the angle between the perihelion and some fixed direction. If the latter angle is measured from the ascending node, it is called the *argument of perihelion* ω.

Another frequently used quantity is the *longitude of perihelion* ϖ, defined as

$$\varpi = \Omega + \omega. \tag{3.24}$$

This is measured partly along the ecliptic, partly along the orbital plane. If the inclination is zero, the direction of the ascending node becomes indeterminate, but the longitude of the perihelion remains a well defined quantity.

The length of the vector $\boldsymbol{e}$ is simply the *eccentricity* of the orbit.

The equation of a conic section is

$$r = \frac{p}{1 + e\cos\phi}, \tag{3.25}$$

where the *semi-latus rectum* or *parameter* p of elliptic and hyperbolic orbits can be expressed in terms of the semi-major axis a and eccentricity e:

$$p = a|1 - e^2|. \tag{3.26}$$

We saw previously that $p = k^2/\mu$, and thus the semi-major axis is

$$a = \frac{k^2}{\mu|1 - e^2|}. \tag{3.27}$$

This gives the angular momentum as a function of the orbital elements:

$$k = \sqrt{a\mu|1 - e^2|}. \tag{3.28}$$

An important relation between the size of the orbit and the energy integral h is obtained by expressing k in terms of h according to (3.17):

$$k^2 = \mu^2(e^2 - 1)/2h. \tag{3.29}$$

Substitution into (3.27) gives

$$a = \frac{\mu(e^2 - 1)}{2h|1 - e^2|} \tag{3.30}$$

or

$$a = \begin{cases} -\mu/2h & \text{if } 0 \leq e < 1 \text{ (ellipse)} \\ \mu/2h & \text{if } e > 1 \text{ (hyperbola).} \end{cases} \tag{3.31}$$

Because a must be positive, we see immediately that h is negative for elliptic orbits and positive for hyperbolic orbits. A parabolic orbit is a limiting case between these, its energy integral being zero.

The five elements Ω, ω, ι, a and e determine the position, size and shape of the orbit. Again, it remains to fix the position of the planet. For this purpose we can use the previously defined time of perihelion τ.

In the Jacobi representation of the three-body problem, the system is divided into a binary orbit and a third-body orbit around the centre of mass of the binary. These are instantaneous or *osculating* two-body orbits; even though both the inner and the outer orbits are described by six orbital elements, these elements do not remain fixed during the evolution of the system. However, it is useful to specify the current state of the system using the osculating orbital elements which would remain constant in purely two-body situations.

Because of the scale-free property of the three-body orbit solutions, one may always choose the semi-major axis of the inner binary as the distance unit. Similarly, the binary mass $m_1 + m_2$ or, alternatively, μ may be put equal to unity. Also, one may choose the orbital plane of the binary as the plane of reference; then the two angles describing the orientation of this plane, Ω and ι for the inner binary are not needed. Then we are left with only three orbital elements for the inner binary, as well as the mass ratio m_1/m_2 of the binary components, as the free parameters of the problem. The outer orbit requires the full complement of six orbital elements plus the third-body mass. Thus the number of free parameters is 11 as it should be in the three-body problem (Section 2.12). Figure 3.5 illustrates the two two-body orbits in their osculating planes A and B.

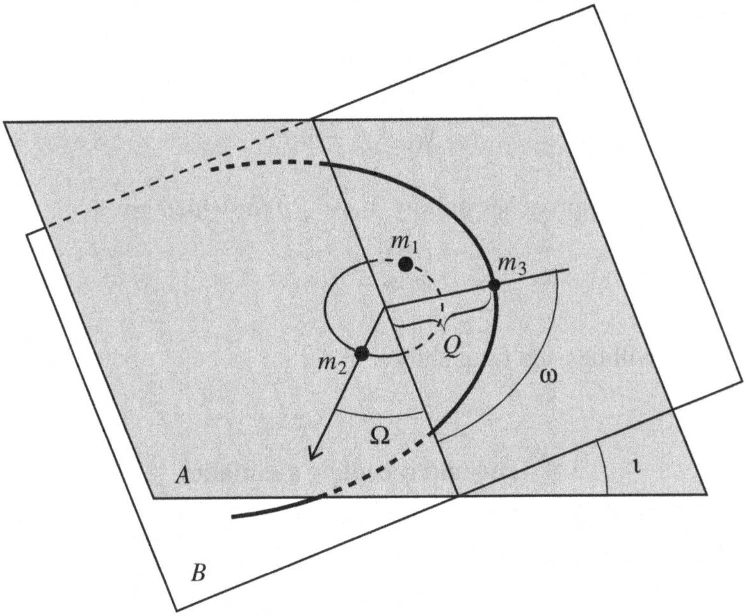

Figure 3.5 The osculating orbital planes A and B for inner and outer binary orbits, respectively, in the Jacobi representation of the three-body problem. The binary has components of masses m_1 and m_2, and the third-body mass m_3. The two planes are inclined by angle ι and the line of nodes is at the angle Ω with respect to a fixed direction in plane A. The pericentre of the outer orbit is at the angle ω with respect to the line of nodes. The pericentre distance of the outer orbit is Q. The pericentre distance of the outer orbit is usually normalised to the semi-major axis of the inner binary.

3.7 Orbital velocity

Assume first that the orbit is an ellipse. From the expressions of the energy integral

$$h = \frac{1}{2}v^2 - \frac{\mu}{r},$$
$$h = -\frac{\mu}{2a},$$

we get

$$-\frac{\mu}{2a} = \frac{1}{2}v^2 - \frac{\mu}{r},$$

from which we can solve the velocity

$$v = \sqrt{\mu\left(\frac{2}{r} - \frac{1}{a}\right)}. \tag{3.32}$$

For a hyperbolic orbit $h = \mu/2a$, and thus

$$v = \sqrt{\mu\left(\frac{2}{r} + \frac{1}{a}\right)}. \tag{3.33}$$

When the distance r approaches infinity, $v \to \sqrt{\mu/a}$, which gives

$$a = \frac{\mu}{v^2}, \tag{3.34}$$

a relation which will be used frequently.

3.8 True and eccentric anomalies

We now know that the planet is in perihelion at time τ. Next we will study how to use the orbital elements to find the position of the planet for an arbitrary instant of time.

The equation of an ellipse can be written as

$$r = \frac{a(1 - e^2)}{1 + e\cos\phi}, \tag{3.35}$$

where ϕ is the true anomaly of the planet. The problem is to find ϕ as a function of time.

We begin this by defining an auxiliary quantity, the *eccentric anomaly* E, as the angle $\angle QCP'$ of Fig. 3.6. We then derive transformation equations between ϕ and E. From Fig. 3.6 we get the following equations for various distances:

$$\begin{aligned}
|CQ| &= |CF| + |FQ|, \\
|CF| &= ae, \\
|CQ| &= a\cos E, \\
|FQ| &= r\cos\phi.
\end{aligned}$$

From these we obtain the length of FQ in two different ways:

$$|FQ| = r\cos\phi = a\cos E - ae$$

or

$$\frac{a(1 - e^2)}{1 + e\cos\phi}\cos\phi = a\cos E - ae.$$

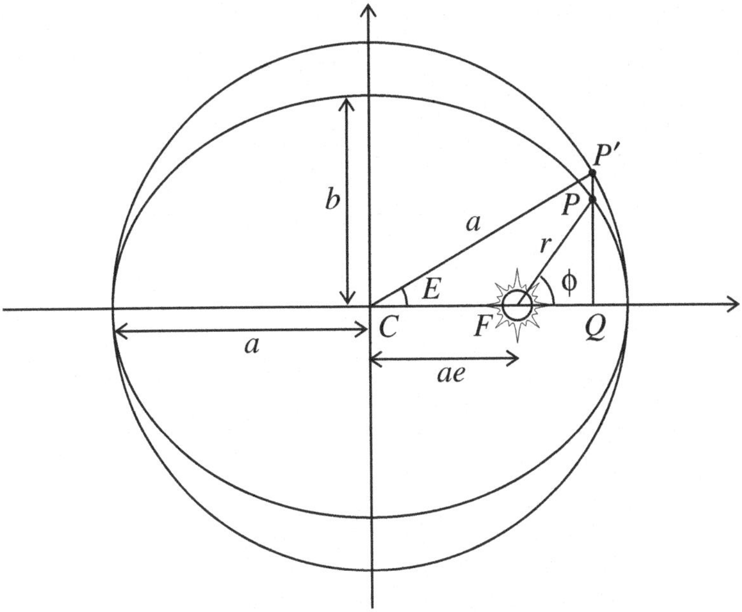

Figure 3.6 Definitions of the true anomaly ϕ and eccentric anomaly E. Point P' lies on a circle centered on C while P is the corresponding point on an ellipse which is obtained from the circle by contraction in the vertical direction by a factor b/a.

From this we can solve for either $\cos\phi$ or $\cos E$:

$$\cos\phi = \frac{\cos E - e}{1 - e\cos E},$$
$$\cos E = \frac{\cos\phi + e}{1 + e\cos\phi}.$$

To get an unambiguous relation between ϕ and E we also need expressions for $\sin\phi$ and $\sin E$. These are obtained by expressing $|QP|$ in two different ways:

$$|QP| = \frac{b}{a}|QP'| = \frac{b}{a}(a\sin E)$$
$$= a\sqrt{1-e^2}\sin E = r\sin\phi = \frac{a(1-e^2)}{1+e\cos\phi}\sin\phi.$$

This gives

$$\sin E = \sqrt{1-e^2}\,\frac{\sin\phi}{1+e\cos\phi}$$

and using the previous expression of $\cos\phi$ in terms of E:

$$\sin\phi = \sqrt{1-e^2}\,\frac{\sin E}{1-e\cos E}.$$

Thus we have the following equations for the angles E and ϕ:

$$\begin{aligned}\sin\phi &= \sqrt{1-e^2}\,\frac{\sin E}{1-e\cos E},\\ \cos\phi &= \frac{\cos E - e}{1-e\cos E},\end{aligned} \tag{3.36}$$

$$\begin{aligned}\sin E &= \sqrt{1-e^2}\,\frac{\sin\phi}{1+e\cos\phi},\\ \cos E &= \frac{\cos\phi + e}{1+e\cos\phi}.\end{aligned} \tag{3.37}$$

3.9 Mean anomaly and Kepler's equation

We continue by defining yet another anomaly. The *mean anomaly* M is defined as the angle between the perihelion and the radius vector assuming that the planet moves at a constant angular velocity:

$$M = 2\pi\frac{t-\tau}{P} = n(t-\tau). \tag{3.38}$$

Here P is the orbital period of the planet and $n = 2\pi/P$ is called the *mean motion*, i.e. the mean angular velocity. According to Kepler's second law the surface velocity is constant. Thus the area of the shaded region in Fig. 3.7 is

$$A = \pi ab\frac{t-\tau}{P}. \tag{3.39}$$

But this area is also

$$\begin{aligned} A &= \frac{b}{a}(\text{area of } FP'X)\\ &= \frac{b}{a}(\text{area of the sector } CP'X - \text{area of the triangle } CP'F)\\ &= \frac{b}{a}\left(\frac{1}{2}a(aE) - \frac{1}{2}(ae)(a\sin E)\right)\\ &= \frac{1}{2}\frac{b}{a}(a^2E - a^2e\sin E)\\ &= \frac{1}{2}ab(E - e\sin E). \end{aligned} \tag{3.40}$$

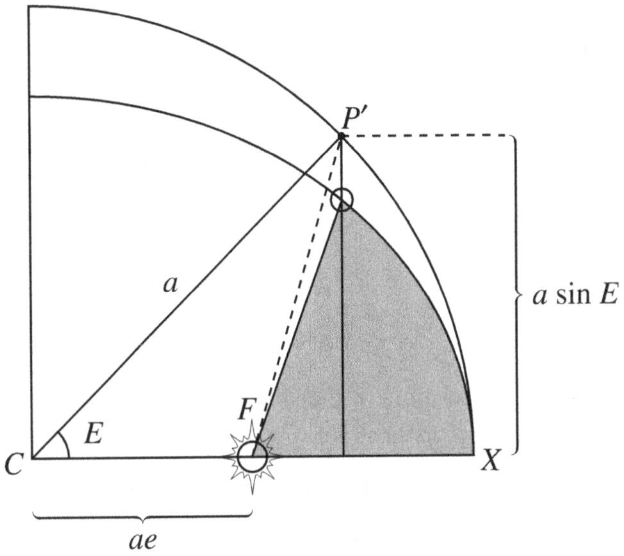

Figure 3.7 To derive Kepler's equation the shaded area is expressed in terms of the mean and eccentric anomalies.

The two expressions (3.39) and (3.40) of the area give the equation

$$A = \frac{1}{2}ab(E - e\sin E) = \pi ab\frac{t-\tau}{P} = \frac{1}{2}abn(t-\tau)$$
$$= \frac{1}{2}abM,$$

or

$$E - e\sin E = M. \tag{3.41}$$

This is called *Kepler's equation*. It gives the connection between the eccentric anomaly and the mean anomaly which increases at a constant rate with time.

3.10 Solution of Kepler's equation

The mean anomaly is easily obtained since it is directly proportional to time. The eccentric anomaly must be solved from the transcendental Kepler equation. Because of the nature of the equation there is no finite expression giving the position as a function of time. The solution must either be found numerically or be evaluated from a series expansion. Next we give a simple method for solving the equation by iteration. Series expansions will be studied later in Section 3.15.

Unless the eccentricity is very high the eccentric anomaly is not very different from the mean anomaly. Therefore, we write it as

$$E = M + x,$$

where x is a small correction. Substituting this into Kepler's equation we get

$$M + x - e\sin(M + x) = M,$$

or

$$x = e\sin(M + x). \tag{3.42}$$

We now apply the addition formula of the sines and replace $\sin x$ and $\cos x$ by the first terms of their Taylor expansions ($\sin x \approx x$, $\cos x \approx 1$):

$$\begin{aligned} x &= e(\cos M \sin x + \sin M \cos x) \\ &\approx ex\cos M + e\sin M. \end{aligned}$$

From this we can solve the first approximation for x:

$$x^{(1)} = \frac{e\sin M}{1 - e\cos M}.$$

We substitute this in (3.42), and solve for an improved approximation for x:

$$x^{(2)} = e\sin\left(M + x^{(1)}\right).$$

We continue this iteration until the consecutive values of x do not change more than the required accuracy. The whole algorithm can be summarized as follows:

$$\begin{aligned} x^{(1)} &= \frac{e\sin M}{1 - e\cos M}, \\ x^{(2)} &= e\sin\left(M + x^{(1)}\right), \\ x^{(3)} &= e\sin\left(M + x^{(2)}\right), \\ &\vdots \\ x^{(n)} &= e\sin\left(M + x^{(n-1)}\right). \end{aligned} \tag{3.43}$$

The iteration can be terminated when $\left|x^{(n)} - x^{(n-1)}\right| < \epsilon$, where ϵ is the accuracy needed. It is absolutely necessary to express the angles in radians, otherwise the result is nonsense. For small eccentricities this method converges very rapidly, and the values do not change noticeably after a couple of iterations.

Example 3.1 The mean motion of asteroid (1221) Amor is $n = 0.371°/\mathrm{d}$ and eccentricity $e = 0.4346$. Find its mean, eccentric and true anomalies 300 days after the perihelion passage.

The orbital period is

$$P = \frac{360^\circ}{0.371^\circ/\mathrm{d}} = 970.35\ \mathrm{d},$$

and thus the mean anomaly is

$$M = 360^\circ \frac{300}{970.35} = 111.30^\circ = 1.9426\ \mathrm{rad}.$$

Next we solve the eccentric anomaly from Kepler's equation. Since the eccentricity is relatively large, the convergence is not very rapid:

$$\begin{aligned}
x^{(1)} &= \frac{e \sin M}{1 - e \cos M} = 0.3497,\\
x^{(2)} &= e \sin\left(M + x^{(1)}\right) = 0.3263,\\
x^{(3)} &= e \sin\left(M + x^{(2)}\right) = 0.3329,\\
&\vdots\\
x^{(7)} &= e \sin\left(M + x^{(6)}\right) = 0.3315.
\end{aligned}$$

After this the values do not change, and thus the solution is

$$E = M + x^{(6)} = 2.2740\ \mathrm{rad} = 130.29^\circ.$$

To find the true anomaly we evaluate

$$\begin{aligned}
\sin\phi &= \sqrt{1 - e^2}\frac{\sin E}{1 - e \cos E} = 0.5362,\\
\cos\phi &= \frac{\cos E - e}{1 - e \cos E} = -0.8441,
\end{aligned}$$

which give

$$\phi = 2.5756\ \mathrm{rad} = 147.57^\circ.$$

Note that due to the large eccentricity of its orbit the asteroid has moved over a much greater angle than the direction given by the mean anomaly.

3.11 Kepler's third law

The last problem is to find the mean anomaly, which requires that we know the orbital period P. The surface velocity law can be expressed in the form

$$\mathrm{d}A = \frac{1}{2} k\, \mathrm{d}t.$$

We integrate this over one orbital period:

$$\int_{\text{orbital ellipse}} \mathrm{d}A = \frac{1}{2}k\int_0^P \mathrm{d}t.$$

Here the left hand side is the area of the ellipse, which in terms of the orbital elements is $\pi a^2\sqrt{1-e^2}$. Thus we get

$$\pi a^2\sqrt{1-e^2} = \frac{1}{2}kP.$$

Substituting the expression (3.28) for k we get

$$P = \frac{2\pi}{\sqrt{\mu}}a^{3/2}. \tag{3.44}$$

This is the exact form of Kepler's third law. Note that it depends also on the mass of the planet. In the case of the planets of the solar system this effect is rather small, and thus it remained unnoticed by Kepler.

From the third law we get an expression for the mean motion:

$$n = \frac{2\pi}{P} = \sqrt{\mu}a^{-3/2}. \tag{3.45}$$

3.12 Position and speed as functions of eccentric anomaly

We set up a rectangular $\xi\eta$ coordinate frame in the orbital plane of the planet in such a way that the origin is in the Sun and the ξ axis points to the perihelion (Fig. 3.8). The unit vectors parallel to the coordinate axes are denoted by $\hat{\boldsymbol{e}}_\xi$ and $\hat{\boldsymbol{e}}_\eta$. The radius vector of the planet is then

$$\boldsymbol{r} = a(\cos E - e)\hat{\boldsymbol{e}}_\xi + b\sin E\hat{\boldsymbol{e}}_\eta. \tag{3.46}$$

The time derivative of this gives the velocity

$$\dot{\boldsymbol{r}} = -a\dot{E}\sin E\hat{\boldsymbol{e}}_\xi + b\dot{E}\cos E\hat{\boldsymbol{e}}_\eta. \tag{3.47}$$

To find the time derivative of E we take Kepler's equation in the form

$$E - e\sin E = \sqrt{\mu}a^{-3/2}(t-\tau)$$

and differentiate with respect to time:

$$\dot{E}(1 - e\cos E) = \sqrt{\mu}a^{-3/2}.$$

This gives

$$\dot{E} = \frac{\sqrt{\mu}a^{-3/2}}{1 - e\cos E}. \tag{3.48}$$

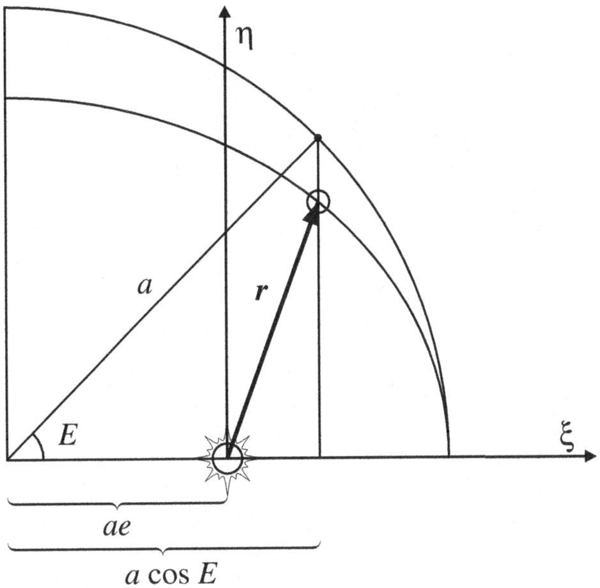

Figure 3.8 The position of the planet in terms of the eccentric anomaly E is $\boldsymbol{r} = a(\cos E - e)\hat{\boldsymbol{e}}_\xi + b \sin E \hat{\boldsymbol{e}}_\eta$.

Using the expressions (3.46) and (3.47) we obtain the lengths of the radius and velocity vectors:

$$\begin{aligned} r^2 = \boldsymbol{r} \cdot \boldsymbol{r} &= a^2(\cos E - e)^2 + b^2 \sin^2 E \\ &= a^2(\cos^2 E - 2e \cos E + e^2 + (1 - e^2) \sin^2 E) \\ &= a^2(1 - 2e \cos E + e^2 \cos^2 E) \\ &= a^2(1 - e \cos E)^2, \end{aligned}$$

from which

$$r = a(1 - e \cos E). \tag{3.49}$$

The speed is calculated in a similar manner:

$$\begin{aligned} v^2 = \dot{\boldsymbol{r}} \cdot \dot{\boldsymbol{r}} &= a^2 \dot{E}^2 (\sin^2 E + (1 - e^2) \cos^2 E) \\ &= \frac{a^2 \mu a^{-3}(1 - e^2 \cos^2 E)}{(1 - e \cos E)^2} \\ &= \frac{\mu}{a} \frac{1 + e \cos E}{1 - e \cos E}, \end{aligned}$$

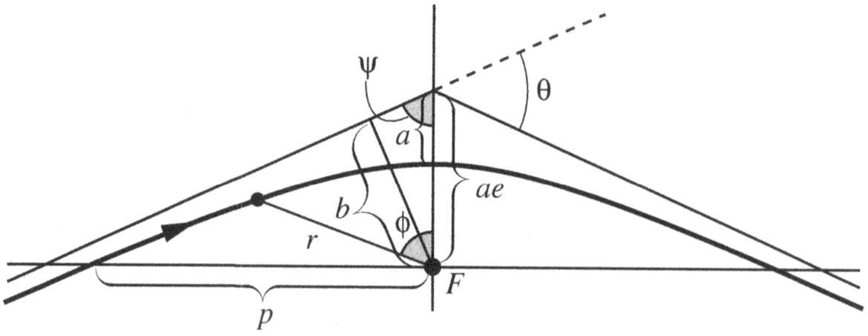

Figure 3.9 A planet (or a comet) approaches the Sun at F from a large distance. The impact distance is b, i.e. the distance by which the planet would miss the Sun at its initial (nearly) straight line orbit. This straight line makes an angle ψ with the major axis (vertical line). The semi-latus rectum p is also marked, as well as the scattering angle θ.

whence

$$v = \sqrt{\frac{\mu}{a}\frac{1 + e\cos E}{1 - e\cos E}}. \tag{3.50}$$

3.13 Hyperbolic orbit

Figure 3.9 illustrates a hyperbolic orbit about the focal point F. The equation of the orbit is (Eq. (3.25))

$$r = \frac{p}{1 + e\cos\phi}. \tag{3.51}$$

If we put $r = \infty$, it follows that $1 + e\cos\phi = 0$ and the asymptotic true anomaly $\phi = \pi - \psi$, $\cos\phi = -\cos\psi$, and

$$\psi = \pm\arccos\left(\frac{1}{e}\right). \tag{3.52}$$

Therefore the incoming and outgoing directions are at the angle ψ relative to the major axis.

Without the orbital curvature the orbit would pass the focus at the minimum distance of

$$ae\sin\psi = a\sqrt{e^2 - 1} = b. \tag{3.53}$$

This is called the *impact parameter* of the orbit. We easily see that

$$b^2 = ap \tag{3.54}$$

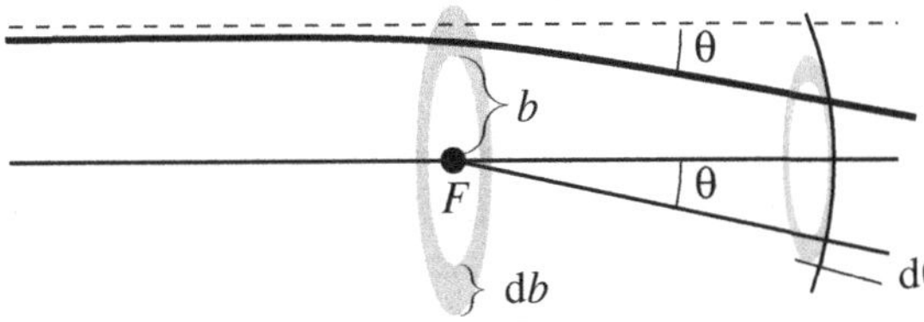

Figure 3.10 A particle approaches a body at F initially along a horizontal path which is subsequently scattered by an angle θ. Orbits whose impact distances lie within the ring of width db are scattered into the solid angle between θ and $\theta + \mathrm{d}\theta$.

where (Eq. (3.26))

$$p = a(e^2 - 1) \tag{3.55}$$

is the semi-latus rectum of the orbit, also called the parameter. Because of the curvature, the incoming and outgoing asymptotic directions differ by an angle

$$\theta = \pi - 2\psi \tag{3.56}$$

which is called the *scattering angle*. Since

$$\sin\frac{\theta}{2} = \sin\left(\frac{\pi}{2} - \psi\right) = \cos\psi = \frac{1}{e},$$

it follows that

$$b^2 = a^2(e^2 - 1) = a^2 \cot^2\frac{\theta}{2}. \tag{3.57}$$

Consider a steady uniform stream of particles approaching the body at point F. What is the probability that a particle is scattered to the solid angle dΩ between θ and $\theta + \mathrm{d}\theta$? In order that particles scatter into this interval, their impact parameters must lie within a corresponding interval b, $b + \mathrm{d}b$ (see Fig. 3.10). To find out how many particles scatter into this solid angle we define the cross-section $\sigma(\theta)\,\mathrm{d}\Omega$ for scattering to the solid angle

$$\mathrm{d}\Omega = 2\pi \sin\theta\, \mathrm{d}\theta$$

as the area of the 'impact ring' $2\pi b\, \mathrm{d}b$:

$$\sigma(\theta) 2\pi \sin\theta\, \mathrm{d}\theta = -2\pi\, b\, \mathrm{d}b. \tag{3.58}$$

The minus sign tells us that increasing b results in smaller θ.

From Eq. (3.57) we get

$$2\pi b\, \mathrm{d}b = 2\pi a^2 \cot\frac{\theta}{2} \mathrm{d}\left(\cot\frac{\theta}{2}\right) = -2\pi \frac{a^2}{4} \frac{\sin\theta\, \mathrm{d}\theta}{\sin^4(\theta/2)}.$$

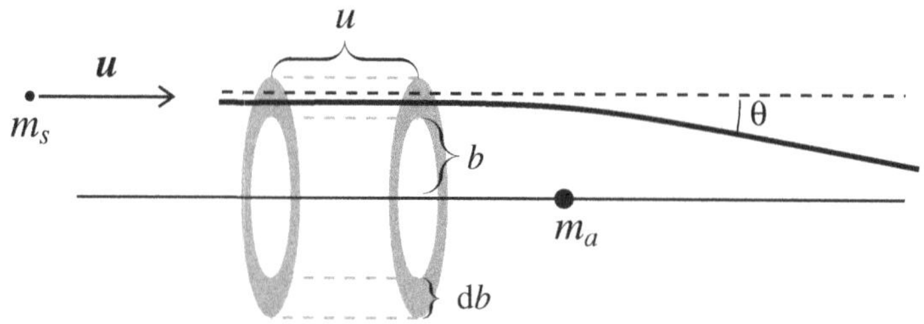

Figure 3.11 Calculating dynamical friction. Small bodies of mass m_s stream past a heavy body of mass m_a with speed u. In unit time the heavy body is influenced by all the bodies in the volume $2\pi b\, \mathrm{d}b\, u$ which are in the impact distance range $[b, b + \mathrm{d}b]$. The number of such bodies is $2\pi b\, \mathrm{d}b\, un$, if n is the number density in the stream.

Therefore

$$\sigma(\theta) = \frac{a^2}{4} \frac{1}{\sin^4(\theta/2)}. \tag{3.59}$$

This is called the *Rutherford differential cross-section* since it was originally derived by Rutherford in connection with scattering of alpha particles by atomic nuclei.

3.14 Dynamical friction

Consider a heavy body moving through a field of light bodies. There is a succession of hyperbolic two-body encounters the net effect of which is to slow down the motion of the heavy body. This slowing down may be described as dynamical friction. In the following we show how dynamical friction is calculated using a succession of hyperbolic two-body encounters (Chandrasekhar 1942, Binney and Tremaine 1987). Let a light body of mass m_s approach a heavy body of mass m_a with the relative speed u at infinity and with impact parameter b. According to Eq. (3.34) the semi-major axis of the orbit is $a = G(m_a + m_s)/u^2$. After the scattering by angle θ there is a change in the velocity of the mass m_a, in the direction parallel to the line of approach, by

$$\Delta V_{||} = \frac{m_s}{m_s + m_a} u(1 - \cos\theta) \tag{3.60}$$

(see Fig. 3.11). In unit time, the heavy body samples a volume of length u and surface area $2\pi b\, \mathrm{d}b$. If n is the number density of light bodies, the heavy body meets

$$2\pi bun\, \mathrm{d}b = \sigma(\theta) 2\pi un \sin\theta\, \mathrm{d}\theta \tag{3.61}$$

light bodies in the impact distance range b, $b + \mathrm{d}b$. Therefore the total change in the speed of the heavy body parallel to its motion through the field of light bodies per time interval Δt is:

$$\frac{\Delta V_{\|}}{\Delta t} = 2\pi \frac{m_s}{m_s + m_a} u^2 n \frac{a^2}{4} \int_{\theta_{\min}}^{\pi} (1 - \cos\theta) \frac{1}{\sin^4(\theta/2)} \sin\theta \, \mathrm{d}\theta.$$

The integral is equal to

$$\int_{\theta_{\min}}^{\pi} 8 \cot\left(\frac{\theta}{2}\right) \mathrm{d}\left(\frac{\theta}{2}\right) = \Big|_{\theta_{\min}}^{\pi} 8 \ln\left(\sin\frac{\theta}{2}\right)$$
$$= -8 \ln\left(\frac{1}{\sqrt{1 + \cot^2(\theta_{\min}/2)}}\right).$$

The minimum deflection angle $\theta_{\min}$ is connected to the maximum impact distance $b_{\max}$:

$$b_{\max}^2 = a^2 \cot^2\left(\frac{\theta_{\min}}{2}\right). \tag{3.62}$$

The quantity $b_{\max}$ describes the extent of the medium of light bodies (at the density n), and it is usually possible to give a good estimate for it. Therefore the integral becomes

$$-8 \ln\left(\frac{1}{\sqrt{1 + \dfrac{b_{\max}^2 u^4}{[G(m_a + m_s)]^2}}}\right) = 4 \ln\left(1 + \frac{b_{\max}^2 u^4}{[G(m_a + m_s)]^2}\right)$$

and

$$-\frac{\mathrm{d}u}{\mathrm{d}t} = \frac{\Delta V_{\|}}{\Delta t} = 2\pi \frac{G^2 m_s (m_a + m_s)}{u^2} n \ln\left(1 + \frac{b_{\max}^2 u^4}{[G(m_a + m_s)]^2}\right).$$

The changes of velocity $\Delta V_{\perp}$ perpendicular to the flow cancel out since passages from the opposite sides of the heavy body give contributions of opposite signs. Then the velocity change is purely parallel to the motion and may be written in vector form

$$\frac{\mathrm{d}\boldsymbol{u}}{\mathrm{d}t} = -4\pi G^2 m_s m_a n \ln\Lambda \cdot \frac{\boldsymbol{u}}{u^3} \tag{3.63}$$

where

$$\Lambda = \frac{b_{\max} u^2}{G m_a} \tag{3.64}$$

if m_s is negligible in comparison with m_a. Since generally $\Lambda^2 \gg 1$ we have replaced $\ln(1 + \Lambda^2)$ by $2 \ln \Lambda$. Equation (3.63) is the basic equation for dynamical friction.

3.15 Series expansions

Since planetary orbits are nearly circular, we can approximate their orbits as circles modified with small corrections. We now study briefly how some quantities related to elliptic orbits can be expressed as series expansions. The derivations given here follow Kovalevsky (1967), which gives more such expansions.

Because the eccentricities are usually small, it is practical to express these quantities as power series of the eccentricity. The calculations are not usually very involved mathematically but may be somewhat laborious. Computer programs for symbolic manipulation can be very useful in this work.

3.15.1 Series expansion of the eccentric anomaly

When computing the position in the orbit we have to solve the transcendental equation $M = E - e \sin E$. If the eccentricity is small, the difference $E - M = e \sin E$ will also be small. We will now derive a series expansion for this difference in terms of the mean anomaly.

We begin by developing $e \sin E$ as a Fourier series with respect to M. The constant term of the series is

$$a_0 = \frac{1}{2\pi} \int_0^{2\pi} e \sin E \, \mathrm{d}M, \tag{3.65}$$

where $\mathrm{d}M$ is obtained by differentiating Kepler's equation:

$$\mathrm{d}M = (1 - e \cos E) \, \mathrm{d}E. \tag{3.66}$$

We take the eccentric anomaly as the new integration variable. This will not change the integration limits. Thus the constant term is

$$\begin{aligned} a_0 &= \frac{1}{2\pi} \int_0^{2\pi} e \sin E (1 - e \cos E) \, \mathrm{d}E \\ &= \frac{1}{2\pi} \int_0^{2\pi} e \sin E \, \mathrm{d}E - \frac{1}{2\pi} \int_0^{2\pi} e^2 \sin E \cos E \, \mathrm{d}E = 0. \end{aligned} \tag{3.67}$$

This shows that over a long period of time the mean and eccentric anomalies are the same on average, as they should be according to their definitions.

Coefficients of the cosine terms are

$$\begin{aligned} a_k &= \frac{1}{\pi}\int_0^{2\pi} e\sin E\cos kM\,\mathrm{d}M \\ &= \frac{1}{\pi}\int_{-\pi}^{\pi} e\sin E\cos kM\,\mathrm{d}M \\ &= 0, \end{aligned} \tag{3.68}$$

since the integrand is an odd function. Thus the only remaining terms are the sine terms

$$b_k = \frac{1}{\pi}\int_0^{2\pi} e\sin E\sin kM\,\mathrm{d}M. \tag{3.69}$$

This can be integrated by parts. In the formula

$$\int u\,\mathrm{d}v = uv - \int v\,\mathrm{d}u$$

we set

$$\begin{aligned} u &= e\sin E, \\ \mathrm{d}v &= \sin kM\,\mathrm{d}M, \end{aligned}$$

which gives

$$\begin{aligned} \mathrm{d}u &= e\cos E\,\mathrm{d}E, \\ v &= -\frac{1}{k}\cos kM = -\frac{1}{k}\cos k(E - e\sin E). \end{aligned}$$

The expression of the coefficient b_k is then

$$\begin{aligned} b_k = &-\frac{1}{\pi k}\bigg|_0^{2\pi}(e\sin E\cos k(E - e\sin E)) \\ &+\frac{1}{\pi k}\int_0^{2\pi}\cos k(E - e\sin E)e\cos E\,\mathrm{d}E. \end{aligned}$$

The substitution term is obviously zero, and to the integrand we can apply the formula

$$\cos x\cos y = \frac{1}{2}(\cos(x+y) + \cos(x-y)),$$

to get

$$\begin{aligned} b_k &= \frac{e}{2\pi k}\int_0^{2\pi} \cos[(k+1)E - ke\sin E]\,\mathrm{d}E \\ &\quad + \frac{e}{2\pi k}\int_0^{2\pi} \cos[(k-1)E - ke\sin E]\,\mathrm{d}E \\ &= \frac{e}{k}(J_{k+1}(ke) + J_{k-1}(ke)), \end{aligned} \tag{3.70}$$

where $J_k(x)$ is the Bessel function

$$J_k(x) = \frac{1}{2\pi}\int_0^{2\pi} \cos(kt - x\sin t)\,\mathrm{d}t. \tag{3.71}$$

Using the property of the Bessel functions

$$J_k(x) = \frac{x}{2k}\,[J_{k-1}(x) + J_{k+1}(x)] \tag{3.72}$$

we have

$$b_k = \frac{2}{k}J_k(ke). \tag{3.73}$$

Thus the eccentric anomaly is

$$E = M + \sum_{k=1}^{\infty} \frac{2J_k(ke)}{k}\sin kM. \tag{3.74}$$

The Bessel functions can be evaluated from the following series expansions (see e.g. Arfken 1970):

$$\begin{aligned} J_0(x) &= 1 - \left(\frac{x}{2}\right)^2 + \frac{1}{4}\left(\frac{x}{2}\right)^4 - \cdots + \frac{(-1)^n}{(n!)^2}\left(\frac{x}{2}\right)^{2n} + \cdots, \\ J_k(x) &= \left(\frac{x}{2}\right)^k \frac{1}{k!}\left[1 - \frac{1}{k+1}\left(\frac{x}{2}\right)^2 + \cdots + \right. \\ &\quad \left. \frac{(-1)^n}{n!(k+1)(k+2)\cdots(k+n)}\left(\frac{x}{2}\right)^{2n} + \cdots\right]. \end{aligned} \tag{3.75}$$

We see that x^k is the lowest power of x contained in $J_k(x)$. If we want to include terms up to the kth power of the eccentricity, it suffices to evaluate the coefficients $b_1, \ldots, b_k$.

The first few Bessel functions are

$$\begin{aligned}
J_0(x) &= 1 - \frac{x^2}{4} + \frac{x^4}{64} + \mathcal{O}(x^6),\\
J_1(x) &= \frac{x}{2} - \frac{x^3}{16} + \mathcal{O}(x^5),\\
J_2(x) &= \frac{x^2}{8} - \frac{x^4}{96} + \mathcal{O}(x^6),\\
J_3(x) &= \frac{x^3}{48} + \mathcal{O}(x^5),\\
J_4(x) &= \frac{x^4}{384} + \mathcal{O}(x^6),\\
J_5(x) &= \mathcal{O}(x^5).
\end{aligned} \tag{3.76}$$

Using these we get the following coefficients for the Fourier series of $E - M$

$$\begin{aligned}
b_1 &= 2J_1(e) = e - \frac{e^3}{8} + \mathcal{O}(e^5),\\
b_2 &= \frac{2J_2(2e)}{2} = \frac{e^2}{2} + \frac{e^4}{6} + \mathcal{O}(e^6),\\
b_3 &= \frac{2J_3(3e)}{3} = \frac{3e^3}{8} + \mathcal{O}(e^5),\\
b_4 &= \frac{2J_4(4e)}{4} = \frac{e^4}{3} + \mathcal{O}(e^6),\\
b_5 &= \frac{2J_5(5e)}{5} = \mathcal{O}(e^5).
\end{aligned} \tag{3.77}$$

Neglecting terms higher than e^3 we have the approximation

$$\begin{aligned}
E &= M + \left(e - \frac{e^3}{8}\right) \sin M + \frac{e^2}{2} \sin 2M + \frac{3e^3}{8} \sin 3M\\
&= M + e \sin M + \frac{e^2}{2} \sin 2M + \frac{e^3}{8}(-\sin M + 3 \sin 3M).
\end{aligned} \tag{3.78}$$

Note that the coefficients of this Fourier series are only approximate. Actually the coefficients are infinite power series. If greater accuracy is needed, we have to use more Fourier terms and also evaluate the coefficients by including higher powers of the eccentricity.

3.15.2 Series of $\sin nE$ and $\cos nE$

Next, we will derive an auxiliary result needed for other expansions. We consider first the Fourier series of $\cos nE$ in terms of M.

The constant term is

$$a_0 = \frac{1}{2\pi}\int_0^{2\pi} \cos nE \, \mathrm{d}M = \frac{1}{2\pi}\int_0^{2\pi} \cos nE(1 - e\cos E)\,\mathrm{d}E. \tag{3.79}$$

When $n > 1$, this integral vanishes. When $n = 1$, the term $e\cos^2 E$ will give a non-zero result. Thus the constant term is

$$a_0 = -\frac{e}{2}\delta_{n1}. \tag{3.80}$$

The coefficients of the sine terms are

$$b_k = \frac{1}{\pi}\int_0^{2\pi} \cos nE \sin kM \, \mathrm{d}M. \tag{3.81}$$

The integrand is odd as a product of odd and even functions. Hence all the coefficients b_k are zero.

The coefficients of the cosine terms are

$$\begin{aligned}
a_k &= \frac{1}{\pi}\int_0^{2\pi} \cos nE \cos kM \, \mathrm{d}M \\
&= \frac{1}{k\pi}\int_0^{2\pi} \cos nE \frac{\mathrm{d}\sin kM}{\mathrm{d}M}\,\mathrm{d}M \\
&= \frac{1}{k\pi}\Big|_0^{2\pi} \cos nE\ \sin kM - \frac{1}{k\pi}\int_0^{2\pi} \sin kM \frac{\mathrm{d}\cos nE}{\mathrm{d}M}\,\mathrm{d}M \\
&= \frac{n}{k\pi}\int_0^{2\pi} \sin nE \sin k(E - e\sin E)\,\mathrm{d}E \\
&= \frac{n}{2k\pi}\int_0^{2\pi} \cos((k-n)E - ke\sin E)\,\mathrm{d}E \\
&\quad - \frac{n}{2k\pi}\int_0^{2\pi} \cos((k+n)E - ke\sin E)\,\mathrm{d}E \\
&= \frac{n}{k}\left[J_{k-n}(ke) - J_{k+n}(ke)\right].
\end{aligned} \tag{3.82}$$

Thus we have the following expansion for $\cos nE$

$$\cos nE = -\frac{e}{2}\delta_{n1} + \sum_{k=1}^{\infty} \frac{n}{k}\left[J_{k-n}(ke) - J_{k+n}(ke)\right]\cos kM. \tag{3.83}$$

In the same manner we find

$$\sin nE = \sum_{k=1}^{\infty} \frac{n}{k} \left[J_{k-n}(ke) + J_{k+n}(ke) \right] \sin kM. \tag{3.84}$$

After some work we get the following approximate expressions by neglecting all terms proportional to the fourth and higher powers of the eccentricity:

$$\begin{aligned}
\cos E &= -\frac{e}{2} + \left(1 - \frac{3e^2}{8}\right) \cos M + \left(\frac{e}{2} - \frac{e^3}{2}\right) \cos 2M \\
&\quad + \frac{3e^2}{8} \cos 3M + \frac{e^3}{3} \cos 4M, \\
\cos 2E &= \left(-e + \frac{e^3}{12}\right) \cos M + (1 - e^2) \cos 2M \\
&\quad + \left(e - \frac{9e^3}{8}\right) \cos 3M + e^2 \cos 4M + \frac{25e^3}{24} \cos 5M, \\
\sin E &= \left(1 - \frac{e^2}{8}\right) \sin M + \left(\frac{e}{2} - \frac{e^3}{8}\right) \sin 2M \\
&\quad + \frac{3e^2}{8} \sin 3M + \frac{e^3}{3} \sin 4M, \\
\sin 2E &= \left(-e + \frac{e^3}{6}\right) \sin M + (1 - e^2) \sin 2M \\
&\quad + \left(e - \frac{9e^3}{8}\right) \sin 3M + e^2 \sin 4M + \frac{25e^3}{24} \sin 5M.
\end{aligned} \tag{3.85}$$

3.15.3 Distance as a function of time

Next we find a series for the length r of the radius vector as a function of the mean anomaly (and hence time). We know that

$$r = a(1 - e \cos E). \tag{3.86}$$

We can use here the previously found series for $\cos E$:

$$\begin{aligned}
\frac{r}{a} &= 1 - e \cos E \\
&= 1 + \frac{e^2}{2} - \sum_{k=1}^{\infty} \frac{e}{k} (J_{k-1}(ke) - J_{k+1}(ke)) \cos kM \\
&= 1 + \frac{e^2}{2} - \sum_{k=1}^{\infty} \frac{2e}{k^2} \frac{\mathrm{d} J_k(ke)}{\mathrm{d}e} \cos kM.
\end{aligned} \tag{3.87}$$

Keeping only the terms up to e^3 we get

$$\frac{r}{a} = 1 + \frac{e^2}{2} + \left(-e + \frac{3e^3}{8}\right)\cos M - \frac{e^2}{2}\cos 2M - \frac{3e^3}{8}\cos 3M. \tag{3.88}$$

The inverse value of the distance is also easily obtained:

$$\begin{aligned} \frac{a}{r} &= \frac{1}{1 - e\cos E} = \frac{1}{\mathrm{d}M/\mathrm{d}E} = \frac{\mathrm{d}E}{\mathrm{d}M} \\ &= 1 + \sum_{k=1}^{\infty} 2J_k(ke)\cos kM. \end{aligned} \tag{3.89}$$

The beginning of this series is

$$\frac{a}{r} = 1 + \left(e - \frac{e^3}{8}\right)\cos M + e^2\cos 2M + \frac{9e^3}{8}\cos 3M. \tag{3.90}$$

Using these results we can evaluate all powers of r. Since $(r/a)^n = (1 - e\cos E)^n$, we get first an expression containing powers of $\cos E$ up to $\cos^n E$. These can be reduced to expressions containing only cosines of E and its multiples. Finally, these can be expressed as series of M. For example

$$\begin{aligned} \left(\frac{r}{a}\right)^2 &= (1 - e\cos E)^2 = 1 - 2e\cos E + e^2\cos^2 E \\ &= 1 - 2e\cos E + \frac{e^2}{2}(1 + \cos 2E). \end{aligned} \tag{3.91}$$

Substituting the series of $\cos E$ and $\cos 2E$ we get

$$\begin{aligned} \left(\frac{r}{a}\right)^2 = 1 &+ \frac{3e^2}{2} + \left(-2e - \frac{5e^3}{4}\right)\cos M \\ &- \frac{e^2}{2}\cos 2M - \frac{e^3}{4}\cos 3M. \end{aligned} \tag{3.92}$$

3.15.4 Legendre polynomials

Legendre polynomials will be needed in later chapters. We will introduce them here, although they are not directly related to the two-body case. See e.g. Arfken (1970) for a thorough treatment of these polynomials and related functions.

Consider the situation of Fig. 3.12. The potential at P due to the mass at Q is proportional to $1/|\boldsymbol{r} - \boldsymbol{r}'|$. We want to express this in terms of the lengths r and r'

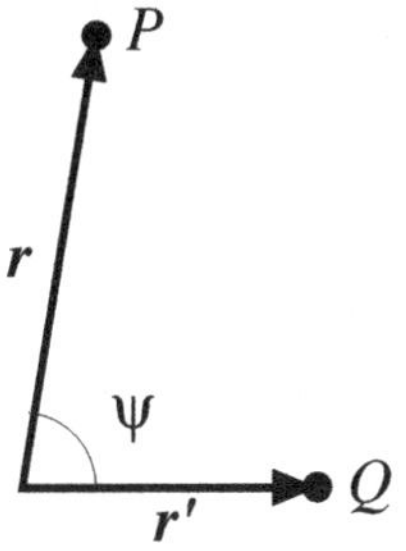

Figure 3.12 The potential at P due to the mass at Q can be expressed as a power series of $t = r'/r$.

and the angle ψ between the vectors $\boldsymbol{r}$ and $\boldsymbol{r}'$. From the cosine formula we get

$$\begin{aligned}\frac{1}{|\boldsymbol{r}-\boldsymbol{r}'|} &= (r^2 + r'^2 - 2rr'\cos\psi)^{-1/2} \\ &= \frac{1}{r}\left(1 + \left(\frac{r'}{r}\right)^2 - 2\frac{r'}{r}\cos\psi\right)^{-1/2}.\end{aligned} \tag{3.93}$$

For brevity, we denote $t = r'/r$ and $x = \cos\psi$. The expression can now be developed as a power series in t:

$$\frac{1}{|\boldsymbol{r}-\boldsymbol{r}'|} = \frac{1}{r}(1 + t^2 - 2tx)^{-1/2} = \frac{1}{r}\sum_{n=0}^{\infty} t^n P_n(x). \tag{3.94}$$

The coefficients $P_n(x)$ appearing here are functions of x, and they constitute a set of special functions we will need. Their expressions can be found from the binomial expansion

$$\begin{aligned}\frac{1}{|\boldsymbol{r}-\boldsymbol{r}'|} &= \frac{1}{r}\sum_{n=0}^{\infty}\frac{(2n)!}{2^{2n}(n!)^2}(2tx - t^2)^n \\ &= \frac{1}{r}\sum_{n=0}^{\infty}\sum_{k=0}^{n}\frac{(-1)^k(2n)!}{2^{2n}n!k!(n-k)!}(2x)^{n-k}t^{n+k}, \\ &= \frac{1}{r}\sum_{n=0}^{\infty}\sum_{k=0}^{\lfloor n/2\rfloor}\frac{(-1)^k(2n-2k)!}{2^{2n-2k}k!(n-k)!(n-2k)!}(2x)^{n-2k}t^n.\end{aligned} \tag{3.95}$$

Comparing this with (3.94) we can pick up the expressions of the functions P_n:

$$P_n(x) = \sum_{k=0}^{\lfloor n/2\rfloor}\frac{(-1)^k(2n-2k)!}{2^{2n}k!(n-k)!(n-2k)!}x^{n-2k}. \tag{3.96}$$

These functions are called the Legendre polynomials, the first ones being

$$\begin{aligned} P_0(x) &= 1, \\ P_1(x) &= x, \\ P_2(x) &= \frac{1}{2}(3x^2 - 1), \\ P_3(x) &= \frac{1}{2}(5x^2 - 3x), \\ P_4(x) &= \frac{1}{8}(35x^4 - 30x^2 + 3). \end{aligned} \tag{3.97}$$

Problems

Problem 3.1 Show that the true and eccentric anomalies are related by

$$\tan\frac{\phi}{2} = \sqrt{\frac{1+e}{1-e}}\tan\frac{E}{2}.$$

Hint: show first that

$$\tan^2\frac{x}{2} = \frac{1-\cos x}{1+\cos x}.$$

Problem 3.2 At the perihelion the distance of a comet from the Sun is 0.5 AU and its velocity is 58 km/s. What is the type of the orbit: ellipse, parabola or hyperbola? Find the semi-major axis, parameter and eccentricity of the orbit.

Problem 3.3 The Earth was at perihelion in 4 January 1995 at 11 UTC. Find the mean, eccentric and true anomalies of the Earth on 12 February 1995 at 12 UTC. Orbital elements of the Earth are $a = 1.000$ AU, $e = 0.0167$.

Problem 3.4 A comet moves in an elliptic orbit with $a = 5.0$ AU and $e = 0.8$. Find the true anomaly of the comet and distance from the Sun, when the time elapsed since the perihelion passage is $P/4$, where P is the period of the comet.

Problem 3.5 An object moves in an elliptic orbit. What is the maximum value of its radial velocity and when is this velocity attained? Apply the results to the Earth.

Problem 3.6 (a) Find the time average of the distance for a body moving in an elliptic orbit:

$$\langle r \rangle = \frac{1}{P}\int_0^P r\,\mathrm{d}t.$$

Hint: use the eccentric anomaly. Why is this different from a?

(b) Find the mean distance averaged with respect to the true anomaly:

$$\bar{r} = \frac{1}{2\pi} \int_0^{2\pi} r \, \mathrm{d}\phi.$$

(c) The semi-major axis is sometimes called the mean distance. What kind of a 'mean' distance could it mean? (This leads to an integral that is somewhat harder to evaluate than the two previous ones.)

Problem 3.7 Show that the solution of a two-body problem can be written in the form

$$\boldsymbol{r}(\tau) = \boldsymbol{r}_0 f(\tau, \boldsymbol{r}_0, \boldsymbol{r}'_0) + \boldsymbol{r}'_0 g(\tau, \boldsymbol{r}_0, \boldsymbol{r}'_0),$$

where

$$\tau = \sqrt{\mu}(t - t_0).$$

The prime (′) means a derivative with respect to τ, and subscript zero refers to the value at $t = t_0$. Here f and g are power series in τ which only contain the values of the radius vector and its first derivative at the initial moment t_0. Find the first few terms of the series f and g.

Problem 3.8 Assume the Earth moves around the Sun in a circular orbit. Also, assume that the mass of the Earth is negligible. Write a program to apply expansions (2.53) and (2.55) to calculate Earth's trajectory using only terms up to Δt. Even if the initial velocity is not quite correct, the Earth should return to its initial position after each revolution. Experiment with different time steps to find the accuracy of the method.

4
Hamiltonian mechanics

The formalism of the previous chapters used a rather arbitrary coordinate system. In the Hamiltonian formalism the coordinates are chosen in quite a different way to reflect more deeply the dynamical properties of the system. In this chapter we derive the Hamiltonian equations of motion. The results of this chapter are later needed mainly to derive some standard results that are the starting point for further studies. The same results can also be obtained in a more traditional way, but the Hamiltonian approach makes the calculations considerably shorter and more straightforward.

Hamiltonian mechanics and its applications to mechanics in general are explained more extensively in many books on theoretical mechanics. This chapter is based mainly on Goldstein (1950).

4.1 Generalised coordinates

We have this far used ordinary Euclidean rectangular coordinates to describe positions and velocities of the objects. They are purely geometric quantities that describe the system in a very simple and understandable way. However, they do not tell us anything about the dynamic properties of the system nor do they utilise any specific features of the system. We now want to find a different kind of description in terms of quantities which do not have these problems.

Motions of bodies may be constrained in various ways. For example, two points of a solid body must always be at the same distance from each other. If the constraint can be expressed as a function of the radius vectors $\boldsymbol{r}_i$ and time t in the form

$$f(\boldsymbol{r}_1, \ldots, \boldsymbol{r}_n, t) = 0, \tag{4.1}$$

the constraint is called holonomic. In order to describe the motion of n bodies we need $3n$ coordinates, but if there are constraints, all coordinates are not independent. Each independent holonomic constraint can be used to eliminate one coordinate.

Let us now assume that the actual number of degrees of freedom of our system is m. For the full description of such a system we will need m quantities $q_1, \ldots, q_m$. The original coordinates are obtained from these using the transformation

$$\begin{aligned} \boldsymbol{r}_1 &= \boldsymbol{r}_1(q_1, \ldots, q_m, t), \\ &\vdots \\ \boldsymbol{r}_n &= \boldsymbol{r}_n(q_1, \ldots, q_m, t). \end{aligned} \tag{4.2}$$

The quantities q_i are called generalised coordinates. They need not be any geometric quantities. They can be distances, angles, areas as well as angular momenta, temperatures etc.

Whatever the coordinates q_i are, the state of the system can be considered as a point $S = S(q_1, \ldots, q_m)$ in a Euclidean m-dimensional configuration space, and the q_i are rectangular coordinates of this space. The configuration space is an abstract space that has nothing to do with the actual geometry of the system.

4.2 Hamiltonian principle

Dynamic properties of the system can be described by the Lagrangian function L

$$L = L(q_1, \ldots, q_m, \dot{q}_1, \ldots, \dot{q}_m, t) = T - V, \tag{4.3}$$

where T is the kinetic energy and V the potential energy (Lagrange 1811). We now consider the changes in the system in some finite interval of time $t \in [t_1, t_2]$. The points S_1 and S_2 of the configuration space correspond to the state of the system at the times t_1 and t_2, respectively. In this time interval the system evolves along a path C in the configuration space from S_1 to S_2. If the forces are conservative, the system will evolve in the time interval $[t_1, t_2]$ in such a way that the line integral

$$\int_C L \, \mathrm{d}t \tag{4.4}$$

obtains its extremum. This extremum principle is called the *Hamiltonian principle* (Hamilton 1834).

The Hamiltonian principle is but one of various principles that in one way or another tell that a dynamic system always tends to evolve in the most 'practical' or 'economical' way. Mechanics can be based on several other similar principles, which cannot be directly derived from Newton's laws. We could here adopt some other principle, which might be slightly more understandable intuitively, and use it to derive the Hamiltonian principle, but the advantage would not be worth the trouble.

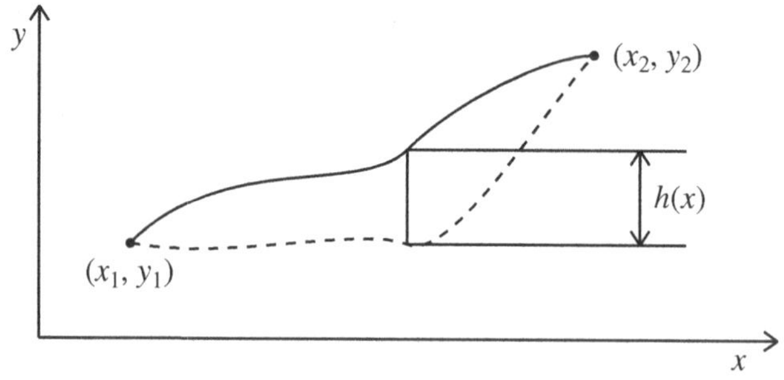

Figure 4.1 Variational calculus is used to find a path giving an extremum to a functional (solid line). The function $h = h(x)$ represents an arbitrary deviation from the optimal path.

4.3 Variational calculus

To use the Hamiltonian principle we need some mathematical tools. Consider first the following simple situation: we have to find a function $y = y(x)$ in the range $x \in [x_1, x_2]$ such that the integral

$$J = \int_{x_1}^{x_2} f(y, y', x)\,\mathrm{d}x \tag{4.5}$$

of a known function $f = f(y, y', x)$, where $y' = \mathrm{d}y/\mathrm{d}x$, attains its extremum value. We can further assume that the points $y_1 = y(x_1)$ and $y_2 = y(x_2)$ are given in advance.

First we have to parametrise the different possibilities. Assume that $y(x, 0)$ is the function sought for. Other possible functions can then be expressed as

$$y(x, a) = y(x, 0) + ah(x), \tag{4.6}$$

where h is an arbitrary function that vanishes at the endpoints of the interval $[x_1, x_2]$ (Fig. 4.1).

Now the value of J obviously depends on the parameter a:

$$J(a) = \int f(y(x, a), y'(x, a), x)\,\mathrm{d}x \tag{4.7}$$

and if J is to have an extremum when $a = 0$, the derivative of J with respect to a must vanish at $a = 0$:

$$\left.\frac{\partial J}{\partial a}\right|_{a=0} = 0. \tag{4.8}$$

Now we just have to calculate this derivative of J:

$$\frac{\partial J}{\partial a} = \int_{x_1}^{x_2} \left(\frac{\partial f}{\partial y}\frac{\partial y}{\partial a} + \frac{\partial f}{\partial y'}\frac{\partial y'}{\partial a} \right) \mathrm{d}x. \tag{4.9}$$

Let us first study the second term:

$$\begin{aligned}
&\int_{x_1}^{x_2} \frac{\partial f}{\partial y'} \frac{\partial y'}{\partial a} \mathrm{d}x \\
&= \int_{x_1}^{x_2} \frac{\partial f}{\partial y'} \frac{\partial^2 y}{\partial x \partial a} \mathrm{d}x \\
&= \left| _{x_1}^{x_2} \frac{\partial f}{\partial y'} \frac{\partial y}{\partial a} - \int_{x_1}^{x_2} \frac{\mathrm{d}}{\mathrm{d}x} \left(\frac{\partial f}{\partial y'} \right) \frac{\partial y}{\partial a} \mathrm{d}x. \right.
\end{aligned}$$

Here $\partial y/\partial a = h(x)$, which vanishes at the ends of the interval. Thus the substitution term is zero, and the derivative of J is

$$\frac{\partial J}{\partial a} = \int_{x_1}^{x_2} \left(\frac{\partial f}{\partial y} - \frac{\mathrm{d}}{\mathrm{d}x} \frac{\partial f}{\partial y'} \right) \frac{\partial y}{\partial a} \mathrm{d}x. \tag{4.10}$$

We will denote the variation of X by δX. This is defined as

$$\delta X = \left. \frac{\partial X}{\partial a} \right|_{a=0} \mathrm{d}a. \tag{4.11}$$

Remember that δX is just an abbreviation for this expression and not a real derivative operator.

Next we multiply (4.10) by da and evaluate both sides at $a = 0$:

$$\delta J = \int_{x_1}^{x_2} \left(\frac{\partial f}{\partial y} - \frac{\mathrm{d}}{\mathrm{d}x} \frac{\partial f}{\partial y'} \right) \delta y \, \mathrm{d}x. \tag{4.12}$$

If f is to yield an extremum, this must vanish. Since $\delta y = h(x) \, \mathrm{d}a$ is arbitrary, the integral can vanish only if the integrand is everywhere zero:

$$\frac{\partial f}{\partial y} - \frac{\mathrm{d}}{\mathrm{d}x} \frac{\partial f}{\partial y'} = 0. \tag{4.13}$$

This is known as the *Lagrange* (or *Euler–Lagrange*) *equation*. Its solution is the function y that makes the integral (4.5) attain its extremum value.

If there are more than one y functions, the generalisation is obvious. The integral is now:

$$J = \int_{x_1}^{x_2} f(y_1(x), \ldots, y_n(x), y_1'(x), \ldots, y_n'(x), x) \, \mathrm{d}x. \tag{4.14}$$

Again we regard J as a function of a parameter a, where a parametrises the different possibilities:

$$\begin{aligned} y_1(x,a) &= y_1(x,0) + ah_1(x), \\ &\vdots \\ y_n(x,a) &= y_n(x,0) + ah_n(x). \end{aligned} \tag{4.15}$$

The variation δJ of the integral is calculated as before. We get

$$\delta J = \int_{x_1}^{x_2} \sum_{i=1}^{n} \left(\frac{\partial f}{\partial y_i} - \frac{\mathrm{d}}{\mathrm{d}x} \frac{\partial f}{\partial y_i'} \right) \delta y_i \, \mathrm{d}x. \tag{4.16}$$

The functions h_i are fully arbitrary independent functions, and so are the y_i. Thus for every i we must have

$$\frac{\partial f}{\partial y_i} - \frac{\mathrm{d}}{\mathrm{d}x} \frac{\partial f}{\partial y_i'} = 0, \quad i = 1, \ldots, n. \tag{4.17}$$

Example 4.1 What is the shortest route between the points (x_1, y_1) and (x_2, y_2)?

If the equation of the path joining the two points is $y = y(x)$, the length of the line element is

$$\mathrm{d}s = \sqrt{\mathrm{d}x^2 + \mathrm{d}y^2} = \sqrt{1 + y'^2}\, \mathrm{d}x$$

and the length of the whole path is

$$s = \int_{x_1}^{x_2} \sqrt{1 + y'^2}\, \mathrm{d}x.$$

This is already of the form (4.5). The function f is the square root in the integral. Now the derivatives needed for the Lagrange equation are readily found:

$$\begin{aligned} \frac{\partial f}{\partial y} &= 0, \\ \frac{\partial f}{\partial y'} &= \frac{y'}{\sqrt{1+y'^2}}, \\ \frac{\mathrm{d}}{\mathrm{d}x}\frac{\partial f}{\partial y'} &= \frac{y''}{\sqrt{1+y'^2}} - \frac{y'^2 y''}{(1+y'^2)^{3/2}}. \end{aligned}$$

Thus the Lagrange equation is simply

$$y'' = 0.$$

The solution of this is obviously $y = ax + b$, where the constants a and b are determined by the requirement that the path must pass through the given points. But this is the equation of a straight line, and so we have managed to prove the profound result that in Euclidean space the shortest paths (or geodesics) are straight lines.

4.4 Lagrangian equations of motion

The Hamiltonian principle requires that the integral

$$\int L(q_1, \ldots, q_m, \dot{q}_1, \ldots, \dot{q}_m, t)\,\mathrm{d}t \tag{4.18}$$

must attain an extremum. The integral is suitably of the form (4.5) when we replace x with the time t. Thus the Lagrange equations give immediately

$$\frac{\partial L}{\partial q_i} - \frac{\mathrm{d}}{\mathrm{d}t}\frac{\partial L}{\partial \dot{q}_i} = 0, \qquad i = 1, \ldots, m. \tag{4.19}$$

These equations are called the *Lagrangian equations of motion.*

As an example consider the planar two-body problem. The Lagrangian is (omitting a constant factor, the mass of the planet)

$$L = T - V = \frac{1}{2}v^2 + \frac{\mu}{r} = \frac{1}{2}(\dot{x}^2 + \dot{y}^2) + \frac{\mu}{\sqrt{x^2 + y^2}}, \tag{4.20}$$

and hence

$$\begin{aligned}
\frac{\partial L}{\partial x} &= -\frac{x\mu}{(x^2 + y^2)^{3/2}}, \\
\frac{\partial L}{\partial y} &= -\frac{y\mu}{(x^2 + y^2)^{3/2}}, \\
\frac{\partial L}{\partial \dot{x}} &= \dot{x}, \\
\frac{\partial L}{\partial \dot{y}} &= \dot{y}.
\end{aligned}$$

The equations of motion are thus

$$\begin{aligned}
-\frac{x\mu}{(x^2 + y^2)^{3/2}} - \ddot{x} = 0, \\
-\frac{y\mu}{(x^2 + y^2)^{3/2}} - \ddot{y} = 0.
\end{aligned} \tag{4.21}$$

This is just the Newtonian equation of motion in component form.

This whole business may not look very useful since we get just the same equations as before. However, now we can easily select a more suitable coordinate frame. In our example the potential depends on the distance r only, which makes polar coordinates the natural choice. The Lagrangian is then

$$L = \frac{1}{2}(\dot{r}^2 + r^2\dot{\phi}^2) + \frac{\mu}{r}. \tag{4.22}$$

Again we find the derivatives:

$$\begin{aligned}
\frac{\partial L}{\partial r} &= r\dot{\phi}^2 - \frac{\mu}{r^2}, \\
\frac{\partial L}{\partial \phi} &= 0, \\
\frac{\partial L}{\partial \dot{r}} &= \dot{r}, \\
\frac{\partial L}{\partial \dot{\phi}} &= r^2\dot{\phi}.
\end{aligned}$$

The equations of motion are

$$\begin{aligned}
&r\dot{\phi}^2 - \mu/r^2 - \ddot{r} = 0, \\
&\frac{\mathrm{d}}{\mathrm{d}t}(r^2\dot{\phi}) = 0.
\end{aligned} \tag{4.23}$$

This already shows some advantages of the Lagrangian formalism. The equations in polar coordinates were as easily obtained as the ordinary equations in rectangular coordinates. Moreover, the latter equation tells immediately that the surface velocity is constant. Thus we obtained Kepler's second law as a by-product without any extra effort.

Obviously the Lagrangian formalism is in some ways better than the usual Newtonian formalism. It applies well to many problems of mechanics. In celestial mechanics its superiority is not quite as pronounced, and for us it is just an intermediate step.

Example 4.2 Derive the equation of motion for a pendulum with a mass m and length l.

In this case we need only the angle ϕ between the vertical and the length l of the shaft of the pendulum to describe the state of the system. The kinetic energy is $T = \frac{1}{2}ml^2\dot{\phi}^2$ and the potential energy $V = mgl(1 - \cos\phi)$, assuming that the gravitational field is constant and taking the lowest point of the pendulum to correspond to zero potential energy. The Lagrangian is now

$$L = \frac{1}{2}ml^2\dot{\phi}^2 - mgl(1 - \cos\phi),$$

from which

$$\frac{\partial L}{\partial \phi} = -mgl \sin \phi,$$
$$\frac{\partial L}{\partial \dot{\phi}} = ml^2 \dot{\phi}.$$

The equation of motion of the pendulum is then

$$-mgl \sin \phi - ml^2 \ddot{\phi} = 0$$

or

$$\ddot{\phi} + \frac{g}{l} \sin \phi = 0.$$

If the angle of oscillation is small, $\sin \phi \approx \phi$, and the equation becomes the familiar equation of a simple harmonic oscillator:

$$\ddot{\phi} + \frac{g}{l} \phi = 0.$$

4.5 Hamiltonian equations of motion

Assume we have a system with m degrees of freedom and a Lagrangian

$$L = L(q_1, \ldots, q_m, \dot{q}_1, \ldots, q_m, t). \tag{4.24}$$

It can happen that one of the coordinates does not appear in the Lagrangian. Such a coordinate is called *cyclic*. Unfortunately the corresponding velocity can still appear in the Lagrangian, and the number of equations is not reduced. We would like somehow to eliminate the equations corresponding to cyclic coordinates. This is what happens in the Hamiltonian formalism.

A momentum has a deeper relation to the dynamics of a system than a mere velocity. So we replace the generalised velocities of the Lagrangian formalism with generalised momenta. In a rectangular frame we have

$$\frac{\partial L}{\partial \dot{x}} = \frac{\partial T}{\partial \dot{x}} = \frac{\partial}{\partial \dot{x}} \frac{m}{2} (\dot{x}^2 + \dot{y}^2 + \dot{z}^2) = m\dot{x} = p_x. \tag{4.25}$$

We now generalise this and define *generalised momenta* p_i as

$$p_i = \frac{\partial L}{\partial \dot{q}_i}. \tag{4.26}$$

The Lagrangian must be replaced by an entity which depends on the variables q_i and p_i. This is accomplished by the *Legendre transformation*. Let $f = f(x, y)$

be a function of two variables x and y. Its total differential is

$$\mathrm{d}f = \frac{\partial f}{\partial x}\,\mathrm{d}x + \frac{\partial f}{\partial y}\,\mathrm{d}y = u\,\mathrm{d}x + v\,\mathrm{d}y \tag{4.27}$$

where u and v are the partial derivatives of f. Now we define a function g in the following way:

$$g = ux - f(x, y). \tag{4.28}$$

The total differential of this is

$$\mathrm{d}g = u\,\mathrm{d}x + x\,\mathrm{d}u - u\,\mathrm{d}x - v\,\mathrm{d}y = x\,\mathrm{d}u - v\,\mathrm{d}y.$$

Because this is a total differential, we must have

$$x = \frac{\partial g}{\partial u}, \qquad v = -\frac{\partial g}{\partial y}$$

or

$$dg = \frac{\partial g}{\partial u}\mathrm{d}u + \frac{\partial g}{\partial y}\,\mathrm{d}y.$$

So we can consider g as a function of u and y.

In a similar manner the Lagrangian can be transformed to obtain the *Hamiltonian* H of the system:

$$H = H(\boldsymbol{q}, \boldsymbol{p}, t) = \sum_{i=1}^{m} \dot{q}_i p_i - L(\boldsymbol{q}, \dot{\boldsymbol{q}}, t). \tag{4.29}$$

Here we have denoted $\boldsymbol{q} = (q_1, \ldots, q_m)$ etc. In what follows, all sums are assumed to be from 1 to m if the limits are not shown explicitly. We now find the total differential of the Hamiltonian:

$$\begin{aligned}
\mathrm{d}H &= \sum \dot{q}_i\,\mathrm{d}p_i + \sum p_i\,\mathrm{d}\dot{q}_i - \sum \frac{\partial L}{\partial q_i}\,\mathrm{d}q_i - \sum \frac{\partial L}{\partial \dot{q}_i}\,\mathrm{d}\dot{q}_i - \frac{\partial L}{\partial t}\,\mathrm{d}t \\
&= \sum \dot{q}_i\,\mathrm{d}p_i + \sum p_i\,\mathrm{d}\dot{q}_i - \sum \frac{\partial L}{\partial q_i}\,\mathrm{d}q_i - \sum p_i\,\mathrm{d}\dot{q}_i - \frac{\partial L}{\partial t}\,\mathrm{d}t.
\end{aligned}$$

According to the Lagrangian equations we have

$$\frac{\partial L}{\partial q_i} = \frac{\mathrm{d}}{\mathrm{d}t}\frac{\partial L}{\partial \dot{q}_i} = \frac{\mathrm{d}}{\mathrm{d}t}p_i = \dot{p}_i,$$

and so

$$\mathrm{d}H = \sum \dot{q}_i\,\mathrm{d}p_i - \sum \dot{p}_i\,\mathrm{d}q_i - \frac{\partial L}{\partial t}\,\mathrm{d}t. \tag{4.30}$$

By the definition of the total differential we have

$$dH = \sum \frac{\partial H}{\partial q_i} dq_i + \sum \frac{\partial H}{\partial p_i} dp_i + \frac{\partial H}{\partial t} dt. \tag{4.31}$$

These two expressions (4.30) and (4.31) for dH have to be identical. So we must have

$$\begin{aligned} \frac{\partial H}{\partial p_i} &= \dot{q}_i, \\ \frac{\partial H}{\partial q_i} &= -\dot{p}_i, \\ \frac{\partial H}{\partial t} &= -\frac{\partial L}{\partial t}. \end{aligned} \tag{4.32}$$

These are called the *Hamiltonian canonical equations of motion*. We have replaced m second order equations by $2m$ first order equations (the last one of (4.32) is not really an equation of motion) in which coordinates and momenta must be considered as independent variables. They are called *canonical coordinates and momenta*. They are generalised coordinates and momenta, and their dimension can be almost anything. However, the product of the corresponding coordinate and momentum must always have the dimension of time × energy. The variables q_i and p_i are called *conjugate variables*, and although they are independent, they describe quantities that are related in a certain way. If we have fixed, say, q_i then p_i cannot be chosen arbitrarily. If q_i is distance, p_i is the corresponding momentum; if q_i is an angle, p_i is the angular momentum; if q_i is time, p_i is energy, etc.

4.6 Properties of the Hamiltonian

If q_m does not appear in the Lagrangian, it is not present in the Hamiltonian either. Then the partial derivative of H with respect to q_m vanishes, and it is seen immediately from the equations of motion that the corresponding momentum p_m is constant. Thus we can use symmetries of the system to simplify its equations of motion.

Next, we will study how the Hamiltonian changes with time. We take the time derivative of the Hamiltonian:

$$\dot{H} = \sum \left(\frac{\partial H}{\partial q_i} \dot{q}_i + \frac{\partial H}{\partial p_i} \dot{p}_i \right) + \frac{\partial H}{\partial t}. \tag{4.33}$$

We use the equations of motion to replace $\dot{q}_i$ and $\dot{p}_i$:

$$\dot{H} = \sum \left(\frac{\partial H}{\partial q_i} \frac{\partial H}{\partial p_i} - \frac{\partial H}{\partial p_i} \frac{\partial H}{\partial q_i} \right) + \frac{\partial H}{\partial t},$$

and hence

$$\frac{\mathrm{d}H}{\mathrm{d}t} = \frac{\partial H}{\partial t}. \tag{4.34}$$

This shows that H will change with time only if time is explicitly present in its expression. If the Hamiltonian does not depend explicitly on time it is a constant of motion.

We show now that H is equal to the total energy if the transformation equations (4.2)

$$\boldsymbol{r}_i = \boldsymbol{r}_i(q_1, \ldots, q_m), \quad i = 1, \ldots, n$$

do not depend explicitly on time and if the potential depends on the coordinates q_i only.

To prove this we need some auxiliary results. First of all, if the transformation equations (4.2) do not contain time explicitly, the kinetic energy T is a quadratic form of the velocities $\dot{q}_i$.

In rectangular coordinates we have

$$T = \sum_{i=1}^{n} \frac{1}{2} m_i v_i^2.$$

We apply the transformation (4.2) to this:

$$\begin{aligned} T &= \sum_{i=1}^{n} \frac{1}{2} m_i \left(\sum_j \frac{\partial \boldsymbol{r}_i}{\partial q_j} \dot{q}_j + \frac{\partial \boldsymbol{r}_i}{\partial t} \right)^2 \\ &= \sum_{i=1}^{n} \frac{1}{2} m_i \left(\frac{\partial \boldsymbol{r}_i}{\partial t} \right)^2 + \sum_j \dot{q}_j \left(\sum_{i=1}^{n} m_i \frac{\partial \boldsymbol{r}_i}{\partial t} \cdot \frac{\partial \boldsymbol{r}_i}{\partial q_j} \right) \\ &\quad + \sum_j \sum_k \dot{q}_j \dot{q}_k \left(\sum_{i=1}^{n} m_i \frac{\partial \boldsymbol{r}_i}{\partial q_j} \cdot \frac{\partial \boldsymbol{r}_i}{\partial q_k} \right). \end{aligned}$$

If the time derivatives of the $\boldsymbol{r}_i$ vanish, only the last sum will remain, and T is of the form

$$T = \sum_j \sum_k a_{jk} \dot{q}_j \dot{q}_k. \tag{4.35}$$

Another auxiliary result is *Euler's theorem*: if f is a homogeneous form of degree n on the variables q_i, then

$$\sum q_i \frac{\partial f}{\partial q_i} = nf. \tag{4.36}$$

In fact we need this only when $n = 2$, but we can as well prove the general case.

Assume that f is of the form

$$f = \sum_{i_1} \cdots \sum_{i_n} a_{i_1 \cdots i_n} q_{i_1} \cdots q_{i_n}.$$

When $n = 1$, we have $f = \sum a_i q_i$ and $\sum q_i(\partial f/\partial q_i) = \sum q_i a_i = f$. Thus the theorem holds.

Assume then that the theorem is valid for all forms of degree n, and let f be of degree $n + 1$

$$\begin{aligned} f &= \sum_{i_1} \cdots \sum_{i_{n+1}} a_{i_1 \cdots i_{n+1}} q_{i_1} \cdots q_{i_{n+1}} \\ &= \sum_{i_{n+1}} q_{i_{n+1}} \left(\sum_{i_1} \cdots \sum_{i_n} a_{i_1 \cdots i_{n+1}} q_{i_1} \cdots q_{i_n} \right) \\ &= \sum_{i_{n+1}} q_{i_{n+1}} f', \end{aligned}$$

where f' is a homogeneous form of degree n. The expression on the left hand side in Euler's theorem is

$$\begin{aligned} \sum_k q_k \frac{\partial f}{\partial q_k} &= \sum q_k \frac{\partial}{\partial q_k} \left(\sum_{i_{n+1}} q_{i_{n+1}} f' \right) \\ &= \sum_k q_k \left(\sum_{i_{n+1}} \frac{\partial q_{i_{n+1}}}{\partial q_k} f' + \sum_{i_{n+1}} q_{i_{n+1}} \frac{\partial f'}{\partial q_k} \right) \\ &= \sum_{i_{n+1}} \left(\sum_k q_k \frac{\partial q_{i_{n+1}}}{\partial q_k} f' + q_{i_{n+1}} \sum_k q_k \frac{\partial f'}{\partial q_k} \right). \end{aligned}$$

We apply the induction assumption to the latter sum:

$$\begin{aligned} \sum_k q_k \frac{\partial f}{\partial q_k} &= \sum_{i_{n+1}} \left(q_{i_{n+1}} f' + q_{i_{n+1}} n f' \right) \\ &= (n+1) \sum_{i_{n+1}} q_{i_{n+1}} f' = (n+1) f. \end{aligned}$$

Thus the theorem holds also for $n + 1$, and hence for all values of n.

Now it is easy to show that H is the total energy. Since the potential V does not depend on the velocities $\dot{q}_i$, we have

$$p_i = \frac{\partial L}{\partial \dot{q}_i} = \frac{\partial T}{\partial \dot{q}_i}, \qquad i = 1, \ldots, m,$$

and thus

$$H = \sum \dot{q}_i p_i - L = \sum \dot{q}_i \frac{\partial T}{\partial \dot{q}_i} - L.$$

Because the kinetic energy is a quadratic form of the $\dot{q}_i$, we can use Euler's theorem ($n = 2$):

$$H = 2T - L = 2T - (T - V) = T + V. \tag{4.37}$$

Thus the Hamiltonian gives the total energy. However, we have to remember that this may not be true if for example the transformations (4.2) depend explicitly on time.

4.7 Canonical transformations

If the Hamiltonian is cyclic in one coordinate the corresponding equation of motion has a trivial solution. Therefore it is advantageous to search for a coordinate frame in which as many coordinates as possible become cyclic. Writing the corresponding Hamiltonian can be difficult, and therefore we have to study how to transform from one frame to another.

Since the momenta p_i are variables independent of the coordinates, we have to find both the new coordinates Q_i and the new momenta P_i as functions of the old ones:

$$\begin{aligned} Q_i &= Q_i(\boldsymbol{q}, \boldsymbol{p}, t), \\ P_i &= P_i(\boldsymbol{q}, \boldsymbol{p}, t). \end{aligned} \tag{4.38}$$

A nice feature of the Hamiltonian formalism is that if we use proper transformations the new variables will also be canonical and the new equations of motion will have exactly the same form as in the original frame:

$$\dot{Q}_i = \frac{\partial K}{\partial P_i}, \qquad -\dot{P}_i = \frac{\partial K}{\partial Q_i}, \tag{4.39}$$

where K is the new Hamiltonian. A transformation preserving the form of the equations is called canonical. Now we have to find what kind of transformations are canonical.

We begin by expressing the Hamiltonian principle in terms of the old and new coordinates:

$$\delta \int_{t_1}^{t_2} L \, \mathrm{d}t = \begin{cases} \delta \int_{t_1}^{t_2} \left(\sum p_i \dot{q}_i - H\right) \mathrm{d}t = 0 \\ \delta \int_{t_1}^{t_2} \left(\sum P_i \dot{Q}_i - K\right) \mathrm{d}t = 0. \end{cases} \tag{4.40}$$

These integrals can differ only by something whose variation vanishes. A function that depends on the end points t_1 and t_2 only is such a function, because the variation always vanishes at both ends of the interval. Thus the integrands can differ by a total time derivative of some function, since in that case

$$\int_{t_1}^{t_2} \frac{\mathrm{d}F}{\mathrm{d}t}\,\mathrm{d}t = F(t_2) - F(t_1). \tag{4.41}$$

(In fact the integrands can differ also by a constant factor, but that means just a change of units.) The function F is the *generating function* of the transformation. It has to depend on both the old and new variables in a suitable way. It can have any of the following four forms:

$$\begin{aligned} F &= F_1(\boldsymbol{q}, \boldsymbol{Q}, t), \\ F &= F_2(\boldsymbol{q}, \boldsymbol{P}, t), \\ F &= F_3(\boldsymbol{p}, \boldsymbol{Q}, t), \\ F &= F_4(\boldsymbol{p}, \boldsymbol{P}, t). \end{aligned} \tag{4.42}$$

The transformation is canonical if the difference of the integrands in Eq. (4.40) equals the time derivative of the generating function:

$$\sum p_i \dot{q}_i - H = \sum P_i \dot{Q}_i - K + \frac{\mathrm{d}F}{\mathrm{d}t}. \tag{4.43}$$

If the generating function is of the form F_1, this condition is

$$\sum p_i \dot{q}_i - H = \sum P_i \dot{Q}_i - K + \sum \frac{\partial F_1}{\partial q_i}\dot{q}_i + \sum \frac{\partial F_1}{\partial Q_i}\dot{Q}_i + \frac{\partial F_1}{\partial t}. \tag{4.44}$$

Since the q and Q are independent, this equation holds identically if and only if

$$\begin{aligned} p_i &= \frac{\partial F_1}{\partial q_i}, \\ P_i &= -\frac{\partial F_1}{\partial Q_i}, \\ K &= H + \frac{\partial F_1}{\partial t}. \end{aligned} \tag{4.45}$$

At least in principle the Q_i can be solved from the expressions of the p_i and be substituted into the expressions of the P_i to give the transformation equations

$$\begin{aligned} Q_i &= Q_i(\boldsymbol{q}, \boldsymbol{p}, t), \\ P_i &= P_i(\boldsymbol{q}, \boldsymbol{p}, t). \end{aligned} \tag{4.46}$$

The corresponding equations for a generating function F_2 can be obtained using a suitable Legendre transformation to change variables.

We begin by writing the differential of the generating function F_1:

$$\mathrm{d}F_1 = \sum \frac{\partial F_1}{\partial q_i}\,\mathrm{d}q_i + \sum \frac{\partial F_1}{\partial Q_i}\,\mathrm{d}Q_i + \frac{\partial F_1}{\partial t}\,\mathrm{d}t.$$

We substitute here the previously found derivatives of F_1:

$$\mathrm{d}F_1 = \sum p_i\,\mathrm{d}q_i - \sum P_i\,\mathrm{d}Q_i + \frac{\partial F_1}{\partial t}\,\mathrm{d}t.$$

Next, apply the Legendre transform

$$F_2 = F_1 - \sum \frac{\partial F_1}{\partial Q_i} Q_i = F_1 + \sum P_i Q_i, \tag{4.47}$$

to get the differential of F_2

$$\begin{aligned} \mathrm{d}F_2 &= \sum p_i\,\mathrm{d}q_i - \sum P_i\,\mathrm{d}Q_i + \sum P_i\,\mathrm{d}Q_i + \sum Q_i\,\mathrm{d}P_i + \frac{\partial F_1}{\partial t}\,\mathrm{d}t \\ &= \sum p_i\,\mathrm{d}q_i + \sum Q_i\,\mathrm{d}P_i + \frac{\partial F_1}{\partial t}\,\mathrm{d}t. \end{aligned} \tag{4.48}$$

By definition the differential of F_2 is

$$\mathrm{d}F_2 = \sum \frac{\partial F_2}{\partial q_i}\,\mathrm{d}q_i + \sum \frac{\partial F_2}{\partial P_i}\,\mathrm{d}P_i + \frac{\partial F_2}{\partial t}\,\mathrm{d}t. \tag{4.49}$$

Because the expressions (4.48) and (4.49) must be identical, we get

$$\begin{aligned} p_i &= \frac{\partial F_2}{\partial q_i}, \\ Q_i &= \frac{\partial F_2}{\partial P_i}. \end{aligned} \tag{4.50}$$

The generating functions F_3 and F_4 can be handled similarly by changing variables with the Legendre transform. We get the following transformation equations

$$\begin{aligned} &F = F_1(\boldsymbol{q}, \boldsymbol{Q}, t) && p_i = \frac{\partial F_1}{\partial q_i} && P_i = -\frac{\partial F_1}{\partial Q_i} \\ &F = F_2(\boldsymbol{q}, \boldsymbol{P}, t) && p_i = \frac{\partial F_2}{\partial q_i} && Q_i = \frac{\partial F_2}{\partial P_i} \\ &F = F_3(\boldsymbol{p}, \boldsymbol{Q}, t) && q_i = -\frac{\partial F_3}{\partial p_i} && P_i = -\frac{\partial F_3}{\partial Q_i} \\ &F = F_4(\boldsymbol{p}, \boldsymbol{P}, t) && q_i = -\frac{\partial F_4}{\partial p_i} && Q_i = \frac{\partial F_4}{\partial P_i}. \end{aligned} \tag{4.51}$$

In all cases the Hamiltonian is transformed in the same way

$$K = H + \frac{\partial F}{\partial t}. \tag{4.52}$$

4.8 Examples of canonical transformations

Consider the generating function

$$F = F_2(\boldsymbol{q}, \boldsymbol{P}) = \sum q_i P_i.$$

Then we have

$$p_i = \frac{\partial F}{\partial q_i} = P_i, \quad Q_i = \frac{\partial F}{\partial P_i} = q_i.$$

Thus the transformation is an identity.

A simple but interesting transformation is obtained using the generating function

$$F = F_1(\boldsymbol{q}, \boldsymbol{Q}) = \sum q_i Q_i.$$

Now

$$p_i = \frac{\partial F}{\partial q_i} = Q_i, \quad P_i = -\frac{\partial F}{\partial Q_i} = -q_i.$$

The new coordinates are identical to the old momenta and the new momenta are the old coordinates with the sign changed. Thus there is no real distinction between coordinates and momenta.

An ordinary geometric transformation, which replaces the old coordinates q_i with new ones Q_i, is generated by

$$F = F_2(\boldsymbol{q}, \boldsymbol{P}) = \sum f_i(\boldsymbol{q}, t) P_i, \tag{4.53}$$

where the f_i are arbitrary functions. Then

$$Q_i = \frac{\partial F}{\partial P_i} = f_i(\boldsymbol{q}, t). \tag{4.54}$$

Since the f_i can be any functions, all such *point transformations* are canonical. If we find a set of canonical geometric coordinates q_i, we can keep the equations of motion canonical in all geometric transformations by choosing the momenta properly.

4.9 The Hamilton–Jacobi equation

We noticed that if the coordinate q_i does not appear in the Hamiltonian the corresponding conjugate momentum p_i is constant. Similarly, if the Hamiltonian does not contain p_i then q_i is constant. Obviously it is worthwhile to look more closely how to transform the Hamiltonian to a form which contains as few variables as possible.

The ideal case would be to have a Hamiltonian vanishing identically, in which case the solution of the equations of motion is trivial. We now try to find a suitable canonical transformation, denoted traditionally by S. We require that S is of the form F_2, or

$$S = S(q_1, \ldots, q_m, P_1, \ldots, P_m, t). \tag{4.55}$$

The old momenta are obtained as derivatives of the generating function:

$$p_i = \frac{\partial S}{\partial q_i}. \tag{4.56}$$

The old and new Hamiltonians are related by

$$H + \frac{\partial S}{\partial t} = K = 0. \tag{4.57}$$

We now express the momenta in H as partial derivatives of the generating function. Then the equation (4.57) becomes

$$H\left(q_1, \ldots, q_m, \frac{\partial S}{\partial q_1}, \ldots, \frac{\partial S}{\partial q_m}, t\right) + \frac{\partial S}{\partial t} = 0. \tag{4.58}$$

This is the *Hamilton–Jacobi equation.* Its solution is the function S which depends on m coordinates and time. Since there are $m + 1$ variables, the solution will contain $m + 1$ constants of integration $\alpha_1, \ldots, \alpha_{m+1}$. Equation (4.58) contains only derivatives of S, and therefore S will be determined only up to an additive constant, i.e. $S + C$, where C is a constant, is also a valid solution. This constant is one of the integration constants; let it be α_{m+1}. This additive constant can be neglected because it does not affect the transformation for which we are searching. Thus the generating function is of the form

$$S = S(q_1, \ldots, q_m, \alpha_1, \ldots, \alpha_m, t). \tag{4.59}$$

It seems that S does not tell how the momenta are transformed. We know, though, that the new momenta are constants and appear as arguments of S. Thus it is natural to take the constants α_i as the new momenta:

$$P_i = \alpha_i, \quad i = 1, \ldots, m. \tag{4.60}$$

The constant values of the new coordinates are obtained as derivatives of the generating function with respect to the momenta P_i or the constants α_i:

$$Q_i = \frac{\partial S(q_1, \ldots, q_m, \alpha_1, \ldots, \alpha_m, t)}{\partial \alpha_i} = \beta_i. \tag{4.61}$$

If the values of the variables q_i and p_i are known for a certain initial instant of time t_0, they can be substituted into the expressions for the momenta (4.56):

$$p_i = \frac{\partial S(q_1, \ldots, q_m, \alpha_1, \ldots, \alpha_m, t)}{\partial q_i}. \tag{4.62}$$

From these we can solve the constants α_i. Substitution of these constants and coordinates q_i at t_0 into (4.61) then gives the values of the new coordinates β_i.

In practice, the work is not quite that simple. Unfortunately we have replaced the simple Hamiltonian equations of motion with a complicated partial differential equation, and there is no guarantee that it can be solved. There is no general theory for solving such equations. We now return to the familiar two-body problem and try to find a way to solve the Hamilton–Jacobi equation.

4.10 Two-body problem in Hamiltonian mechanics: two dimensions

We start with the simple planar problem. However, the results obtained here can be used later when dealing with the general case.

The kinetic energy of the planet of mass m with respect to the Sun in polar coordinates is

$$T = \frac{1}{2} m(\dot{r}^2 + r^2\dot{\phi}^2), \tag{4.63}$$

and its potential energy is

$$V = -\mu m/r, \quad \mu = G(m + M_\odot). \tag{4.64}$$

The momenta corresponding to the coordinates ϕ and r are

$$\begin{aligned} p_\phi &= \frac{\partial T}{\partial \dot{\phi}} = mr^2\dot{\phi}, \\ p_r &= \frac{\partial T}{\partial \dot{r}} = m\dot{r}. \end{aligned} \tag{4.65}$$

We take the following relations as known:

$$\begin{aligned} h &= \frac{1}{2}v^2 - \frac{\mu}{r}, \\ k &= r^2\dot{\phi} = \sqrt{a\mu(1-e^2)}, \\ h &= -\mu/2a. \end{aligned} \tag{4.66}$$

The Hamiltonian in terms of the coordinates r and ϕ and their conjugate momenta is

$$H = T + V = \frac{1}{2m}\left(p_r^2 + \frac{p_\phi^2}{r^2}\right) - \frac{\mu m}{r}. \tag{4.67}$$

Since h is the total energy per unit mass, we have also

$$H = mh. \tag{4.68}$$

Because the gravitational potential depends on the distance only, it is natural to use polar coordinates. Then the Hamiltonian is cyclic in ϕ. We could use this result immediately. However, we will carry out all calculations in detail to see how the cyclic coordinates behave in the solution of the Hamilton–Jacobi equation.

We try to find a coordinate system in which the Hamiltonian is identically zero. Let the generating function of this transformation be

$$S = S(r, \phi, P_1, P_2, t), \tag{4.69}$$

where P_1 and P_2 are the new momenta, which must be constants. The Hamilton–Jacobi equation is

$$H\left(r, \frac{\partial S}{\partial r}, \frac{\partial S}{\partial \phi}\right) + \frac{\partial S}{\partial t} = 0$$

or

$$\frac{1}{2m}\left(\left(\frac{\partial S}{\partial r}\right)^2 + \frac{1}{r^2}\left(\frac{\partial S}{\partial \phi}\right)^2\right) - \frac{\mu m}{r} + \frac{\partial S}{\partial t} = 0. \tag{4.70}$$

There is only one rather general method for solving such partial differential equations: separation of variables. We assume that the solution can be expressed as

$$S(r, \phi, t) = S_r(r) + S_\phi(\phi) + S_t(t), \tag{4.71}$$

where S_r depends only on the distance r, S_ϕ only on the angle ϕ and S_t only on the time t. Now we substitute this trial solution into our Eq. (4.70):

$$\frac{1}{2m}\left(\left(\frac{\mathrm{d}S_r}{\mathrm{d}r}\right)^2 + \frac{1}{r^2}\left(\frac{\mathrm{d}S_\phi}{\mathrm{d}\phi}\right)^2\right) - \frac{\mu m}{r} = -\frac{\mathrm{d}S_t}{\mathrm{d}t}. \tag{4.72}$$

Here the left hand side is a function of r and ϕ only and the right hand side depends only on time. The equation can hold only if both sides have the same constant value. We denote this value by α_1. Now the original equation can be split into two simpler equations:

$$\frac{\mathrm{d}S_t}{\mathrm{d}t} = -\alpha_1,$$

$$\frac{1}{2m}\left(\left(\frac{\mathrm{d}S_r}{\mathrm{d}r}\right)^2 + \frac{1}{r^2}\left(\frac{\mathrm{d}S_\phi}{\mathrm{d}\phi}\right)^2\right) - \frac{\mu m}{r} = \alpha_1.$$

In the latter equation only the term $(\mathrm{d}S_\phi/\mathrm{d}\phi)^2$ depends on ϕ, and so it must be another constant, α_2. Thus we have three ordinary differential equations to determine S:

$$\begin{aligned}\frac{\mathrm{d}S_t}{\mathrm{d}t} &= -\alpha_1,\\ \frac{\mathrm{d}S_\phi}{\mathrm{d}\phi} &= \alpha_2,\\ \frac{\mathrm{d}S_r}{\mathrm{d}r} &= \sqrt{2m\left(\alpha_1+\frac{\mu m}{r}\right)-\frac{\alpha_2^2}{r^2}}.\end{aligned} \tag{4.73}$$

The generating function is then

$$S = -\alpha_1 t + \alpha_2\phi + \int^r \mathrm{d}r\sqrt{2m\left(\alpha_1+\frac{\mu m}{r}\right)-\frac{\alpha_2^2}{r^2}}. \tag{4.74}$$

We will need only the derivatives of this, so there is no need to evaluate the integral.

When we take the integration constants α_1 and α_2 as the new momenta we get

$$\begin{aligned}P_1 &= \alpha_1,\\ P_2 &= \alpha_2,\\ Q_1 &= \frac{\partial S}{\partial P_1} = \frac{\partial S}{\partial \alpha_1},\\ Q_2 &= \frac{\partial S}{\partial P_2} = \frac{\partial S}{\partial \alpha_2}.\end{aligned} \tag{4.75}$$

We now have to determine the constants α_1 and α_2. Since

$$H + \frac{\partial S}{\partial t} = H - \alpha_1 = 0,$$

we have

$$\alpha_1 = H = mh. \tag{4.76}$$

Hence α_1 (and the momentum P_1) is the total energy in the heliocentric coordinate frame. Using the properties of the generating function we have

$$\frac{\partial S}{\partial \phi} = p_\phi,$$

and we saw previously that $\partial S/\partial\phi = \alpha_2$. Therefore

$$\alpha_2 = p_\phi = mr^2\dot{\phi} = mk = m\sqrt{a\mu(1-e^2)}. \tag{4.77}$$

This is the length of the angular momentum vector of the planet. The new momenta are now

$$\begin{aligned} P_1 &= \alpha_1 = mh = -\frac{m\mu}{2a}, \\ P_2 &= \alpha_2 = m\sqrt{a\mu(1-e^2)}. \end{aligned} \tag{4.78}$$

It remains to find the new coordinates Q_1 and Q_2. They are determined by the generating function:

$$\begin{aligned} Q_1 &= \frac{\partial S}{\partial \alpha_1} = -t + \int \frac{m\,\mathrm{d}r}{\sqrt{2m(\alpha_1 + \mu m/r) - (\alpha_2/r)^2}} \\ &= -t + I_1, \\ Q_2 &= \frac{\partial S}{\partial \alpha_2} = \phi - \frac{\alpha_2}{m}\int \frac{m\,\mathrm{d}r}{r^2\sqrt{2m(\alpha_1 + \mu m/r) - (\alpha_2/r)^2}} \\ &= \phi - \frac{\alpha_2}{m} I_2. \end{aligned} \tag{4.79}$$

Evaluating the integrals I_1 and I_2 is the most tedious part of the problem. We begin by substituting the values of α_1 and α_2 obtained from (4.78):

$$\begin{aligned} I_1 &= \int \frac{m\,\mathrm{d}r}{\sqrt{2m\,(-m\mu/(2a) + m\mu/r) - m^2 a\mu(1-e^2)/r^2}} \\ &= \frac{1}{\sqrt{\mu}}\int \frac{r\,\mathrm{d}r}{\sqrt{-r^2/a + 2r - a(1-e^2)}}, \\ I_2 &= \int \frac{m\,\mathrm{d}r}{r^2\sqrt{2m\,(-m\mu/(2a) + m\mu/r) - m^2 a\mu(1-e^2)/r^2}} \\ &= \frac{1}{\sqrt{\mu}}\int \frac{\mathrm{d}r}{r\sqrt{-r^2/a + 2r - a(1-e^2)}}. \end{aligned} \tag{4.80}$$

The usual trick is to use the eccentric anomaly, which often simplifies such integrals considerably. We replace r with E:

$$\begin{aligned} r &= a(1 - e\cos E), \\ \mathrm{d}r &= ae\sin E\,\mathrm{d}E. \end{aligned} \tag{4.81}$$

The integral I_1 is now easily evaluated:

$$\begin{aligned} I_1 &= \frac{1}{\sqrt{\mu}} \int \frac{a(1 - e\cos E) ae \sin E \,\mathrm{d}E}{\sqrt{-a(1 - e\cos E)^2 + 2a(1 - e\cos E) - a(1 - e^2)}} \\ &= \frac{a^{3/2}}{\sqrt{\mu}} \int \frac{e \sin E(1 - e\cos E)\,\mathrm{d}E}{\sqrt{-1 + 2e\cos E - e^2\cos^2 E + 2 - 2e\cos E - 1 + e^2}} \\ &= \frac{a^{3/2}}{\sqrt{\mu}} \int \frac{e\sin E(1 - e\cos E)\,\mathrm{d}E}{e \sin E} \\ &= \frac{a^{3/2}}{\sqrt{\mu}} \int (1 - e\cos E)\,\mathrm{d}E \\ &= \frac{a^{3/2}}{\sqrt{\mu}} (E - e\sin E). \end{aligned} \tag{4.82}$$

The factor $a^{-3/2}\mu^{-1/2}$ is the inverse of the mean motion, and thus

$$Q_1 = -t + I_1 = -t + \frac{1}{n}(E - e\sin E) = -t + \frac{M}{n},$$

and since the mean anomaly M is $M = n(t - \tau)$,

$$Q_1 = -\tau. \tag{4.83}$$

Thus the variable Q_1 is essentially the time of perihelion (with a minus sign).

Calculation of the integral I_2 begins in a similar manner:

$$\begin{aligned} I_2 &= \frac{1}{\sqrt{\mu}} \int \frac{ae\sin E\,\mathrm{d}E}{a(1 - e\cos E)\sqrt{a}e\sin E} \\ &= \frac{1}{\sqrt{a\mu(1 - e^2)}} \int \frac{\sqrt{1 - e^2}\,\mathrm{d}E}{1 - e\cos E}. \end{aligned} \tag{4.84}$$

We need here one more substitution. The integrand resembles the expression (3.36) for $\sin\phi$, where ϕ is the true anomaly. Let us here denote the true anomaly by the symbol f. Then

$$\sin f = \sqrt{1 - e^2}\frac{\sin E}{1 - e\cos E}.$$

Differentiating this we get

$$\begin{aligned} \cos f\,\mathrm{d}f &= \sqrt{1 - e^2}\frac{(1 - e\cos E)\cos E - e\sin^2 E}{(1 - e\cos E)^2}\,\mathrm{d}E \\ &= \sqrt{1 - e^2}\frac{\cos E - e}{(1 - e\cos E)^2}\,\mathrm{d}E \\ &= \frac{\sqrt{1 - e^2}}{1 - e\cos E}\cos f\,\mathrm{d}E, \end{aligned}$$

from which

$$\mathrm{d}f = \frac{\sqrt{1-e^2}}{1-e\cos E}\,\mathrm{d}E.$$

Thus the integrand of I_2 is just $\mathrm{d}f$, and the value of the integral is simply f. The variable Q_2 is

$$Q_2 = \phi - \frac{\alpha_2}{m} I_2 = \phi - \frac{\alpha_2}{m}\frac{f}{\sqrt{a\mu(1-e^2)}} = \phi - f,$$

where the last equality is obtained by substituting the expression (4.77) for α_2. Thus we get

$$f = \phi - Q_2. \tag{4.85}$$

The true anomaly f is measured from the perihelion and ϕ from some arbitrary direction. Thus Q_2 must be the the angular distance of the perihelion from this arbitrary direction. If this direction is the ascending node, Q_2 is just the argument of perihelion ω.

To summarise:

$$\begin{aligned} K &= 0, \\ Q_1 &= -\tau, \\ Q_2 &= \omega, \\ P_1 &= mh, \\ P_2 &= m\sqrt{a\mu(1-e^2)}. \end{aligned} \tag{4.86}$$

We have now found a coordinate frame in which the Hamiltonian vanishes and all coordinates and momenta are constants. In other words, the canonical variables in the equations of motion are also integration constants of the equations!

Since the original Hamiltonian does not depend on time the equation

$$H + \frac{\partial S}{\partial t} = 0$$

shows that S_t must be $-\alpha_1 t$, where α_1 is the constant value of H. Further, because H does not depend on the angle ϕ, it follows from the Hamiltonian equations of motion that the momentum p_ϕ is constant and therefore

$$p_\phi = \frac{\partial S}{\partial \phi} = \frac{\mathrm{d}S_\phi}{\mathrm{d}\phi}$$

is constant, too. Hence S must be $\alpha_2\phi$.

The part of the generating function corresponding to cyclic coordinates can be written directly without having to carry out the separation of variables in the Hamilton–Jacobi equation.

4.11 Two-body problem in Hamiltonian mechanics: three dimensions

The kinetic energy of a planet in spherical coordinates is

$$T = \frac{1}{2}m(\dot{r}^2 + r^2\dot{\theta}^2 + r^2\cos^2\theta\dot{\phi}^2)$$

and the potential energy is as before

$$V = -\mu m/r.$$

The conjugate momenta corresponding to the spherical coordinates are

$$\begin{aligned} p_r &= m\dot{r}, \\ p_\theta &= mr^2\dot{\theta}, \\ p_\phi &= mr^2\cos^2\theta\dot{\phi}. \end{aligned} \tag{4.87}$$

The Hamiltonian is

$$H = \frac{1}{2m}\left(p_r^2 + \frac{p_\theta^2}{r^2} + \frac{p_\phi^2}{r^2\cos^2\theta}\right) - \frac{\mu m}{r} \tag{4.88}$$

and the Hamilton–Jacobi equation is

$$\frac{1}{2m}\left(\left(\frac{\partial S}{\partial r}\right)^2 + \frac{1}{r^2}\left(\frac{\partial S}{\partial \theta}\right)^2 + \frac{1}{r^2\cos^2\theta}\left(\frac{\partial S}{\partial \phi}\right)^2\right) - \frac{\mu m}{r} + \frac{\partial S}{\partial t} = 0. \tag{4.89}$$

Using a trial solution

$$S = S_t(t) + S_r(r) + S_\theta(\theta) + S_\phi(\phi) \tag{4.90}$$

we can separate the Hamilton–Jacobi equation into four equations:

$$\begin{aligned} \frac{\mathrm{d}S_t}{\mathrm{d}t} &= -\alpha_1, \\ \frac{\mathrm{d}S_\phi}{\mathrm{d}\phi} &= \alpha_2, \\ \left(\frac{\mathrm{d}S_\theta}{\mathrm{d}\theta}\right)^2 + \frac{\alpha_2^2}{\cos^2\theta} &= \alpha_3^2, \\ \left(\frac{\mathrm{d}S_r}{\mathrm{d}r}\right)^2 + \frac{\alpha_3^2}{r^2} &= 2m\left(\alpha_1 + \frac{\mu m}{r}\right). \end{aligned} \tag{4.91}$$

Before continuing, we will study the meaning of the constants α_i. As before we have

$$\alpha_1 = mh. \tag{4.92}$$

To find the constant α_2 we use the second equation (4.91) and notice that the derivative of the generating function on the left hand side is the momentum p_ϕ:

$$\alpha_2 = \frac{\partial S}{\partial \phi} = p_\phi = mr^2 \cos^2 \theta \dot{\phi}.$$

Since the projection of the radius vector on the xy plane is $r \cos\theta$ and the projection of the velocity is $r \cos\theta\dot{\phi}$, we suspect that α_2 is a projection of the product of $\boldsymbol{r}$ and $\dot{\boldsymbol{r}}$. Let us find the z-component of the angular momentum $m\boldsymbol{r} \times \dot{\boldsymbol{r}}$. It is obtained by taking the vector product of the projections of $m\boldsymbol{r}$ and $\dot{\boldsymbol{r}}$ on the xy plane. Since only the component of $\dot{\boldsymbol{r}}$ perpendicular to $\boldsymbol{r}$ affects this product, the z-component of the angular momentum is

$$mr \cos\theta r \cos\theta\dot{\phi} = mr^2 \cos^2\theta\dot{\phi} = \alpha_2.$$

Thus the constant α_2 is the z-component of the angular momentum. It can be written also as

$$\alpha_2 = m\sqrt{a\mu(1 - e^2)} \cos \iota. \tag{4.93}$$

It remains to find the third constant:

$$\begin{aligned}\alpha_3 &= \sqrt{\left(\frac{\partial S}{\partial \theta}\right)^2 + \frac{\alpha_2^2}{\cos^2\theta}} = \sqrt{p_\theta^2 + \frac{p_\phi^2}{\cos^2\theta}} \\ &= \sqrt{m^2 r^4 \dot{\theta}^2 + \frac{m^2 r^4 \cos^4\theta\dot{\phi}^2}{\cos^2\theta}} \\ &= mr^2\sqrt{\dot{\theta}^2 + \cos^2\theta\dot{\phi}^2} \\ &= mr^2 \dot{f},\end{aligned} \tag{4.94}$$

where f is the angle measured in the direction of motion (i.e. the true anomaly). But this is the total angular momentum, and hence

$$\alpha_3 = m\sqrt{a\mu(1 - e^2)}. \tag{4.95}$$

Now we know the new momenta:

$$\begin{aligned}P_1 &= \alpha_1 = mh, \\ P_2 &= \alpha_2 = m\sqrt{a\mu(1 - e^2)} \cos \iota, \\ P_3 &= \alpha_3 = m\sqrt{a\mu(1 - e^2)}.\end{aligned} \tag{4.96}$$

Next we have to find the new coordinates. The generating function is

$$\begin{aligned} S = &- t\alpha_1 + \phi\alpha_2 + \int \mathrm{d}\theta \sqrt{\alpha_3^2 - \frac{\alpha_2^2}{\cos^2\theta}} \\ &+ \int \mathrm{d}r \sqrt{2m\left(\alpha_1 + \frac{\mu m}{r}\right) - \frac{\alpha_3^2}{r^2}}. \end{aligned} \tag{4.97}$$

The new coordinates are the derivatives of this:

$$\begin{aligned} Q_1 &= \frac{\partial S}{\partial \alpha_1} = -t + \int \frac{m\,\mathrm{d}r}{\sqrt{2m(\alpha_1 + \mu m/r) - \alpha_3^2/r^2}} = -t + I_1, \\ Q_2 &= \frac{\partial S}{\partial \alpha_2} = \phi - \alpha_2 \int \frac{\mathrm{d}\theta}{\cos^2\theta\sqrt{\alpha_3^2 - \alpha_2^2/\cos^2\theta}} = \phi - \alpha_2 I_3, \\ Q_3 &= \frac{\partial S}{\partial \alpha_3} = \alpha_3 \int \frac{\mathrm{d}\theta}{\sqrt{\alpha_3^2 - \alpha_2^2/\cos^2\theta}} \\ &\quad - \frac{\alpha_3}{m} \int \frac{m\,\mathrm{d}r}{r^2\sqrt{2m(\alpha_1 + \mu m/r) - \alpha_3^2/r^2}} \\ &= \alpha_3(I_4 - I_2/m), \end{aligned} \tag{4.98}$$

where I_1 and I_2 are the integrals evaluated in the previous section.

We still have to calculate the integrals I_3 and I_4. We begin with $\alpha_3 I_4$. We replace α_2 and α_3 by their expressions in terms of the orbital elements (4.93) and (4.95):

$$\begin{aligned} \alpha_3 I_4 = \alpha_3 \int \frac{\mathrm{d}\theta}{\sqrt{\alpha_3^2 - \alpha_2^2/\cos^2\theta}} &= \int \frac{\mathrm{d}\theta}{\sqrt{1 - (\alpha_2/\alpha_3)^2/\cos^2\theta}} \\ = \int \frac{\mathrm{d}\theta}{\sqrt{1 - \cos^2\iota/\cos^2\theta}} &= \int \frac{\cos\theta\,\mathrm{d}\theta}{\sqrt{\cos^2\theta - \cos^2\iota}}. \end{aligned} \tag{4.99}$$

We replace the angle θ by an angle η measured from the ascending node (Fig. 4.2). From the sine formula of spherical trigonometry we get

$$\frac{\sin\theta}{\sin\iota} = \frac{\sin\eta}{\sin\pi/2}$$

or

$$\sin\theta = \sin\iota\sin\eta. \tag{4.100}$$

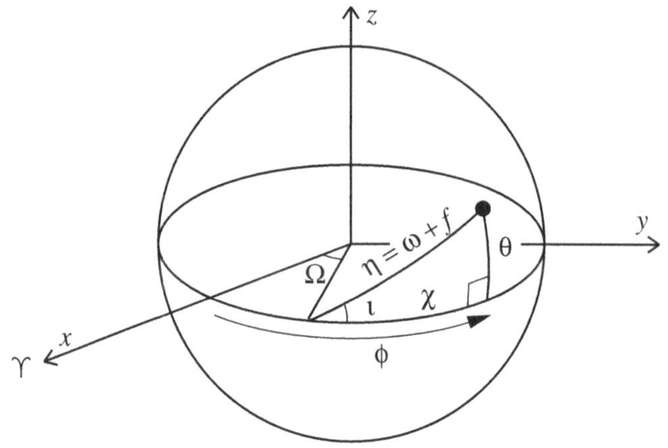

Figure 4.2 Quantities appearing in the solution of the three-dimensional Hamilton–Jacobi equation.

This gives the formulae needed to transform the integral (4.99):

$$\begin{aligned} \cos\theta &= \sqrt{1-\sin^2\iota\sin^2\eta}, \\ \cos\theta\,\mathrm{d}\theta &= \sin\iota\cos\eta\,\mathrm{d}\eta. \end{aligned} \tag{4.101}$$

Substitution into (4.99) gives

$$\begin{aligned} \alpha_3 I_4 &= \int \frac{\sin\iota\cos\eta\,\mathrm{d}\eta}{\sqrt{1-\sin^2\iota\sin^2\eta-\cos^2\iota}} \\ &= \int \frac{\sin\iota\cos\eta\,\mathrm{d}\eta}{\sqrt{\sin^2\iota(1-\sin^2\eta)}} = \int \mathrm{d}\eta = \eta. \end{aligned} \tag{4.102}$$

Thus we get

$$\alpha_3 I_4 = \eta. \tag{4.103}$$

The remaining integral is evaluated in a similar manner. We begin by eliminating the constants α_2 and α_3:

$$\begin{aligned} \alpha_2 I_3 &= \alpha_2 \int \frac{\mathrm{d}\theta}{\cos^2\theta\sqrt{\alpha_3^2-\alpha_2^2/\cos^2\theta}} \\ &= \frac{\alpha_2}{\alpha_3}\int \frac{\mathrm{d}\theta}{\cos^2\theta\sqrt{1-(\alpha_2/\alpha_3)^2/\cos^2\theta}} \\ &= \int \frac{\cos\iota\,\mathrm{d}\theta}{\cos^2\theta\sqrt{1-\cos^2\iota/\cos^2\theta}}. \end{aligned} \tag{4.104}$$

From spherical trigonometry we get (Fig. 4.2)

$$\tan\theta \cot\iota = \sin\chi, \tag{4.105}$$

and hence

$$\begin{aligned}\frac{\mathrm{d}\theta}{\cos^2\theta} &= \cos\chi \tan\iota\, \mathrm{d}\chi,\\ \frac{1}{\cos^2\theta} &= 1 + \tan^2\theta = 1 + \sin^2\chi \tan^2\iota.\end{aligned} \tag{4.106}$$

Substitution into the integral (4.104) gives

$$\begin{aligned}\alpha_2 I_3 &= \int \frac{\cos\iota \cos\chi \tan\iota\, \mathrm{d}\chi}{\sqrt{1-(1+\sin^2\chi \tan^2\iota)\cos^2\iota}}\\ &= \int \frac{\sin\iota \cos\chi\, \mathrm{d}\chi}{\sqrt{1-\cos^2\iota - \sin^2\chi \sin^2\iota)}}\\ &= \int \frac{\sin\iota \cos\chi\, \mathrm{d}\chi}{\sqrt{\sin^2\iota(1-\sin^2\chi)}}\\ &= \int \mathrm{d}\chi = \chi.\end{aligned} \tag{4.107}$$

In summary, the new coordinates are

$$\begin{aligned}Q_1 &= -t + I_1 = -t + M/n,\\ Q_2 &= \phi - \alpha_2 I_3 = \phi - \chi,\\ Q_3 &= \alpha_3 I_4 - \frac{\alpha_3}{m} I_2 = \eta - \frac{\alpha_3}{m\sqrt{a\mu(1-e^2)}} f = \eta - f.\end{aligned} \tag{4.108}$$

As before, $Q_1 = -\tau$. From Fig. 4.2 we can see that $\phi - \chi$ is the longitude of the ascending node and $\eta - f$ the argument of perihelion. Thus the canonical coordinates can be expressed in terms of the familiar orbital elements in a very simple way:

$$\begin{aligned}Q_1 &= -\tau,\\ Q_2 &= \Omega,\\ Q_3 &= \omega.\end{aligned} \tag{4.109}$$

The corresponding momenta are

$$\begin{aligned}P_1 &= mh = -\frac{m\mu}{2a},\\ P_2 &= m\sqrt{a\mu(1-e^2)}\cos\iota,\\ P_3 &= m\sqrt{a\mu(1-e^2)}.\end{aligned} \tag{4.110}$$

We have now completely solved the two-body Hamilton–Jacobi equation. All the canonical variables are constants, but just like the standard orbital elements they can be used to determine unambiguously the position of the planet.

4.12 Delaunay's elements

We have found a way to describe the two-body system in a manner which in a certain sense is the simplest possible: we have found a coordinate frame in which the Hamiltonian vanishes and all variables in the equations of motion are themselves integration constants of the equations. Before leaving the two-body problem we will introduce a slightly different set of canonical variables that is frequently used in celestial mechanics.

The mass of the planet appears only as a multiplicative constant in the Hamiltonian and canonical momenta, and can as well be omitted. Thus we get the following simplified canonical variables:

$$\begin{aligned} q_1 &= -\tau, \\ q_2 &= \Omega, \\ q_3 &= \omega, \\ p_1 &= -\frac{\mu}{2a}, \\ p_2 &= \sqrt{a\mu(1-e^2)}\cos\iota, \\ p_3 &= \sqrt{a\mu(1-e^2)}. \end{aligned} \tag{4.111}$$

Except for the first one the coordinates q_i are angles. To make the set of coordinates more uniform we replace q_1 with the mean anomaly $M = n(t - \tau)$ keeping the other coordinates intact. We require that

$$\begin{aligned} h &= q_2 = \Omega, \\ g &= q_3 = \omega, \\ l &= n(t + q_1), \\ H &= p_2 = \sqrt{a\mu(1-e^2)}\cos\iota, \\ G &= p_3 = \sqrt{a\mu(1-e^2)}, \\ L &= ? \end{aligned} \tag{4.112}$$

where h, g and l are the new coordinates and H, G and L the corresponding momenta. The new coordinates are traditionally labelled by these symbols, which should not be confused with the previously used h (energy per unit mass), H (Hamiltonian), G (gravitational constant) and L (angular momentum).

We have to find the new Hamiltonian K and the momentum L in such a way that the set of variables remains canonical. It is quite easy to guess that the function

$$F = \left(nL - \frac{3\mu}{2a}\right)(t + q_1) + q_2 H + q_3 G \tag{4.113}$$

will generate the correct transformation. The constant in the first term is needed to make the new variables as simple as possible. We get

$$p_1 = \frac{\partial F}{\partial q_1} = -\frac{3\mu}{2a} + nL,$$

from which

$$L = \frac{1}{n}\left(-\frac{\mu}{2a} + \frac{3\mu}{2a}\right) = \frac{\mu}{an} = \frac{\mu}{a\sqrt{\mu}a^{-3/2}} = \sqrt{a\mu}.$$

The new Hamiltonian is (since the old Hamiltonian is zero)

$$K = \frac{\partial F}{\partial t} = -\frac{3\mu}{2a} + \frac{\mu}{a} = -\frac{\mu}{2a} = -\frac{\mu^2}{2L^2}.$$

The orbit of the planet is now described by the following quantities:

$$\begin{aligned} l &= n(t - \tau) = M, \\ g &= \omega, \\ h &= \Omega, \\ L &= \sqrt{a\mu}, \\ G &= \sqrt{a\mu(1 - e^2)}, \\ H &= \sqrt{a\mu(1 - e^2)} \cos \iota, \\ K &= -\frac{\mu^2}{2L^2}. \end{aligned} \tag{4.114}$$

These are called *Delaunay's elements*. They are an example of the more general action-angle variables.

4.13 Hamiltonian formulation of the three-body problem

For a three-body problem it is usually convenient to use the Jacobi form and the Delaunay elements to describe both the inner orbit (subscript i, see Fig. 4.3) and

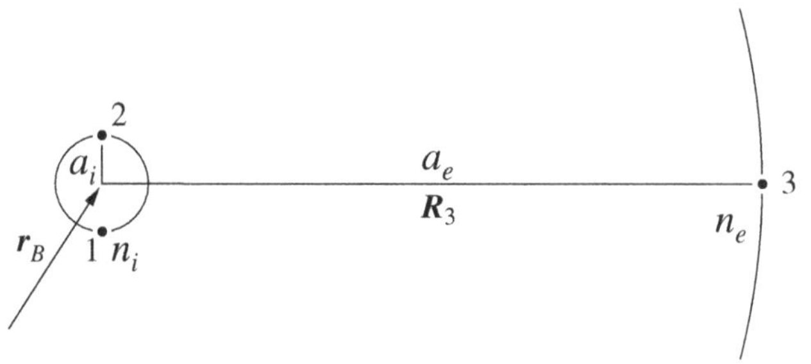

Figure 4.3 Internal (i) and external (e) systems.

the external orbit (subscript e). Using the osculating elements we may write

$$\begin{aligned} l_i &= M_i & L_i &= n_i a_i^2 \\ g_i &= \omega_i & G_i &= n_i a_i^2 \sqrt{(1-e_i^2)} \\ h_i &= \Omega_i & H_i &= n_i a_i^2 \sqrt{(1-e_i^2)} \cos \iota_i \\ l_e &= M_e & L_e &= n_e a_e^2 \\ g_e &= \omega_e & G_e &= n_e a_e^2 \sqrt{(1-e_e^2)} \\ h_e &= \Omega_e & H_e &= n_e a_e^2 \sqrt{(1-e_e^2)} \cos \iota_e. \end{aligned} \tag{4.115}$$

The osculating elements are calculated from the initial values as follows:

$$\begin{aligned} a_i &= \left(\frac{2}{r} - \frac{v^2}{Gm_B}\right)^{-1}, \\ a_e &= \left(\frac{2}{R_3} - \frac{V_3^2}{GM}\right)^{-1}, \\ n_i &= \sqrt{Gm_B/a_i^3}, \\ n_e &= \sqrt{GM/a_e^3}, \\ e_i &= \sqrt{1 + \frac{(\boldsymbol{r}\cdot\dot{\boldsymbol{r}})^2 - r^2 v^2}{Gm_B a_i}}, \\ e_e &= \sqrt{1 + \frac{(\boldsymbol{R}_3\cdot\dot{\boldsymbol{R}}_3)^2 - R_3^2 V_3^2}{GMa_e}}, \end{aligned} \tag{4.116}$$

etc. The first two values are found by solving the semi-major axis from the formula (3.32) for the orbital velocity. The mean motions n_i and n_e are given by (3.45). Finally, the eccentricities follow from the relation $k = \sqrt{a\mu(1-e^2)} = r\sqrt{v^2 - \dot{r}^2}$.

The Hamiltonian is

$$H = -\frac{\mu^2}{2L_i^2} - \frac{m^2}{2L_e^2} + Gm_3\left(\frac{m_B}{R_3} - \frac{m_1}{r_{13}} - \frac{m_2}{r_{23}}\right), \tag{4.117}$$

and the Hamiltonian equations of motion are

$$\begin{aligned}
\dot{L}_i &= -\frac{\partial H}{\partial l_i} & \dot{L}_e &= -\frac{\partial H}{\partial l_e} \\
\dot{G}_i &= -\frac{\partial H}{\partial g_i} & \dot{G}_e &= -\frac{\partial H}{\partial g_e} \\
\dot{H}_i &= -\frac{\partial H}{\partial h_i} & \dot{H}_e &= -\frac{\partial H}{\partial h_e} \\
\dot{l}_i &= \frac{\partial H}{\partial L_i} & \dot{l}_e &= \frac{\partial H}{\partial L_e} \\
\dot{g}_i &= \frac{\partial H}{\partial G_i} & \dot{g}_e &= \frac{\partial H}{\partial G_e} \\
\dot{h}_i &= \frac{\partial H}{\partial H_i} & \dot{h}_e &= \frac{\partial H}{\partial H_e}.
\end{aligned} \tag{4.118}$$

In order to be able to solve these equations, we have to express quantities R_3, r_{13} and r_{23} in the Hamiltonian in terms of the Delaunay elements. This is not an easy task, and makes the use of the Hamiltonian method more difficult than it appears at first.

The three components of the total angular momentum can be written with the aid of Delaunay elements as follows:

$$\boldsymbol{L} = (K_i \sin h_i + K_e \sin h_e, -K_i \cos h_i - K_e \cos h_e, H_i + H_e) \tag{4.119}$$

where

$$K_i = \sqrt{G_i^2 - H_i^2}, \quad K_e = \sqrt{G_e^2 - H_e^2} \tag{4.120}$$

(Problem 4.6).

4.14 Elimination of nodes

The Hamiltonian equations of motion may be simplified by choosing an optimal coordinate system. We take the invariable plane (perpendicular to the total angular momentum **L**) as the reference plane (Figs. 4.4 and 4.5). Then the components of

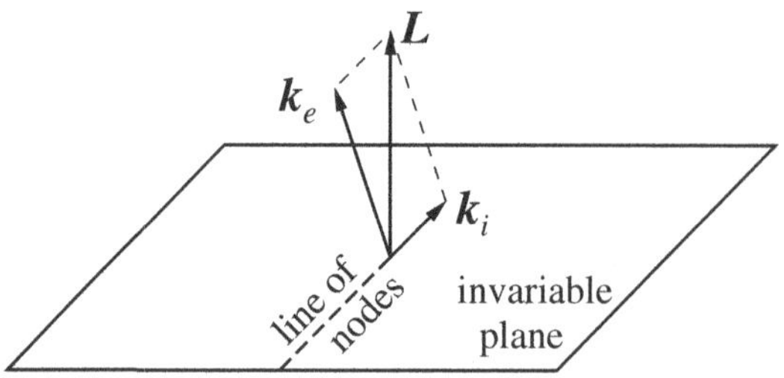

Figure 4.4 The invariable plane is perpendicular to the angular momentum vector $\boldsymbol{L}$. The line of nodes is perpendicular to the plane spanned by the vectors $\boldsymbol{k}_e$ and $\boldsymbol{k}_i$.

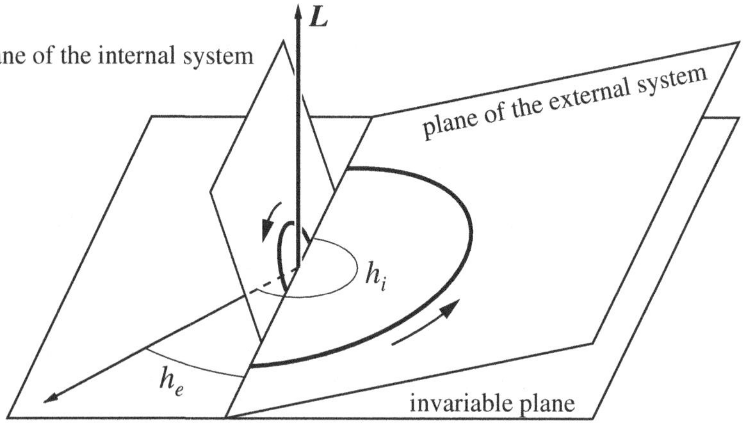

Figure 4.5 Elimination of nodes. In the system defined by the invariable plane the longitudes of the ascending nodes h_e and h_i differ by π.

$\boldsymbol{L}$ are (0, 0, L), or

$$\begin{aligned} K_i &= K_e \qquad G_i^2 - H_i^2 = G_e^2 - H_e^2, \\ h_i + \pi &= h_e \qquad H_i + H_e = L. \end{aligned} \tag{4.121}$$

From here we solve for H_i and H_e:

$$\begin{aligned} H_i &= \left(L^2 + G_i^2 - G_e^2\right)/2L, \\ H_e &= \left(L^2 - G_i^2 + G_e^2\right)/2L. \end{aligned} \tag{4.122}$$

The quantities H_i and H_e are constants of motion which we infer as follows. The energy of the system, i.e. its Hamiltonian, does not depend on h_e since h_e only specifies the orientation of the three-body system in space which is of no consequence as regard to the total energy. The difference $h_e - h_i$ would generally matter, but in our special coordinate system this difference is equal to π (Fig. 4.5). Consequently the Hamiltonian cannot depend on h_i either. Since the Hamiltonian does not depend explicitly on either h_e or h_i, the conjugate momenta H_e and H_i must be constants of motion.

This is an important result. For the two binaries it tells us that

$$\begin{aligned} \sqrt{m_B a_i \left(1 - e_i^2\right)} \cos \iota_i &= \text{constant}, \\ \sqrt{M a_e \left(1 - e_e^2\right)} \cos \iota_e &= \text{constant}. \end{aligned} \tag{4.123}$$

The values of the constants are determined at some initial moment of time, and Eq. (4.123) governs the subsequent evolution of the system.

This choice of a coordinate system is called 'elimination of nodes'. From the set of equations (4.118) this choice eliminates those which relate h_i to H_i and h_e to H_e.

4.15 Elimination of mean anomalies

Often we are interested in the evolution of the three-body system over a long period of time. In particular, we would like to know whether there are, and under what conditions, long term trends of evolution, so-called *secular evolution*. In a hierarchical triple, the long term corresponds to calculating the trends over many orbital periods of both the inner and the outer binaries. In the lowest order, when the size of the inner binary orbit is much smaller than the outer orbit, the two semi-major axes remain constant, i.e. they do not show secular evolution (see Chapter 9 for a more thorough discussion). Then one may average the Hamiltonian over both orbital cycles, and the averaged Hamiltonian can be used in place of the original Hamiltonian (Marchal 1990). Now the new Hamiltonian does not depend explicitly on the mean anomalies l_i and l_e, and the corresponding canonical momenta L_i and L_e are constants, as they should be since this means that the semi-major axes of the inner and the outer orbits remain constant.

It is rather obvious that the semi-major axes can remain constant only if the system is sufficiently hierarchical, i.e. $a_e/a_i \gg 1$. In other words, the two binary orbits should not perturb each other too much. We need methods of perturbation theory to look for a suitable Hamiltonian to carry out the elimination of mean anomalies.

Problems

Problem 4.1 Show that the equation

$$\frac{\partial f}{\partial x} - \frac{\mathrm{d}}{\mathrm{d}x}\left(f - y'\frac{\partial f}{\partial y'}\right) = 0$$

is equivalent to the Lagrange equation (4.13).

Problem 4.2 A body with a mass M moves along the x axis in a potential $-kx$ ($k > 0$). Write its Lagrangian and Hamiltonian equations of motion.

Problem 4.3 The rectangular coordinates of a body are x, y and z and the corresponding momenta p_x, p_y and p_z. Apply a transformation with a generating function

$$f = (x - at)P_X + (y - bt)P_Y + (z - ct)P_Z,$$

where a, b and c are constants and t is time. Express the new coordinates X, Y and Z and the momenta P_X, P_Y and P_Z as functions of the old ones. What is the new Hamiltonian if the original one is

$$H = \frac{p_x^2 + p_y^2 + p_z^2}{2m}?$$

What is the physical interpretation of the transformation?

Problem 4.4 The rectangular coordinates of a body are x and y and the momenta p_x and p_y. The Hamiltonian is

$$H = \frac{p_x^2 + p_y^2}{2m} - \frac{k}{\sqrt{x^2 + y^2}},$$

where $k > 0$ is a constant. Find a generating function for transformation to polar coordinates, the new momenta and the new Hamiltonian.

Problem 4.5 Delaunay's elements of an object on 1 April 1983 were $l = 30°$, $g = 10°, h = 100°, L = 0.03\ \mathrm{AU}^2/\mathrm{d}, G = 0.029\ \mathrm{AU}^2/\mathrm{d}, H = 0.025\ \mathrm{AU}^2/\mathrm{d}$. Find its ordinary orbital elements.

Problem 4.6 Verify Eq. (4.119).

5

The planar restricted circular three-body problem and other special cases

After the two-body problem, the next more complicated system consists of three bodies. Let us call these bodies the Sun, planet and asteroid. Some further assumptions are made to keep the system as simple as possible. The word *restricted* means here that the mass of the asteroid is so small that it does not significantly affect the motion of the primaries (the Sun and the planet). The primaries move in *circular* orbits, and the asteroid is assumed to move in the *same plane* as the primaries. The perturbations due to the third body can be neglected and the positions of the primaries can be calculated analytically for all times. The problem is now to find the trajectory of the massless body.

The assumption about the mass of the asteroid is a little problematic. If the primaries affect the motion of the asteroid, it must, of course, affect their motions according to Newton's third law. The accuracy required determines whether the third body can actually be considered massless. Discarding Newton's third law has a side effect: total energy is no longer conserved. However, the energy conservation law can be replaced by another similar law.

5.1 Coordinate frames

When studying the restricted circular three body problem, the units are usually chosen in such a way that the properties of the system depend on a single parameter.

- The total mass of the primaries is taken as the unit of mass. The mass of the planet is denoted by μ, whence the mass of the Sun is $1 - \mu$. (This must not be confused with the μ appearing in the two-body problem.)
- The distance between the primaries is the unit of distance. The distances of the planet and the Sun from the centre of mass are then $1 - \mu$ and μ, respectively.
- The unit of time is chosen so that the mean motion of the primaries is $n = 1$. Hence the mean anomaly equals the time.

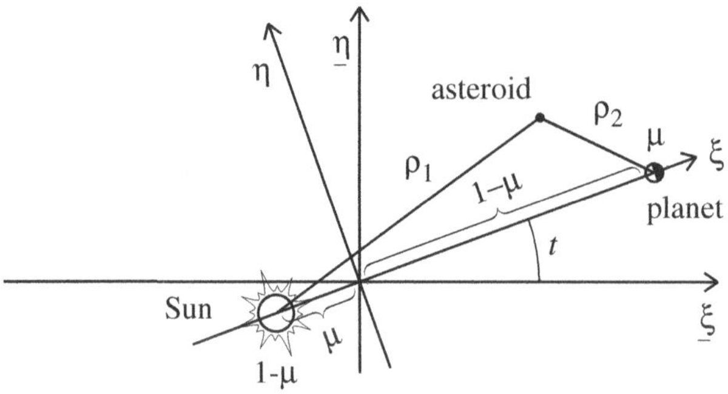

Figure 5.1 The inertial sidereal $\underline{\xi\eta}$ and rotating synodic $\xi\eta$ coordinate frames of the three-body problem.

From these it follows that the gravitational constant is unity. The only remaining parameter is μ.

To describe the orbits one needs, first of all, a fixed sidereal frame $(\underline{\xi}, \underline{\eta})$, the origin of which is at the centre of mass of the system (Fig. 5.1). In this frame, the positions of the primaries as functions of time are

$$\begin{aligned} \text{Sun:} & \begin{cases} \underline{\xi} = -\mu \cos t \\ \underline{\eta} = -\mu \sin t \end{cases} \\ \text{planet:} & \begin{cases} \underline{\xi} = (1-\mu)\cos t \\ \underline{\eta} = (1-\mu)\sin t. \end{cases} \end{aligned} \tag{5.1}$$

Things look simpler in a frame where the primaries are stationary. Therefore we will also use a synodic $\xi\eta$ frame, the origin of which is also at the centre of mass, but with the positive ξ axis pointing in the direction of the planet.

The synodic frame rotates with an angular velocity of $n = 1$. In this frame the coordinates of the Sun are $(-\mu, 0)$ and those of the planet are $(1-\mu, 0)$. Denote the position of the asteroid by (ξ, η) and the distances of the asteroid from the Sun and the planet by ρ_1 and ρ_2, respectively.

5.2 Equations of motion

The Newtonian equations of motion in the rotating frame are now required. Beginning with the Hamiltonian equations in the inertial sidereal frame, a canonical transformation is employed to get the equations in the rotating (synodic) frame.

Denoting the conjugate momenta corresponding to the $\underline{\xi}$- and $\underline{\eta}$-coordinates by $p_{\underline{\xi}}$ and $p_{\underline{\eta}}$, the Hamiltonian of the asteroid is

$$\underline{H} = \frac{1}{2}\left(p_{\underline{\xi}}^2 + p_{\underline{\eta}}^2\right) - \frac{1-\mu}{\rho_1} - \frac{\mu}{\rho_2}, \tag{5.2}$$

where

$$\begin{aligned}\rho_1 &= \left((\underline{\xi} + \mu\cos t)^2 + (\underline{\eta} + \mu\sin t)^2\right)^{1/2},\\ \rho_2 &= \left((\underline{\xi} - (1-\mu)\cos t)^2 + (\underline{\eta} - (1-\mu)\sin t)^2\right)^{1/2}.\end{aligned} \tag{5.3}$$

The $\underline{\xi}$, $\underline{\eta}$ and ξ, η coordinates are related by the equations

$$\begin{aligned}\underline{\xi} &= \xi\cos t - \eta\sin t,\\ \underline{\eta} &= \xi\sin t + \eta\cos t.\end{aligned} \tag{5.4}$$

Transformation to the rotating $\xi\eta$ frame can be accomplished using a generating function

$$\begin{aligned}F &= F(p_{\underline{\xi}}, p_{\underline{\eta}}, \xi, \eta)\\ &= -(\xi\cos t - \eta\sin t)p_{\underline{\xi}} - (\xi\sin t + \eta\cos t)p_{\underline{\eta}}.\end{aligned} \tag{5.5}$$

It is easy to see that this gives just the equations (5.4) for the coordinates:

$$\begin{aligned}\underline{\xi} &= -\frac{\partial F}{\partial p_{\underline{\xi}}} = \xi\cos t - \eta\sin t,\\ \underline{\eta} &= -\frac{\partial F}{\partial p_{\underline{\eta}}} = \xi\sin t + \eta\cos t.\end{aligned}$$

The new momenta are

$$\begin{aligned}P_\xi &= -\frac{\partial F}{\partial \xi} = p_{\underline{\xi}}\cos t + p_{\underline{\eta}}\sin t,\\ P_\eta &= -\frac{\partial F}{\partial \eta} = -p_{\underline{\xi}}\sin t + p_{\underline{\eta}}\cos t.\end{aligned} \tag{5.6}$$

The transformation is just a rotation, and thus the sum of the squares of the momenta remains unchanged, which is of course easy to check using the equations (5.6):

$$p_{\underline{\xi}}^2 + p_{\underline{\eta}}^2 = P_\xi^2 + P_\eta^2.$$

The new Hamiltonian is

$$H = \underline{H} + \frac{\partial F}{\partial t} = \frac{1}{2}\left(P_\xi^2 + P_\eta^2\right) - \frac{1-\mu}{\rho_1} - \frac{\mu}{\rho_2}$$
$$+ \xi p_{\underline{\xi}} \sin t + \eta p_{\underline{\xi}} \cos t - \xi p_{\underline{\eta}} \cos t + \eta p_{\underline{\eta}} \sin t.$$

The last four terms can be written as

$$\xi(p_{\underline{\xi}} \sin t - p_{\underline{\eta}} \cos t) + \eta(p_{\underline{\xi}} \cos t + p_{\underline{\eta}} \sin t)$$
$$= \xi(-P_\eta) + \eta P_\xi.$$

Thus we have

$$H = \frac{1}{2}\left(P_\xi^2 + P_\eta^2\right) + P_\xi \eta - P_\eta \xi - \frac{1-\mu}{\rho_1} - \frac{\mu}{\rho_2}, \tag{5.7}$$

where

$$\begin{aligned} {\rho_1}^2 &= (\xi + \mu)^2 + \eta^2, \\ {\rho_2}^2 &= (\xi - (1-\mu))^2 + \eta^2. \end{aligned} \tag{5.8}$$

The equations of motion of the asteroid in the synodic frame are now

$$\begin{aligned} \dot{\xi} &= \frac{\partial H}{\partial P_\xi} = P_\xi + \eta, \\ \dot{\eta} &= \frac{\partial H}{\partial P_\eta} = P_\eta - \xi, \\ \dot{P}_\xi &= -\frac{\partial H}{\partial \xi} = \frac{\partial}{\partial \xi}\left(\frac{1-\mu}{\rho_1} + \frac{\mu}{\rho_2}\right) + P_\eta, \\ \dot{P}_\eta &= -\frac{\partial H}{\partial \eta} = \frac{\partial}{\partial \eta}\left(\frac{1-\mu}{\rho_1} + \frac{\mu}{\rho_2}\right) - P_\xi. \end{aligned} \tag{5.9}$$

The Newtonian equations of motion are obtained by eliminating the momenta from (5.9). The derivatives of the momenta are found from the first two equations:

$$\begin{aligned} \dot{P}_\xi &= \ddot{\xi} - \dot{\eta}, \\ \dot{P}_\eta &= \ddot{\eta} + \dot{\xi}. \end{aligned}$$

Substitution into the two last equations (5.9) gives

$$\begin{aligned} \ddot{\xi} - \dot{\eta} &= \dot{\eta} + \xi + \frac{\partial}{\partial \xi}\left(\frac{1-\mu}{\rho_1} + \frac{\mu}{\rho_2}\right), \\ \ddot{\eta} + \dot{\xi} &= -\dot{\xi} + \eta + \frac{\partial}{\partial \eta}\left(\frac{1-\mu}{\rho_1} + \frac{\mu}{\rho_2}\right). \end{aligned}$$

Thus, the equations of motion are

$$\begin{aligned} \ddot{\xi} - 2\dot{\eta} &= \frac{\partial \Omega}{\partial \xi}, \\ \ddot{\eta} + 2\dot{\xi} &= \frac{\partial \Omega}{\partial \eta}, \end{aligned} \tag{5.10}$$

where Ω can be considered an effective potential:

$$\Omega = \frac{1}{2}\left(\xi^2 + \eta^2\right) + \frac{1-\mu}{\rho_1} + \frac{\mu}{\rho_2}. \tag{5.11}$$

The terms $-2\dot{\eta}$ and $2\dot{\xi}$ in (5.10) are the Coriolis terms, while the term $\frac{1}{2}(\xi^2 + \eta^2)$ in (5.11) represents the centrifugal potential. Sometimes a constant term is added to Ω, since it simplifies some results, but it is not needed here.

Example 5.1 We will see the appearance of the Coriolis terms and the centrifugal potential more clearly if we use the tools of Section 2.14 and rederive Eqs. (5.10) and (5.11) in vector form. Let the angular velocity of the comoving frame relative to the inertial frame be $\boldsymbol{\omega} = \hat{\boldsymbol{e}}_\zeta$, and the position vector of the third body in the comoving frame $\boldsymbol{r}$. Then Eq. (2.76) leads to

$$\ddot{\boldsymbol{r}} = -\nabla U - \boldsymbol{\omega} \times (\boldsymbol{\omega} \times \boldsymbol{r}) - 2\boldsymbol{\omega} \times \dot{\boldsymbol{r}},$$

where

$$U = -\frac{1-\mu}{\rho_1} - \frac{\mu}{\rho_2},$$

and the derivatives are taken in the rotating (comoving) frame. This can be written

$$\ddot{\boldsymbol{r}} + 2\boldsymbol{\omega} \times \dot{\boldsymbol{r}} = \nabla \Omega$$

if

$$\Omega = \frac{1}{2}(\boldsymbol{\omega} \times \boldsymbol{r})^2 - U.$$

In checking this result it is useful to remember the vector expression $\nabla(\boldsymbol{\omega} \times \boldsymbol{r})^2 = -2\boldsymbol{\omega} \times (\boldsymbol{\omega} \times \boldsymbol{r})$. Here we have the generalisation of Eqs. (5.10) and (5.11), with the Coriolis and centrifugal terms clearly shown.

5.3 Jacobian integral

The constants derived earlier for the two-body problem do not apply to the three-body problem. One of the constants, corresponding to the energy integral h, can be found relatively easily. We multiply the first equation of motion by $\dot{\xi}$, the second by $\dot{\eta}$, and add:

$$\dot{\xi}\ddot{\xi} - 2\dot{\xi}\dot{\eta} = \dot{\xi}\frac{\partial\Omega}{\partial\xi},$$
$$\dot{\eta}\ddot{\eta} + 2\dot{\eta}\dot{\xi} = \dot{\eta}\frac{\partial\Omega}{\partial\eta}$$
$$\Rightarrow$$
$$\dot{\xi}\ddot{\xi} + \dot{\eta}\ddot{\eta} = \frac{\partial\Omega}{\partial\xi}\dot{\xi} + \frac{\partial\Omega}{\partial\eta}\dot{\eta}.$$

Since Ω does not depend explicitly on time, the expression on the right is the total time derivative of Ω. The left hand side can be expressed in terms of the derivatives of the velocities:

$$\frac{\mathrm{d}}{\mathrm{d}t}\left(\dot{\xi}^2 + \dot{\eta}^2\right) = 2\frac{\mathrm{d}\Omega}{\mathrm{d}t}.$$

Integration gives

$$\dot{\xi}^2 + \dot{\eta}^2 = 2\Omega - C$$

or

$$C = 2\Omega(\xi, \eta) - v^2, \tag{5.12}$$

where v is the velocity of the asteroid relative to the primaries, and C is a constant, called the *Jacobian integral.*

Because the square of the velocity cannot be negative, the motion of the asteroid is restricted to the region where $v^2 = 2\Omega - C \geq 0$ or

$$\Omega \geq C/2. \tag{5.13}$$

When the position and velocity of the asteroid are known for some initial moment, the Jacobian integral can be evaluated. Since Ω is a function of position only, condition (5.13) tells immediately whether the asteroid can ever reach a given point (ξ, η). This condition does not say anything about the shape of the orbit, it only determines the region where the asteroid could move.

The condition (5.13) shows that the larger the value of C, the smaller the allowed region is. In Fig. 5.2 the shaded areas are forbidden regions for the asteroid $(\mu = 0.3)$.

When C is large, the allowed region consists of three separate areas. The asteroid can orbit either of the primaries, in which case it can never wander far from that primary, or it can move on a very wide orbit around both primaries. It can never move from one allowed region to another if the regions are not connected.

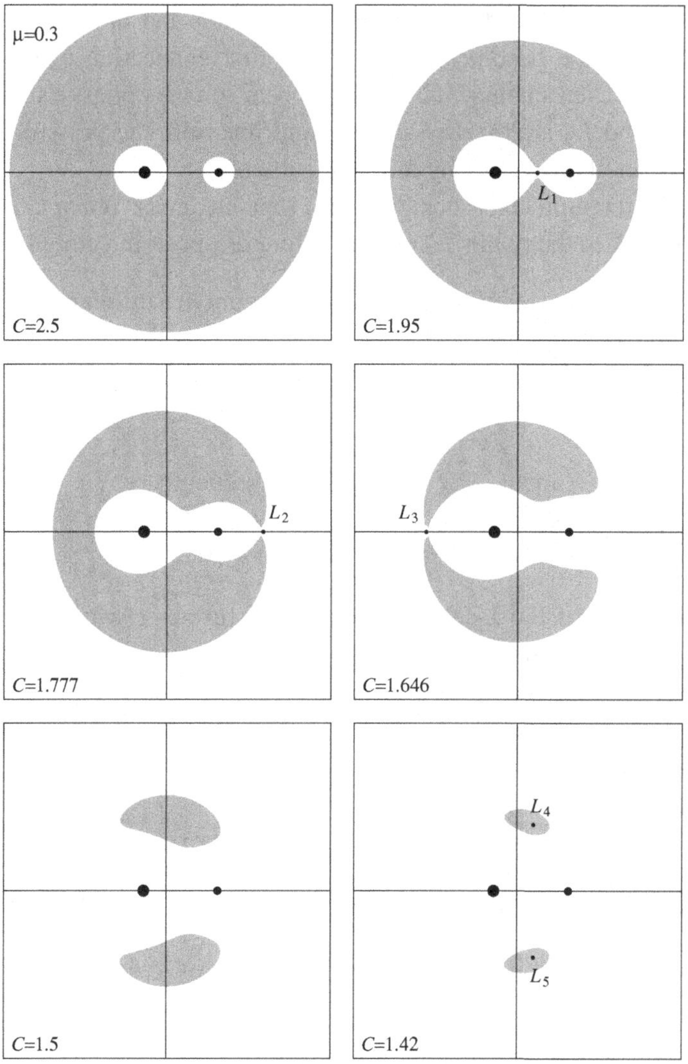

Figure 5.2 Forbidden regions (shaded areas) for different values of the Jacobian integral C in the case $\mu = 0.3$.

When C becomes smaller, a connection opens between the two inner regions, first at the point L_1. A small body orbiting the Sun can then be captured to an orbit around the planet, or the planet can lose a satellite to a solar orbit. However, the small body can never escape from the system. Its orbit is then said to be stable in Hill's sense (Hill 1878). There is no stability of this kind in the general three-body problem where the third body can always escape from the vicinity of the other two. But even in the general three-body problem there are some special motions for which the third body may remain in the vicinity of the other two forever.

When the value of C is further reduced, the outer and inner regions become connected and escape becomes possible. The first connection opens at L_2 and when C becomes even smaller, another one appears on the opposite side at L_3. The points L_1, L_2 and L_3 are on the same straight line with the primaries; they are, however, numbered differently by different authors.

The forbidden region consists now of two separate areas. When C is decreased these areas shrink to the points L_4 and L_5 before disappearing completely.

Example 5.2 We may rederive the Jacobian integral in a different way by writing in vector form (see Example 5.1)

$$\frac{\partial}{\partial t}\left(\frac{1}{2}\dot{\boldsymbol{r}}^2\right) = \dot{\boldsymbol{r}} \cdot \ddot{\boldsymbol{r}} = \dot{\boldsymbol{r}} \cdot \nabla\Omega - 2\dot{\boldsymbol{r}} \cdot (\boldsymbol{\omega} \times \dot{\boldsymbol{r}}) = \frac{\partial \Omega}{\partial t},$$

since $\dot{\boldsymbol{r}} \cdot (\boldsymbol{\omega} \times \dot{\boldsymbol{r}}) = 0$ and $\dot{\boldsymbol{r}} \cdot \nabla\Omega = \partial\Omega/\partial t$. Consequently

$$\Omega - \frac{1}{2}\dot{\boldsymbol{r}}^2 = \text{constant} = \frac{1}{2}C.$$

The significance of the Jacobian integral C is further clarified by writing the specific angular momentum (angular momentum per unit mass)

$$\boldsymbol{k} = \boldsymbol{r} \times \dot{\boldsymbol{r}} + \boldsymbol{r} \times (\boldsymbol{\omega} \times \boldsymbol{r})$$

and the energy per unit mass

$$h = \frac{1}{2}(\dot{\boldsymbol{r}}^2 + 2\dot{\boldsymbol{r}} \cdot (\boldsymbol{\omega} \times \boldsymbol{r}) + (\boldsymbol{\omega} \times \boldsymbol{r})^2) - U$$

in the comoving frame for the third body (Eq. (2.74)). With the help of

$$\boldsymbol{\omega} \cdot \boldsymbol{k} = \boldsymbol{\omega} \cdot (\boldsymbol{r} \times \dot{\boldsymbol{r}}) + \boldsymbol{\omega} \cdot \boldsymbol{r} \times (\boldsymbol{\omega} \times \boldsymbol{r}) = \dot{\boldsymbol{r}} \cdot (\boldsymbol{\omega} \times \boldsymbol{r}) + (\boldsymbol{\omega} \times \boldsymbol{r})^2$$

we get

$$\begin{aligned} h - \boldsymbol{\omega} \cdot \boldsymbol{k} &= \frac{1}{2}\dot{\boldsymbol{r}}^2 + \dot{\boldsymbol{r}} \cdot (\boldsymbol{\omega} \times \boldsymbol{r}) + \frac{1}{2}(\boldsymbol{\omega} \times \boldsymbol{r})^2 + U \\ &\quad - \dot{\boldsymbol{r}} \cdot (\boldsymbol{\omega} \times \boldsymbol{r}) - (\boldsymbol{\omega} \times \boldsymbol{r})^2 \\ &= \frac{1}{2}\dot{\boldsymbol{r}}^2 - \Omega = -\frac{1}{2}C. \end{aligned}$$

Therefore, even though h and $\boldsymbol{k}$ are not individually conserved in the time-varying potential field of the two primaries, the combination

$$h - \boldsymbol{\omega} \cdot \boldsymbol{k} = h - nk_\zeta = -\frac{1}{2}C$$

is conserved. Here k_ζ is the component of the angular momentum perpendicular to the orbital plane of the primaries, i.e. in the ζ-direction. From the above it follows that

$$dh = n\, dk_\zeta .$$

Thus the changes of the ζ-component of the specific angular momentum are tied together with the changes of the specific energy of the third body.

In terms of orbital elements, $h = -1/2a$, $k_\zeta = \sqrt{a(1-e^2)}\cos\iota$, and $n = 1$. Then C is called the Tisserand parameter T:

$$T = C = \frac{1}{a} + 2\sqrt{a(1-e^2)}\cos\iota . \tag{5.14}$$

5.4 Lagrangian points

The function Ω has five interesting special points $L_1, \ldots, L_5$. These are called the Lagrangian points. We now find their positions.

Inspecting the equations of motion (5.10) we can see that they have a trivial solution

$$\dot{\xi} = \dot{\eta} = 0 \tag{5.15}$$

if the derivatives of Ω vanish. Thus the third body can remain at rest at such a point. Since $\partial\Omega/\partial\xi = \partial\Omega/\partial\eta = 0$, these points are also extrema or saddle points of Ω. In the forbidden regions of Fig. 5.2 the values of Ω are smaller than in the allowed regions, and thus the points L_1, L_2 and L_3 are obviously saddle points and L_4 and L_5 are minima. Thus the points $L_1, \ldots, L_5$ are indeed determined by the condition

$$\frac{\partial\Omega}{\partial\xi} = \frac{\partial\Omega}{\partial\eta} = 0. \tag{5.16}$$

Calculating these derivatives we get a pair of equations for the Lagrangian points:

$$\frac{\partial\Omega}{\partial\xi} = \xi - \frac{(1-\mu)(\xi+\mu)}{{\rho_1}^3} - \frac{\mu(\xi-(1-\mu))}{{\rho_2}^3} = 0, \tag{5.17}$$

$$\frac{\partial\Omega}{\partial\eta} = \eta - \frac{(1-\mu)\eta}{{\rho_1}^3} - \frac{\mu\eta}{{\rho_2}^3} = 0. \tag{5.18}$$

The latter has a trivial solution $\eta = 0$. Substituting this into the first equation produces:

$$\xi - \frac{(1-\mu)(\xi+\mu)}{[(\xi+\mu)^2]^{3/2}} - \frac{\mu(\xi-(1-\mu))}{[(\xi-(1-\mu))^2]^{3/2}} = 0. \tag{5.19}$$

Simplifying this requires care with the signs of the square roots. To the right of both primaries, $\xi + \mu$ and $\xi - (1-\mu)$ are positive. Between the primaries the latter is

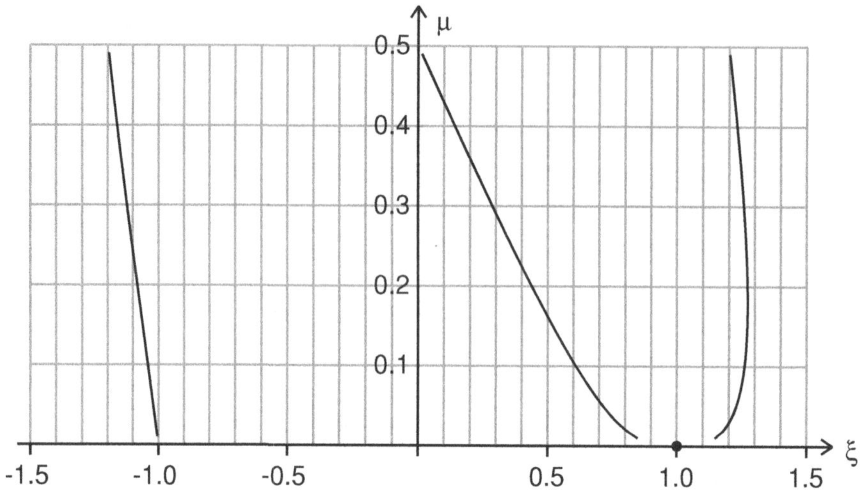

Figure 5.3 ξ-coordinates of the Lagrangian points L_1, L_2 and L_3 for different values of μ.

negative, and the negative branch of the square root is taken. To the left of the primaries, both values are negative. The following three equations therefore arise:

$$\begin{aligned}
&\xi - \frac{1-\mu}{(\xi+\mu)^2} - \frac{\mu}{(\xi-(1-\mu))^2} = 0, && 1-\mu < \xi,\\
&\xi - \frac{1-\mu}{(\xi+\mu)^2} + \frac{\mu}{(\xi-(1-\mu))^2} = 0, && -\mu < \xi < 1-\mu, \qquad (5.20)\\
&\xi + \frac{1-\mu}{(\xi+\mu)^2} + \frac{\mu}{(\xi-(1-\mu))^2} = 0, && \xi < -\mu.
\end{aligned}$$

In principle these are three equations of the fifth degree, but each of them has only one root in the range where it is valid. The equations can easily be solved by numerical methods. Figure 5.3 shows the ξ-coordinates of the points L_1, L_2 and L_3 for different values of μ.

When $\eta \neq 0$ we have from (5.18)

$$1 - \frac{1-\mu}{{\rho_1}^3} - \frac{\mu}{{\rho_2}^3} = 0.$$

Multiply this by $\xi + \mu$ and subtract from (5.17). Then

$$\frac{\mu}{{\rho_2}^3} - \mu = 0.$$

Thus $\rho_2 = 1$.

Similarly, multiply (5.18) by $\xi - (1-\mu)$ to obtain

$$\xi - (1-\mu) - \frac{(1-\mu)(\xi - (1-\mu))}{{\rho_1}^3} - \frac{\mu(\xi - (1-\mu))}{{\rho_2}^3} = 0.$$

When this is subtracted from (5.17), the result is

$$1 - \mu - \frac{1-\mu}{{\rho_1}^3} = 0.$$

It follows that $\rho_1 = 1$. Since the distance between the primaries is unity, these two points with the primaries form two equilateral triangles. Their coordinates satisfy

$$\begin{aligned} (\xi - (1-\mu))^2 + \eta^2 &= 1, \\ (\xi + \mu)^2 + \eta^2 &= 1. \end{aligned} \tag{5.21}$$

From the geometry of the configuration it is obvious that the solution is

$$\begin{aligned} \xi &= \frac{1}{2} - \mu, \\ \eta &= \pm\sqrt{3}/2. \end{aligned} \tag{5.22}$$

5.5 Stability of the Lagrangian points

Perturbing a body near a Lagrangian point is a problem well studied in the classical literature (e.g. Hagihara 1976). Here we follow the analysis given in Szebehely (1967).

Let (ξ_0, η_0) be any of the Lagrangian points and (x, y) the position of the body relative to this point:

$$\begin{aligned} x &= \xi - \xi_0, \\ y &= \eta - \eta_0. \end{aligned} \tag{5.23}$$

Concentrating on a small neighbourhood of the Lagrangian point permits the derivatives of Ω to be replaced by their linearised approximations. At the Lagrangian points

$$\frac{\partial \Omega}{\partial \xi} = \frac{\partial \Omega}{\partial \eta} = 0. \tag{5.24}$$

Then the following approximations can be used in their immediate vicinity:

$$\begin{aligned} \frac{\partial \Omega}{\partial \xi} &\approx x\frac{\partial^2 \Omega}{\partial \xi^2} + y\frac{\partial^2 \Omega}{\partial \xi \partial \eta}, \\ \frac{\partial \Omega}{\partial \eta} &\approx x\frac{\partial^2 \Omega}{\partial \xi \partial \eta} + y\frac{\partial^2 \Omega}{\partial \eta^2}. \end{aligned} \tag{5.25}$$

The linearised equations of motion are then

$$\begin{aligned} \ddot{x} - 2\dot{y} &= x\frac{\partial^2\Omega}{\partial\xi^2} + y\frac{\partial^2\Omega}{\partial\xi\,\partial\eta}, \\ \ddot{y} + 2\dot{x} &= x\frac{\partial^2\Omega}{\partial\xi\,\partial\eta} + y\frac{\partial^2\Omega}{\partial\eta^2}. \end{aligned} \tag{5.26}$$

Consider first the points L_1, L_2 and L_3, where $\eta = 0$. At these points, the derivatives of Ω are

$$\begin{aligned} \frac{\partial^2\Omega}{\partial\xi^2} &= 1 + 2\alpha, \\ \frac{\partial^2\Omega}{\partial\eta^2} &= 1 - \alpha, \\ \frac{\partial^2\Omega}{\partial\xi\,\partial\eta} &= 0, \end{aligned} \tag{5.27}$$

where

$$\alpha = \frac{1-\mu}{{\rho_1}^3} + \frac{\mu}{{\rho_2}^3}. \tag{5.28}$$

The equations of motion become now

$$\begin{aligned} \ddot{x} - 2\dot{y} &= x(1 + 2\alpha), \\ \ddot{y} + 2\dot{x} &= y(1 - \alpha). \end{aligned} \tag{5.29}$$

It is useful to study the trial trajectory

$$\begin{aligned} x &= A\,\mathrm{e}^{\omega t}, \\ y &= B\,\mathrm{e}^{\omega t}, \end{aligned} \tag{5.30}$$

where A, B and ω are constants. If a solution exists for which the real part of ω is non-zero, the coordinates of the body can grow without limit, and the orbit is unstable. But if ω must be purely imaginary, the motion is just oscillation that must remain bounded.

Substitution of the trial orbit into the equations of motion gives

$$\begin{aligned} A\omega^2\,\mathrm{e}^{\omega t} - 2B\omega\,\mathrm{e}^{\omega t} &= A\,\mathrm{e}^{\omega t}(1 + 2\alpha), \\ B\omega^2\,\mathrm{e}^{\omega t} + 2A\omega\,\mathrm{e}^{\omega t} &= B\,\mathrm{e}^{\omega t}(1 - \alpha) \end{aligned}$$

or

$$\begin{aligned} A\omega^2 - 2B\omega &= A(1 + 2\alpha), \\ B\omega^2 + 2A\omega &= B(1 - \alpha). \end{aligned} \tag{5.31}$$

Eliminating A and B, and after some algebra, the result is:

$$\omega^4 + \omega^2(2-\alpha) + (1+2\alpha)(1-\alpha) = 0. \tag{5.32}$$

If ω is to be purely imaginary, there must be two negative solutions for ω^2. The product of these roots must then be positive, so that

$$(1+2\alpha)(1-\alpha) > 0. \tag{5.33}$$

Since the first factor is always positive, then

$$\alpha < 1. \tag{5.34}$$

The Lagrangian points must satisfy Eq. (5.17):

$$\frac{\partial\Omega}{\partial\xi} = \xi - \frac{(1-\mu)(\xi+\mu)}{\rho_1{}^3} - \frac{\mu(\xi-(1-\mu))}{\rho_2{}^3} = 0.$$

Rearranging the terms,

$$\xi - \xi\left[\frac{1-\mu}{\rho_1{}^3} + \frac{\mu}{\rho_2{}^3}\right] - \frac{\mu(1-\mu)}{\rho_1{}^3} + \frac{\mu(1-\mu)}{\rho_2{}^3} = 0$$

or

$$\xi(1-\alpha) - \mu(1-\mu)\left[\frac{1}{\rho_1{}^3} - \frac{1}{\rho_2{}^3}\right] = 0,$$

from which

$$1-\alpha = \frac{\mu(1-\mu)}{\xi}\left[\frac{1}{\rho_1{}^3} - \frac{1}{\rho_2{}^3}\right]. \tag{5.35}$$

It is easy to see that for all Lagrangian points on the ξ axis, the bracketed expression and ξ have opposite signs. Thus the right hand side of the equation is always negative, and

$$\alpha > 1. \tag{5.36}$$

This conflicts with the previous requirement that $\alpha < 1$. Therefore it is impossible to find purely imaginary solutions for ω, and the points L_1, L_2 and L_3 are unstable. If a body in any of these points is disturbed, it will move away. This is quite reasonable physically, for these points are saddle points of Ω.

The coordinates of the point L_4 are $\xi = \frac{1}{2}(1-2\mu)$ and $\eta = \sqrt{3}/2$, so that the linearised equations of motion are

$$\begin{aligned} \ddot{x} - 2\dot{y} &= \frac{3}{4}x + \frac{3\sqrt{3}}{4}(1-2\mu)y, \\ \ddot{y} + 2\dot{x} &= \frac{9}{4}y + \frac{3\sqrt{3}}{4}(1-2\mu)x. \end{aligned} \tag{5.37}$$

Substitution of the trial solution (5.30) gives

$$\omega^4 + \omega^2 + \frac{27}{4}\mu(1-\mu) = 0. \tag{5.38}$$

Because

$$\omega_1^2\omega_2^2 = \frac{27}{4}\mu(1-\mu) > 0,$$

the possible real roots have the same sign. Since also

$$\omega_1^2 + \omega_2^2 = -1 < 0,$$

both roots ω_i^2 must be negative. We require that Eq. (5.38) has real roots and thus ω is imaginary:

$$1 - 4\frac{27}{4}\mu(1-\mu) > 0$$

or

$$27\mu^2 - 27\mu + 1 > 0. \tag{5.39}$$

This is satisfied when

$$\mu < \frac{1}{2} - \sqrt{\frac{23}{108}} \approx 0.0385. \tag{5.40}$$

Thus stable orbits around L_4 and L_5 are possible provided that $\mu < 0.0385$. The Trojan asteroids around the points L_4 and L_5 of the Sun–Jupiter system are a good example of this. They orbit the Sun in the same orbit with Jupiter, but 60° ahead of or behind Jupiter. Actually the asteroids librate around the Lagrangian points, and can move quite far away without escaping. Other planets may have their own Trojans around their L_4 or L_5 points. The first one to be discovered outside the Jupiter system was the Mars Trojan minor planet (5261) Eureka (Mikkola *et al.* 1994).

The above analysis applies only to orbits which start close to the Lagrangian points L_4 or L_5. By numerical orbit integration it is possible to investigate what happens when the displacement is relatively large. Then the small body may librate around L_4 or L_5 on a wide arc centred at the primary. Such orbits are called tadpole orbits (Fig. 5.4). The tadpoles of L_4 and L_5 may also meet in which case the orbit is called a horseshoe orbit. An example of the latter is asteroid (3753) Cruithne, which librates in an inclined horseshoe orbit in the Sun–Earth system and occasionally even becomes a quasi-satellite of the Earth (Wiegert *et al.* 1997, Brasser *et al.* 2004). Venus has a similar quasi-satellite (Fig. 5.5; Mikkola *et al.* 2004). Classical two-dimensional horseshoe orbits are found in 'Janus' moons of Saturn.

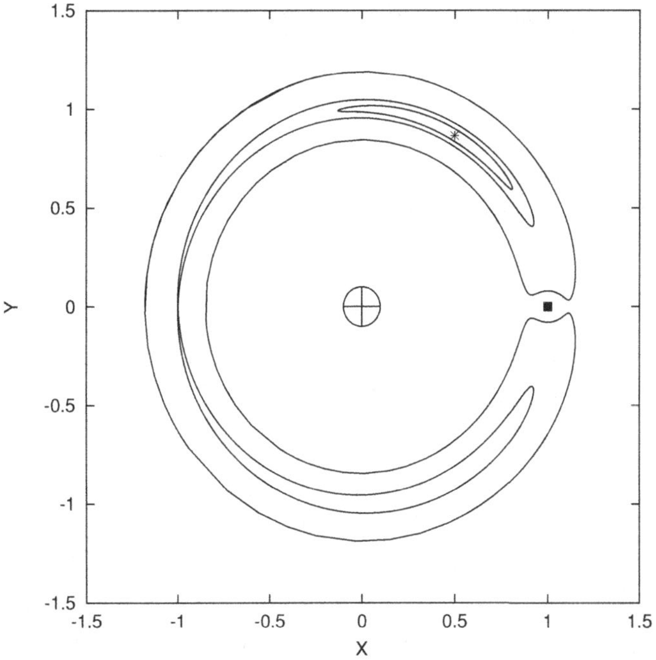

Figure 5.4 Tadpole orbits and horseshoe orbits. The orbits of three asteroids relative to the Sun (⊕), the Earth (■), and the point L_4 ($*$) are drawn. They correspond to slightly different values of the Jacobi constant C. The orbit closest to L_4 is a tadpole orbit, the one going close to the Earth is a horseshoe orbit. Between them is a limiting case where two tadpoles meet to make a horseshoe.

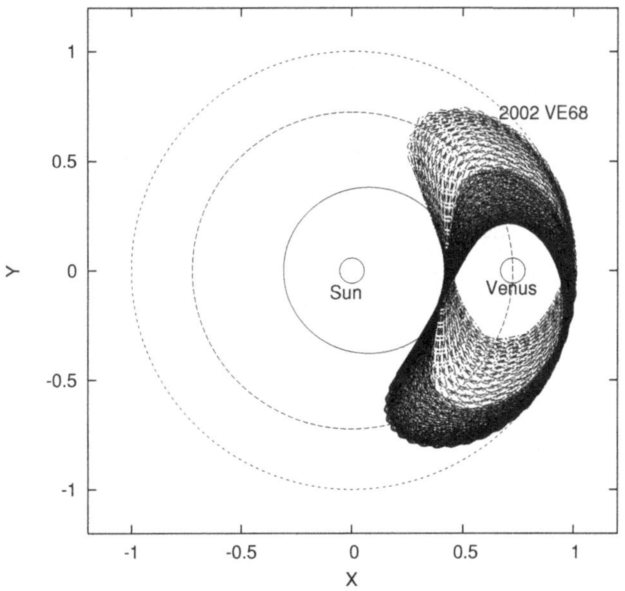

Figure 5.5 The path of asteroid 2002 VE68 which comes close to Venus and from time to time becomes a quasi-satellite of Venus.

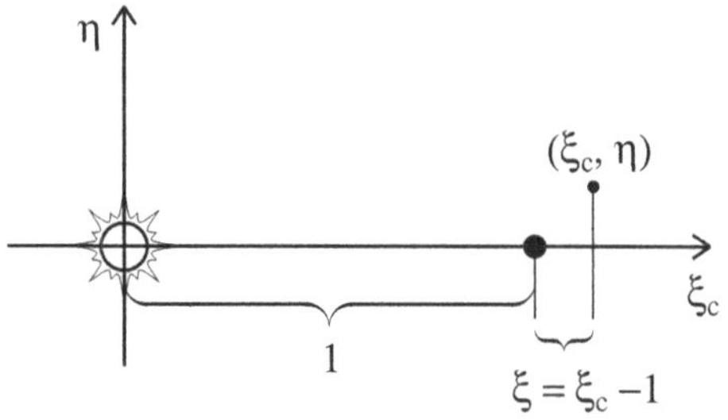

Figure 5.6 A satellite is close to a planet. The coordinates of the planet are $(1, 0)$ and those of the planet (ξ, η). A transformation to coordinates centred at the planet is achieved by replacing $\xi - 1$ by a new coordinate (also called ξ in the text).

5.6 Satellite orbits

We turn now to study the stability of satellite orbits around a planet. These may be, for example, small asteroidal satellites. We neglect effects such as tides or perturbations by other planets, take the planetary orbit to be circular and the satellite orbit to be coplanar with the planetary orbit. Then the theory of the restricted three-body problem may be applied. Let us begin by writing (5.10) as

$$\begin{aligned} \ddot{\xi} - 2\dot{\eta} &= \xi - (1-\mu)\frac{\xi + \mu}{{\rho_1}^3} - \mu\frac{\xi - (1-\mu)}{{\rho_2}^3}, \\ \ddot{\eta} + 2\dot{\xi} &= \eta - \left[\frac{1-\mu}{{\rho_1}^3} + \frac{\mu}{{\rho_2}^3}\right]\eta. \end{aligned} \tag{5.41}$$

To simplify matters, we take the planet/Sun mass ratio μ to be small in which case $1 - \mu \approx 1$. It is useful to consider the orbit of the satellite relative to the planet, and therefore adopt the coordinate system centred on the planet via the transformation $\xi \to 1 + \xi$ (Fig. 5.6). Then

$$\begin{aligned} \ddot{\xi} - 2\dot{\eta} - \xi &= 1 - \frac{1+\xi}{{\rho_1}^3} - \mu\frac{\xi}{{\rho_2}^3}, \\ \ddot{\eta} + 2\dot{\xi} - \eta &= -\frac{\eta}{{\rho_1}^3} - \mu\frac{\eta}{{\rho_2}^3}, \\ {\rho_1}^2 &= (1+\xi)^2 + \eta^2 \approx 1 + 2\xi. \end{aligned}$$

In this approximation, terms of order ξ^2 and η^2 have been neglected, and therefore

$$\rho_1^{-3} \approx (1 + 2\xi)^{-3/2} \approx 1 - 3\xi.$$

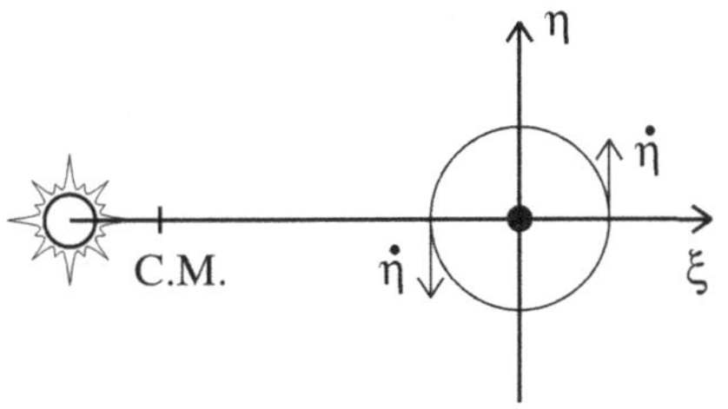

Figure 5.7 Direct circular motion of the satellite around the planet. In retrograde motion the direction of motion is the opposite.

Thus

$$\begin{aligned} \ddot{\xi} - 2\dot{\eta} &= 3\xi - \mu\frac{\xi}{\rho_2^3}, \\ \ddot{\eta} + 2\dot{\xi} &= \mu\frac{\eta}{\rho_2^3}. \end{aligned} \tag{5.42}$$

These are called *Hill's equations*.

It is obvious that the satellite orbit cannot be stable if the Sun causes a greater acceleration on the satellite than the planet. The limiting acceleration, when the satellite crosses the ξ axis, $\ddot{\xi} = 0$, defines a boundary. Consider this boundary in three different situations.

If the satellite motion is radial along the ξ axis (i.e. $\dot{\eta} = 0$), at the stationary point $\dot{\xi} = \ddot{\xi} = 0$ Eq. (5.42) gives $3\xi = \mu\xi/\rho_2^3$, or

$$\rho_2 = (\rho_2)_H = \left(\frac{\mu}{3}\right)^{1/3}. \tag{5.43}$$

A sphere of this radius around the planet is called the *Hill sphere*.

In another case the satellite is moving in a circular orbit around the planet either in direct (+) or retrograde (−) sense (Fig. 5.7). For $\ddot{\xi} = 0$ at $\eta = 0$, (5.42) then gives

$$\ddot{\xi} = 3\xi - \omega^2\xi \pm 2\omega\xi = 0 \tag{5.44}$$

where ω is the angular speed of the satellite around the planet ($\omega^2 = \mu/\rho_2^3$; $\dot{\eta} = \pm\omega\xi$). For direct orbits, this is equivalent to $(\omega - 3)(\omega + 1) = 0$, i.e. $\omega = 3$; for retrograde orbits, $(\omega + 3)(\omega - 1) = 0$, i.e. $\omega = 1$. The corresponding distances are (Innanen 1979, 1980)

$$\begin{aligned} (\rho_2)_d &= \left(\frac{\mu}{9}\right)^{1/3}, \\ (\rho_2)_r &= \mu^{1/3}. \end{aligned} \tag{5.45}$$

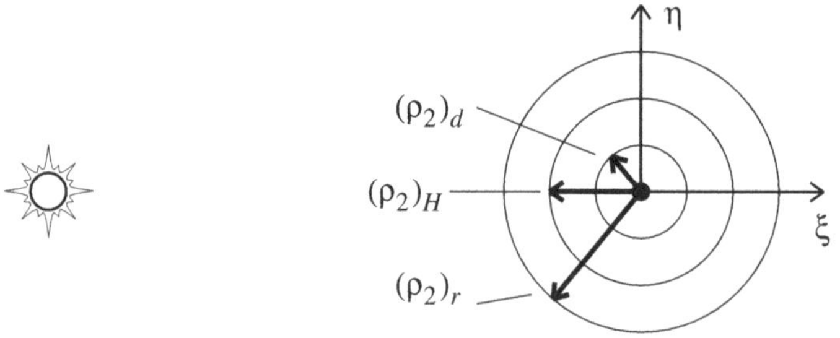

Figure 5.8 Hill sphere of radius $(\rho_2)_H$ limits the region where radial satellite motion is stable. Direct circular orbits are stable within $(\rho_2)_d$ and retrograde orbits within $(\rho_2)_r$ from the primary.

Since the perturbing (tidal) force is strongest when the satellite crosses the ξ axis, the spheres of radii $(\rho_2)_d$ and $(\rho_2)_r$ outline the regions of stability for satellites in direct and retrograde motions, respectively. Note that the Hill sphere is intermediate between these two spheres. The radius of the retrograde sphere of stability is about twice the corresponding radius for direct circular orbits (Fig. 5.8).

The orbits which approach the spheres of stability are modified and are not circular any more. Therefore the limits derived here cannot be taken as absolute boundaries. However, their relation to the zero velocity curves may be studied by looking at the Jacobi constant which in the limit of small ξ, $\rho_1 \approx 1$, $1-\mu \approx 1$ is

$$C \approx 3 + 2\mu/\rho_2 - v^2. \tag{5.46}$$

The velocities at the limits of stability for direct and retrograde circular orbits are $v_d^2 = \mu/(\rho_2)_d$ and $v_r^2 = \mu/(\rho_2)_r$, respectively, and the corresponding values for the Jacobi constants are

$$\begin{aligned} C_d &\approx 3 + \mu/(\rho_2)_d \\ C_r &\approx 3 + \mu/(\rho_2)_r. \end{aligned} \tag{5.47}$$

The Jacobi constant corresponding to the Hill sphere is

$$C_H \approx 3 + 2\mu/(\rho_2)_H. \tag{5.48}$$

Figure 5.9 illustrates the zero velocity curves corresponding to these values of the Jacobi constant. Note that the zero velocity surface corresponding to C_H is not quite a sphere but it has an elongation connecting to the Lagrangian point L_1 where it meets the corresponding surface for the primary (the Sun). It is through this point that satellites in direct circular orbits may escape from the direct influence of the planet, or the Sun. The zero velocity surface corresponding to C_r envelopes both the Sun and the planet and their Hill spheres. This allows asteroids and other bodies

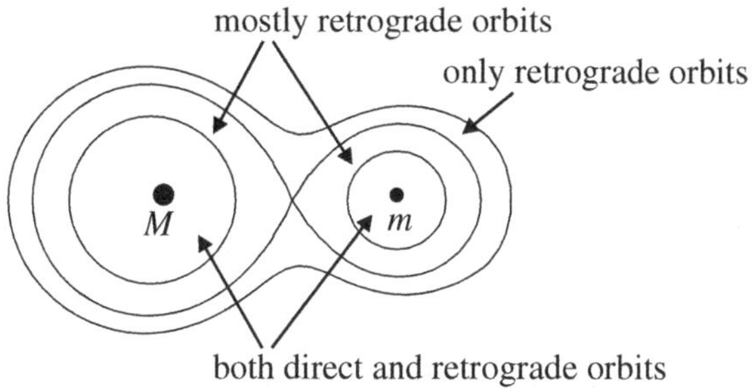

Figure 5.9 Zero velocity curves for three values of Jacobi constant. The primary is labelled M and the secondary (planet) by m. The different regions of stability for satellite orbits around the secondary are indicated.

circling the Sun to become retrograde satellites of planets. Such orbits are only quasi-stable since the reverse process is equally possible. Inside the zero velocity surface corresponding to C_d, satellites of all inclinations are stable. The retrograde satellites of Jupiter, Saturn and Uranus are all well inside the above theoretical limit. The limit derived for retrograde orbits is not to be taken too strictly, since retrograde satellites can exist even at much greater distances (Hénon 1970, Mikkola and Innanen 1997).

Equation (5.46) leads to a useful expression for the speed of approach v to a planet at the distance $(\rho_2)_H$ of the Hill sphere. Substitution of $\rho_2 = (\mu/3)^{1/3}$ gives

$$T = C \approx 3 - v^2 + 2 \times 3^{1/3}\mu^{2/3} \approx 3 - v^2, \tag{5.49}$$

since μ is a small quantity for a planet.

5.7 The Lagrangian equilateral triangle

Lagrange studied a special case of a three-body problem where the three bodies are placed at the corners of an equilateral triangle. He found a solution where the three bodies remain at constant distances from each other while they revolve around their common centre of mass. Later Euler found a solution to the same problem where the lengths of the sides can vary, keeping, however, their ratio constant. In both cases the accelerations of the bodies are toward the centre of mass of the system, and the three bodies rotate around this point in synchronism (Fig. 5.10).

We may use the equations of motion in the Lagrangian form to solve the problem. Since $r_{12} = r_{23} = r_{31}$, $\boldsymbol{W} = 0$ and

$$\ddot{\boldsymbol{r}}_{12} = -GM\boldsymbol{r}_{12}/r_{12}^3. \tag{5.50}$$

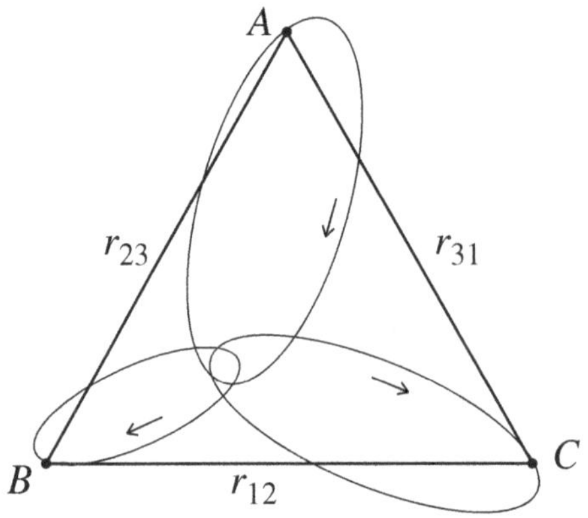

Figure 5.10 Lagrangian equilateral triangle and possible orbits.

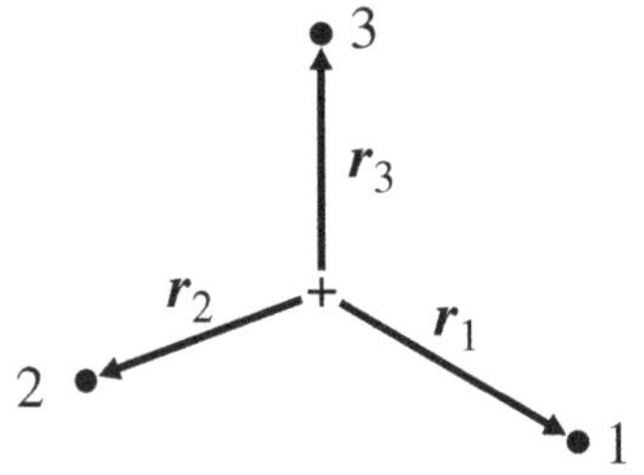

Figure 5.11 The positions of the three bodies in the centre-of-mass system.

This is the equation of motion of the two-body problem. Therefore each one of the three bodies describes an elliptical orbit around their common centre of mass (Fig. 5.10).

To see this more clearly, let us write the equation of motion for body number 1 in the centre of mass system ($\boldsymbol{R} = 0$ in Eq. (2.4), Fig. 5.11). Due to symmetry, the equations of motion of the two other bodies have the same form. By the definition of the centre of mass

$$m_1\boldsymbol{r}_1 + m_2\boldsymbol{r}_2 + m_3\boldsymbol{r}_3 = 0,$$

which may also be written as

$$(m_1 + m_2 + m_3)\boldsymbol{r}_1 + m_2(\boldsymbol{r}_2 - \boldsymbol{r}_1) + m_3(\boldsymbol{r}_3 - \boldsymbol{r}_1) = 0$$

or

$$M\boldsymbol{r}_1 = m_2\boldsymbol{r}_{12} + m_3\boldsymbol{r}_{13}.$$

The square of this is

$$M^2 r_1^2 = m_2^2 r_{12}^2 + m_3^2 r_{13}^2 + 2 m_2 m_3 \boldsymbol{r}_{12} \cdot \boldsymbol{r}_{13}.$$

In an equilateral triangle the angle between the vectors $\boldsymbol{r}_{12}$ and $\boldsymbol{r}_{13}$ is 60°; therefore

$$M^2 r_1^2 = \left(m_2^2 + m_3^2 + m_2 m_3\right) r^2,$$

where $r = r_{12} = r_{23} = r_{31}$.

The equation of motion, Eq. (2.29), becomes

$$\ddot{\boldsymbol{r}}_1 = -G\left(m_2 \frac{\boldsymbol{r}_{12}}{r_{12}^3} + m_3 \frac{\boldsymbol{r}_{13}}{r_{13}^3}\right) = -GM\frac{\boldsymbol{r}_1}{r^3},$$

and if we replace r by r_1 using the equation above:

$$\ddot{\boldsymbol{r}}_1 + GM_1 \frac{\boldsymbol{r}_1}{r_1^3} = 0, \qquad (5.51)$$

where we have defined

$$M_1 = \frac{\left(m_2^2 + m_3^2 + m_2 m_3\right)^{3/2}}{M^2}. \qquad (5.52)$$

This is the equation of the two-body motion (Eq. (3.2)) about the centre of mass when $\mu = GM_1$ (Danby 1962, Roy 2005).

Let us consider the special case of an equal mass (m) Lagrangian equilateral triangle of side length r, with circular orbits. There $M_1 = m/\sqrt{3}$, $r = \sqrt{3} r_1$, orbital speed $v^2 = \mu/r_1 = Gm/r$, kinetic energy $T = \frac{3}{2} m v^2 = \frac{3}{2} Gm^2/r$, potential energy $V = -3Gm^2/r$, total energy $E_0 = -\frac{3}{2} Gm^2/r$, and the total angular momentum relative to the centre of mass $L_0 = 3 m r_1 v = \sqrt{3G} m^{3/2} \sqrt{r}$. In the last relation r may be replaced by $|E_0|$:

$$L_0 = (3/\sqrt{2}) G m^{5/2} / |E_0|^{1/2}. \qquad (5.53)$$

By comparison with Eq. (2.67) we see that L_0 is about 85% of L_{max}, i.e. close to the standard reference value for three-body systems. The zero eccentricity Lagrangian equilateral triangle corresponds to the zero eccentricity two-body orbit, and its angular momentum is maximal for its orbit size. This gives some justification for the use of the standard reference value L_{max} as the upper limit of the range of angular momenta in bound strongly interacting three-body systems in the next chapter. In Chapter 9 we will study systems with $L_0 > L_{max}$, but they are clearly unbound or hierarchical.

The significance of a solution like this depends on its stability. If the initial positions and velocities are varied very slightly, the varied orbit may remain close to the original one at all future times. Then the original solutions are considered stable. If, on the other hand, the slightly varied orbit deviates more and more from the original orbit as time increases, the solution is deemed unstable. In the case of the Lagrangian equilateral triangle, it has been shown that the solution is stable only if one of the bodies dominates the system. The heaviest body has to have more than 95% of the total mass for stability. Figure 5.12 outlines the stable regions as a function of eccentricity e of the orbit and the mass ratio $R = m_2/M$ where m_2 is the middle mass (neither smallest nor largest) and M is the total mass.

5.8 One-dimensional three-body problem

This is a problem where all three bodies are constrained to move on a line. When bodies collide, they bounce back elastically; otherwise Newtonian gravity is assumed to be the only force in the system (Schubart 1956, Hénon 1976b, 1977). This problem is not of much practical significance, and its main value lies in illustrating the general features of three-body orbits which are encountered also in the general three-body problem.

Without loss of generality we may choose the system of coordinates such that the gravitational constant $G = 1$ (this is a common choice in numerical orbit calculations), all the masses are equal to one unit, and the total energy $E = -1$. One could also consider the more general case of the three masses being unequal, but for the present the equal mass case is as good as any other. For the starting configuration, the three bodies are placed at equal distances R from each other. Then, after the speed of one of the outer bodies, say $\dot{r}_1$, is fixed, the speed of the other outer body, say $\dot{r}_2$, is uniquely determined by the known value of the total energy. Even though it may appear that the requirement of the starting distances being equal, $r_1 = r_2 = R$, is a special case, it has been shown not to be so (Mikkola and Hietarinta 1989, 1990, 1991). One may view it such that the equal distances become unequal as soon as the system starts evolving (except when $\dot{r}_1 = -\dot{r}_2$, a special case). By considering the later moments of time in the orbits with symmetric starting distances as new starting configurations, a whole range of starting positions, both symmetric and asymmetric, is covered.

For this reason, the one-dimensional problem can be fully evaluated by studying all possible initial values in a two-dimensional space. One of the dimensions is obviously R; for the second dimension it is useful to make a transformation to another parameter Θ defined as follows (Hietarinta and Mikkola, 1993):

$$\begin{aligned} \dot{r}_1 &= 2\sqrt{T_0}\cos(\Theta - \pi/3), \\ \dot{r}_2 &= 2\sqrt{T_0}\cos(\Theta + \pi/3). \end{aligned} \tag{5.54}$$

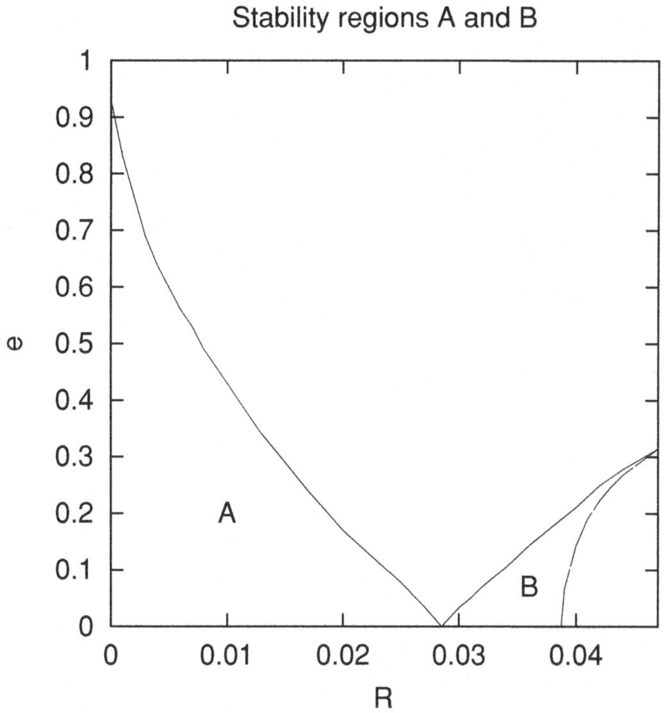

Figure 5.12 Regions of stability (labelled A and B) in the eccentricity–mass ratio plane for elliptic Lagrangian motions (Danby 1964).

Here T_0 is the initial kinetic energy of the system. It is determined by the known value of the total energy and the initial distance R. Therefore Θ is uniquely defined as soon as $\dot{r}_1$ is given, and can be used as the second parameter in place of $\dot{r}_1$. Then the initial value space is the (R,Θ) plane. Even though the one-dimensional three-body problem may appear simple, the solutions require numerical orbit calculations. Figure 5.13 illustrates the time evolution of the coordinate x of each body as a function of time t. It shows the typical feature that the system is composed of a tight binary and a third body which have a close encounter every now and then until one encounter finally leads to an escape of one of the bodies. When this final breakup has happened, the system parameters become stable. An example of such parameters is the semi-major axis of the binary a. In Fig. 5.14 we plot the values of a in the (R,Θ) plane.

Figure 5.14 shows that there are three different regions in the initial value space. At the bottom of the figure there are three semicircular areas where the end result varies continuously with the starting conditions. In these trajectories, there is only one close three-body encounter before the system breaks up. These events would usually be called ‘scattering’. At the top of the figure, in the middle, there is a

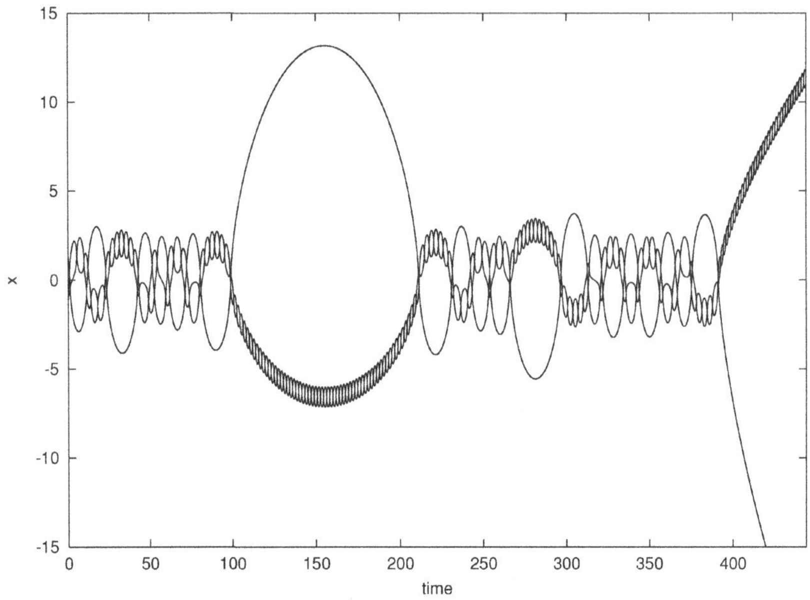

Figure 5.13 The time evolution of the positions of three bodies in the x axis in the one-dimensional three-body problem. Small loops represent binaries, large loops ejections of the third body. In the end (at time $\approx$ 400 units) the third body escapes.

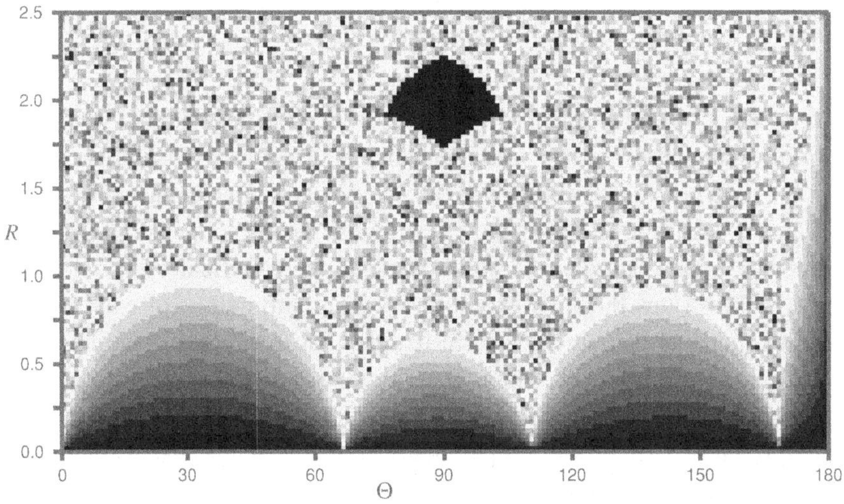

Figure 5.14 A grey scale representation of the semi-major axis a of the binary which remains after an escape in the one-dimensional three-body problem. The largest values of a are represented by white, the smallest ones by black. No escapes happen in the fanlike region in the top middle, where this quantity is not defined. The parameters R (vertical axis) and Θ (horizontal axis) describe the initial conditions. The (R, Θ) plane represents all possible initial values.

region where no escapes take place, i.e. these systems are stable. At the centre of this region there is a stable periodic orbit (Schubart 1956); the stability of this orbit is shown by the fact that the neighbouring orbits are similar to the central orbit. In the remainder of the figure the outcome of the three-body evolution is extremely sensitive to the initial conditions. We may call this region chaotic.

This classification of three-body orbits applies quite generally to the three-body problem. However, one should note that in the present case the region of stable orbits covers an unusually large fraction of the initial value space. There are other periodic solutions in the general problem (e.g. the Lagrangian orbits of the previous section), but they are so rare and difficult to find, that they can usually be neglected as potential astrophysical orbits. But when they are found, they often have the interesting property that in their vicinity there exists an infinite number of quasi-periodic solutions. These orbits fill a set of positive measure in phase space (so-called Kolmogorov–Arnold–Moser theorem, Kolmogorov 1954, Arnold *et al.* 1988). A thorough discussion of periodic orbits is found for example in Marchal (1990).

Problems

Problem 5.1 Derive equations of motion for the three-dimensional restricted three-body problem. The primaries are still confined to the $\xi\eta$ plane, but the massless body can be anywhere. What happens to the Jacobian integral?

Problem 5.2 Assume that the primaries are the Sun and Jupiter and the third body is a comet observed near the Sun. Explain why the following approximations are then valid:

$$\begin{aligned} \dot{\xi}^2 + \dot{\eta}^2 + \dot{\zeta}^2 = v^2 &= \frac{2}{\rho_1} - \frac{1}{a}, \\ \xi\dot{\eta} - \eta\dot{\xi} &= \sqrt{a(1-e^2)}\cos\iota, \\ \frac{1-\mu}{\rho_1} &= \frac{1}{\rho_1}, \\ \frac{\mu}{\rho_2} &= 0. \end{aligned} \tag{5.55}$$

Here a, e and ι are the ordinary orbital elements of the comet. Using these approximations, show that the expression

$$\frac{1}{a} + 2\sqrt{a(1-e^2)}\cos\iota$$

is constant. This is known as *Tisserand's criterion for comets' identity*.

Problem 5.3 Show that in the restricted three-body problem the velocity of the third body is zero on the curve

$$C = (1 - \mu)\left({\rho_1}^2 + \frac{2}{\rho_1}\right) + \mu\left({\rho_2}^2 + \frac{2}{\rho_2}\right).$$

Using this, show that the minimum value of C is 3.

Problem 5.4 Show that the Lagrangian points L_4 and L_5 are minima of the function Ω. Hint: suppose that the function f of two variables has an extremum at (x, y) and its second derivatives exist at this point. The extremum is a minimum if

$$\frac{\partial^2 f}{\partial x^2 \partial}\frac{\partial^2 f}{\partial y^2 \partial} - \left(\frac{\partial^2 f}{\partial x \partial y}\right)^2 > 0. \tag{5.56}$$

Problem 5.5 It has been suggested that the gegenschein (a faint glow in the direction opposite to the Sun) is caused by dust in the Lagrangian point L_2. Find the position of this point and show that it is not within the umbra of the Earth.

Problem 5.6 A binary star has two components of one solar mass orbiting in a circular orbit with a radius of 10 AU. A spacecraft crosses the line joining the stars at a distance of 1 AU from one of the stars with a velocity of 10 km/s. Find the Jacobian integral of the spacecraft. Can it go across the Lagrangian point between the stars? Sketch the region where the motion of the spacecraft is confined as long as it does not use any propulsion.

6

Three-body scattering

Three-body systems tend to be unstable. Often they are only short-lived stages in the evolution of a dynamical system. Typically a body comes from a large distance, meets a binary, and escapes again far away. The meeting could be a distant flyby or a close encounter with one of the binary members. Both types of events are important and will be studied in turn. Here we will study only the latter situation, and limit ourselves to the case where the third body is of low mass in comparison with the binary. The general scattering problem is left to Chapters 8 and 10. As in the two-body problem, the transfer of the third body from one hyperbolic relative orbit to another is called *scattering*.

6.1 Scattering of small fast bodies from a binary

The restricted circular three-body problem deals with the motion of a 'massless' body in the gravitational field of a zero eccentricity binary. What we will now discuss is a similar problem, namely the motion of a low mass body in the binary field. In our problem the low mass body arrives from a large distance with a high speed, scatters from the binary and flies away. The problem is similar to the two-body scattering of Section 3.13. We present the discussion of three-body scattering following Gould (1991).

The limitation to small and fast third bodies allows us to make an important simplifying assumption: (1) the orbital positions of the binary members remain fixed during the encounter. Additionally we assume that (2) the scattering is a two-body scattering from one member of the binary while the influence of the second member is neglected.

Let the third-body mass be m_s and the masses of the binary components m_a and m_b. The relative velocity of the binary components is $\boldsymbol{v}_0 = \dot{\boldsymbol{r}}_a - \dot{\boldsymbol{r}}_b$, and the approach velocity of the third body at large distance from the binary is $\boldsymbol{u}_s$, relative to the centre of mass of the binary. Relative to the binary component of mass m_a

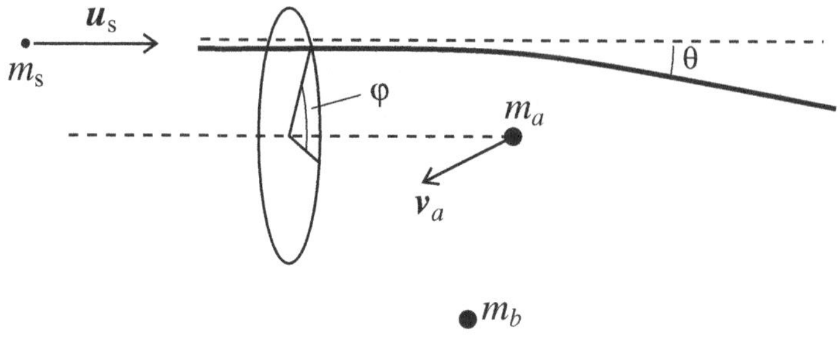

Figure 6.1 The scattering of the third body (mass m_s) from one of the binary components (mass m_a). The asymptotic arrival direction is specified by the orientation angle φ and by the impact distance b which is related to the scattering angle θ. In the centre of mass system of the binary the velocity of the third body is $\boldsymbol{u}_s$ and the velocity of the binary component $\boldsymbol{v}_a = (\mathcal{M}/m_a)\boldsymbol{v}_0$.

the asymptotic approach velocity is

$$\boldsymbol{u} = \boldsymbol{u}_s - \frac{\mathcal{M}}{m_a}\boldsymbol{v}_0. \tag{6.1}$$

Here $\mathcal{M}$ is the reduced mass:

$$\mathcal{M} = \frac{m_a m_b}{m_a + m_b}. \tag{6.2}$$

The scattering of the third body from the body of mass m_a is now calculated assuming that the process is not influenced by the binary component of mass m_b. We then carry out the calculation in the centre of mass frame of bodies with masses m_s and m_a. In that frame the initial speed of the latter body is

$$v_{||}^{(0)} = -\frac{m_s}{m_s + m_a}u. \tag{6.3}$$

After the two-body scattering the same quantity becomes

$$v_{||}^{(1)} = -\frac{m_s}{m_s + m_a}u\cos\theta \tag{6.4}$$

where θ is the deflection angle. The speed in the perpendicular direction is

$$v_{\perp}^{(1)} = -\frac{m_s}{m_s + m_a}u\sin\theta \tag{6.5}$$

(see Fig. 6.1). Therefore the change of speed in the parallel direction is

$$\begin{aligned}\Delta v_{||} = v_{||}^{(1)} - v_{||}^{(0)} &= \frac{m_s}{m_s + m_a}u(1 - \cos\theta) \\ &= \frac{2m_s}{m_s + m_a}u\sin^2\frac{\theta}{2}\end{aligned} \tag{6.6}$$

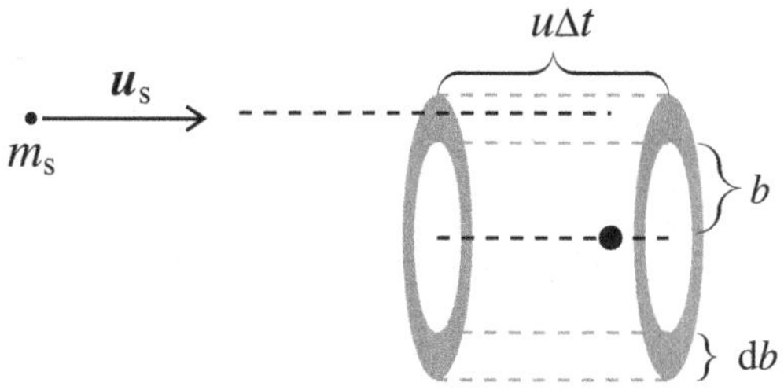

Figure 6.2 A stream of particles enters a cylindrical ring with speed u. The volume $2\pi b\, \mathrm{d}b\, u\Delta t$ includes all the particles which have entered the ring during the time interval Δt. If the rate of entry is $\Delta N/\Delta t$, there are ΔN particles of mass m_s inside the cylindrical ring.

and in the perpendicular direction

$$\Delta v_\perp = v_\perp^{(1)} - 0 = -\frac{m_s}{m_s + m_a} u \sin\theta. \tag{6.7}$$

The total change is

$$\begin{aligned}\Delta v &= \sqrt{\Delta v_\parallel^2 + \Delta v_\perp^2} = \frac{m_s}{m_s + m_a} u \sqrt{\sin^2\theta + (1 - \cos\theta)^2} \\ &= \frac{2m_s}{m_s + m_a} u \sqrt{\frac{1}{2}(1 - \cos\theta)} = \frac{2m_s}{m_s + m_a} u \sin\frac{\theta}{2}.\end{aligned} \tag{6.8}$$

As long as the orbital positions of the binary members do not change, the change in the binding energy $|E_B|$ of the binary comes solely from the change in the kinetic energy:

$$-\Delta|E_B| = \Delta\left(\frac{1}{2}\mathcal{M}v_0^2\right) = \frac{1}{2}\mathcal{M}(\Delta v)^2 + \mathcal{M}\boldsymbol{v}_0 \cdot \Delta\boldsymbol{v} \tag{6.9}$$

where $\Delta\boldsymbol{v} = \Delta\boldsymbol{v}_0$ is the velocity change induced on the binary component. We have made use of the vector formula for two vectors $\boldsymbol{A}$ and $\boldsymbol{B}$,

$$\Delta(\boldsymbol{A} \cdot \boldsymbol{B}) = \Delta\boldsymbol{A} \cdot \boldsymbol{B} + \boldsymbol{A} \cdot \Delta\boldsymbol{B} + \Delta\boldsymbol{A} \cdot \Delta\boldsymbol{B},$$

and have put $\boldsymbol{A} = \boldsymbol{B} = \boldsymbol{v}_0$. The first term is obviously positive, while it turns out that the second term is negative on average. Therefore the two terms are called the gain term and the loss term, respectively.

Let us now consider a stream of 'third' bodies streaming parallel to each other with speed u and going through a ring of impact distances between b and $b + \mathrm{d}b$ (Fig. 6.2). We cut a piece of length $u\Delta t$ from this stream, where Δt is a time interval from the present time backwards. All the bodies which have entered the ring during the interval Δt are now somewhere inside the double walls formed by the two concentric cylinders. The volume inside the double walls is $2\pi b\, \mathrm{d}b\, u\Delta t$.

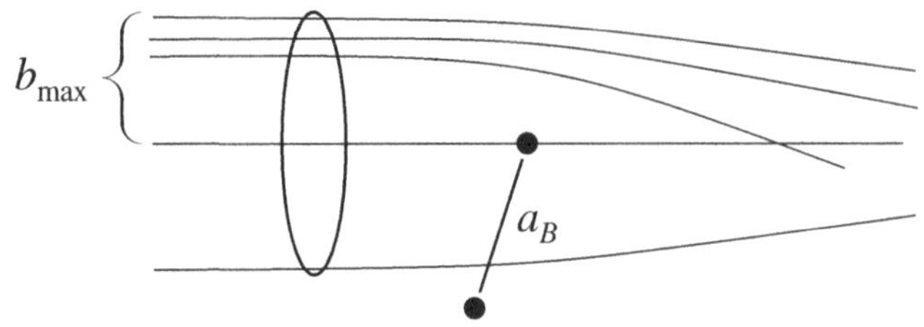

Figure 6.3 A stream of particles entering through a catchment radius $b_{\max}$ is channelled through the binary orbital radius by gravitational focussing.

If the rate of bodies entering the ring is $\Delta N/\Delta t$, then there are ΔN bodies of mass m_s in this volume. Thus the mass density is

$$\rho = \frac{\Delta N m_s}{2\pi b \, \mathrm{d}b \, u \Delta t} \tag{6.10}$$

which gives the rate

$$\frac{\Delta N}{\Delta t} = \frac{2\pi b \, \mathrm{d}b \, u\rho}{m_s}. \tag{6.11}$$

From Eqs. (3.58) and (3.59) we get

$$2\pi b \, \mathrm{d}b = 2\pi \frac{a^2}{4} \frac{\sin\theta \, \mathrm{d}\theta}{\sin^4(\theta/2)} \tag{6.12}$$

where a and b are the semi-major and semi-minor axes of the hyperbolic third-body orbits.

Associated with each body which enters the ring there is the energy gain

$$\frac{1}{2}\mathcal{M}(\Delta v)^2 = \frac{1}{2}\mathcal{M}\left(2\frac{m_s}{m_a + m_s}\right)^2 u^2 \sin^2\frac{\theta}{2}. \tag{6.13}$$

Thus the rate of energy gain over the whole 'applicable' range of θ is

$$\begin{aligned} -\frac{\mathrm{d}|E_B|}{\mathrm{d}t} &= \int_{\theta_{\min}}^{\pi} 2\pi \frac{a^2}{4} \frac{\sin\theta}{\sin^4(\theta/2)} \frac{u\rho}{m_s} \frac{1}{2}\mathcal{M}\left(2\frac{m_s}{m_s + m_a}\right)^2 u^2 \sin^2\frac{\theta}{2} \, \mathrm{d}\theta \\ &= 2\pi \frac{a^2}{4} \frac{u\rho}{m_s} \frac{1}{2}\mathcal{M}\left(2\frac{m_s}{m_s + m_a}\right)^2 u^2 \int_{\theta_{\min}}^{\pi} \frac{\sin\theta}{\sin^4(\theta/2)} \sin^2\frac{\theta}{2} \, \mathrm{d}\theta. \end{aligned} \tag{6.14}$$

The minimum scattering angle $\theta_{\min}$ is related to the maximum distance $b_{\max}$ from which the scattering can take place without undue influence from the other binary member (see Fig. 6.3). The maximum scattering distance is

$$b_{\max} = \frac{a}{\tan(\theta_{\min}/2)} \approx \frac{G m_a/u^2}{\sin(\theta_{\min}/2)}. \tag{6.15}$$

The last step involves the assumption that $\theta_{\min}/2$ is small and thus $\tan(\theta_{\min}/2) \approx \sin(\theta_{\min}/2)$. This is true when $v_0/u \ll 1$ as we will see shortly. The maximum scattering range $b_{\max}$ should be related to the semi-major axis of the binary

$$a_B = \frac{G(m_a + m_b)}{v_0^2} \approx \frac{2Gm_a}{v_0^2} \tag{6.16}$$

if m_a and m_b are not very unequal.

To get a definite value for $b_{\max}$, let us require that $b_{\max} = \frac{1}{2}a_B$, i.e.

$$\frac{Gm_a}{u^2}\frac{1}{\sin(\theta_{\min}/2)} \approx \frac{Gm_a}{v_0^2}$$

which gives

$$\sin\frac{\theta_{\min}}{2} \approx \frac{v_0^2}{u^2}. \tag{6.17}$$

Now we may proceed to evaluate the integral in (6.14):

$$\begin{aligned}\int_{\theta_{\min}}^{\pi} \frac{\sin\theta\,\mathrm{d}\theta}{\sin^2(\theta/2)} &= 4\int_{\theta_{\min}}^{\pi} \cot\left(\frac{\theta}{2}\right)\mathrm{d}\left(\frac{\theta}{2}\right) \\ &= 4\Big|_{\theta_{\min}}^{\pi} \ln\left(\sin\frac{\theta}{2}\right) \approx 4\ln\frac{u^2}{v_0^2}.\end{aligned}$$

Then

$$-\frac{\mathrm{d}|E_B|}{\mathrm{d}t} = 4\pi\frac{G^2 m_s \rho \mathcal{M}}{u}\ln\frac{u^2}{v_0^2}. \tag{6.18}$$

The second binary component makes an equal contribution and doubles the total value:

$$\left(-\frac{\mathrm{d}|E_B|}{\mathrm{d}t}\right)_{\text{gain}} = \frac{8\pi G^2 m_s \rho \mathcal{M}}{u}\ln\frac{u^2}{v_0^2}. \tag{6.19}$$

The calculation of the loss term takes place in the same manner. We start by noting that the $\Delta\boldsymbol{v}$ vector may be divided into the components parallel to $\boldsymbol{u}$, $\Delta\boldsymbol{v}_{\|}$ and perpendicular to $\boldsymbol{u}$, $\Delta\boldsymbol{v}_{\perp}$. Then the $\mathcal{M}\boldsymbol{v}_0 \cdot \Delta\boldsymbol{v}$ term is also divided similarly and it is the sum of $\mathcal{M}\boldsymbol{v}_0 \cdot \Delta\boldsymbol{v}_{\|}$ and $\mathcal{M}\boldsymbol{v}_0 \cdot \Delta\boldsymbol{v}_{\perp}$. The latter term averages to zero when we add results from all different orientation angles φ. For every angle φ there is an equally likely scattering with $\varphi + \pi$, from the opposite side of the binary component. The scatterings from the opposite sides produce velocity changes $\Delta\boldsymbol{v}_{\perp}$ in opposite directions and thus cancel each other (Fig. 6.4).

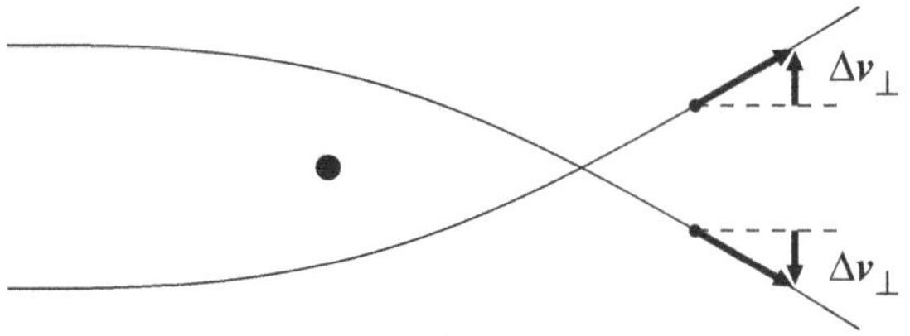

Figure 6.4 For each contribution $\Delta \boldsymbol{v}_\perp$ there is an opposite contribution equal in magnitude.

This approximation is not always justified since the other binary component also influences the scattering. Here we will neglect this influence, but we will return to the problem in Sections 6.6 and 6.8 where this approximation cannot be correct.

We are then left with only the $\mathcal{M}\boldsymbol{v}_0 \cdot \Delta\boldsymbol{v}_{||}$ term which represents frictional drag on the binary component. After making use of $\Delta\boldsymbol{v} = \Delta\boldsymbol{v}_{||} = \Delta v_{||}\boldsymbol{u}/u$, and substituting the expressions for $\boldsymbol{u}$ and $\Delta v_{||}$, we have

$$\begin{aligned}\mathcal{M}(\boldsymbol{v}_0 \cdot \Delta\boldsymbol{v}) &= \mathcal{M}(\boldsymbol{v}_0 \cdot \boldsymbol{u})\frac{\Delta v_{||}}{u} \\ &= \mathcal{M}\left[\boldsymbol{v}_0 \cdot \left(\boldsymbol{u}_s - \frac{\mathcal{M}}{m_a}\boldsymbol{v}_0\right)\right]\frac{2m_s}{m_s + m_a}\sin^2\frac{\theta}{2}.\end{aligned} \tag{6.20}$$

Let us consider first the product $\boldsymbol{v}_0 \cdot \boldsymbol{u}_s$. If the angle between the two vectors is χ, then $\boldsymbol{v}_0 \cdot \boldsymbol{u}_s = v_0 u_s \cos\chi$. During the orbital cycle of a circular binary the angle χ varies between limits $\pi/2 \pm \chi_{\max}$. For every value of $\pi/2 + \chi$ there is an equally likely value $\pi/2 - \chi$; since $\cos(\pi/2 + \chi) = -\cos(\pi/2 - \chi)$, the contributions from the opposite sides of $\chi = \pi/2$ cancel each other, and the average over one orbital cycle $\langle \boldsymbol{v}_0 \cdot \boldsymbol{u}_s \rangle = 0$ (Fig. 6.5).

As a consequence, when $m_s \ll m_a$,

$$\begin{aligned}\mathcal{M}(\boldsymbol{v}_0 \cdot \Delta\boldsymbol{v}) &= -\frac{2m_s}{m_s + m_a}\frac{\mathcal{M}^2}{m_a}v_0^2\sin^2\frac{\theta}{2} \\ &\approx -2m_s\frac{\mathcal{M}^2}{m_a^2}v_0^2\sin^2\frac{\theta}{2}.\end{aligned} \tag{6.21}$$

As far as the θ factors are concerned, this is of the same form as the term $\frac{1}{2}\mathcal{M}(\Delta v)^2$, and we can use the previous result for the integral over θ. The loss term is obtained from the gain term simply by replacing

$$c_1 = \frac{1}{2}\mathcal{M}\left(2\frac{m_s}{m_a}\right)^2 u^2$$

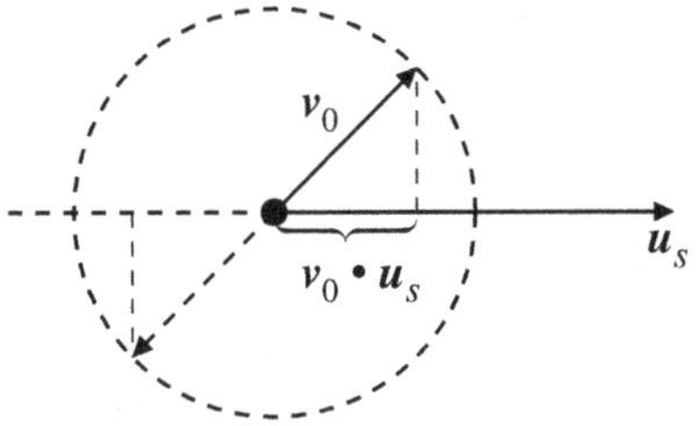

Figure 6.5 In a circular orbit the velocity vector $\boldsymbol{v}_0$ rotates uniformly through 360°. At opposite phases, 180° from each other, the products $\boldsymbol{v}_0 \cdot \boldsymbol{u}_s$ are equal in magnitude but opposite in sign. Therefore $\langle \boldsymbol{v}_0 \cdot \boldsymbol{u}_s \rangle = 0$ over a complete orbital cycle.

by

$$c_2 = -2\frac{\mathcal{M}^2}{m_a^2} m_s v_0^2.$$

It is achieved by multiplying $(\mathrm{d}|E_B|/\mathrm{d}t)_{\text{gain}}$ by c_2/c_1 which gives

$$\left(-\frac{\mathrm{d}|E_B|}{\mathrm{d}t}\right)_{\text{loss}} = -\frac{8\pi G^2 \rho \mathcal{M}^2 v_0^2}{u^3} \ln \frac{u^2}{v_0^2}. \tag{6.22}$$

The net change of the binding energy is the sum of gains and losses, i.e.

$$\frac{\mathrm{d}|E_B|}{\mathrm{d}t} = \frac{16\pi G^2 \rho \mathcal{M}}{u^3} \ln \frac{u^2}{v_0^2} (|E_B| - E_s) \tag{6.23}$$

where

$$|E_B| = \frac{1}{2}\mathcal{M} v_0^2 \tag{6.24}$$

is the binding energy of the binary and

$$E_s = \frac{1}{2} m_s u^2 \approx \frac{1}{2} m_s u_s^2 \tag{6.25}$$

is the kinetic energy of the third body far away from the binary.

It is common terminology to say that a binary is *hard* if $|E_B| > E_s$, and *soft* if $|E_B| < E_s$. The above equation then leads us to the conclusion (so called *Heggie's law*, in fact first discovered by Gurevich and Levin in 1950): Hard binaries become harder, soft binaries become softer. The derivation of Heggie's law in the general three-body problem is more complex since the assumptions (1) and (2) stated above are not generally valid.

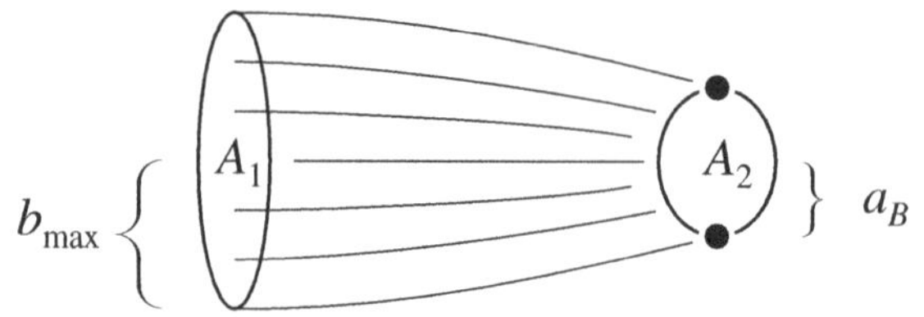

Figure 6.6 The focussing factor ν is the ratio of the two areas, $A_1/A_2 = \pi b_{max}^2/\pi a_B^2$.

When the binary is very hard, $|E_B| \gg E_s$, Eq. (6.23) becomes

$$\frac{d|E_B|}{dt} = \frac{8\pi G^2 \rho \mathcal{M}^2 v_0^2}{u^3} \ln \frac{u^2}{v_0^2}. \tag{6.26}$$

6.2 Evolution of the semi-major axis and eccentricity

The evolution of the binary is often more easily visualised through the evolution of its semi-major axis than its binding energy. Therefore for hard binaries influenced by a constant stream of small bodies the quantity

$$\frac{\dot{a}_B}{a_B} = -\frac{2a_B}{\mathcal{M}G(m_a + m_b)} \frac{d|E_B|}{dt} = R_a \pi a_B G \rho/u \tag{6.27}$$

is studied where R_a is a dimensionless quantity

$$R_a = -\frac{16\mathcal{M}}{m_a + m_b} \frac{v_0^2}{u^2} \ln \frac{u^2}{v_0^2}. \tag{6.28}$$

Before looking at this constant more closely, let us consider the other factors on the right hand side of Eq. (6.27) and their physical significance. Using Eq. (6.16) it may be factorised as follows:

$$\frac{\dot{a}_B}{a_B} = R_a \left(\pi a_B^2\right) \left(\frac{\rho u}{m_a + m_b}\right) \left(\frac{v_0^2}{u^2}\right). \tag{6.29}$$

The factor πa_B^2 is the surface area of the binary orbit while the factor ρu is the mass flow rate of third particles per unit area; the division by $m_a + m_b$ normalises it to the total binary mass. This is necessary since the rate of change of the binary semi-major axis is also normalised to the value of the semi-major axis a_B. The product of these two factors is thus the normalised mass flow rate through the surface area of the binary orbit. The last factor is related to focussing: the stream which goes through the binary orbit is actually captured from a wider cross-section of the stream, and is channelled through the binary area (Fig. 6.6).

The focussing may be described by the focussing factor ν:

$$\nu = \frac{\pi b_{\max}^2}{\pi a_B^2} = \frac{\dfrac{G(m_a + m_b)}{u^2} p_{\max}}{\dfrac{G(m_a + m_b)}{v_0^2} a_B} = \frac{v_0^2}{u^2} \frac{p_{\max}}{a_B}. \tag{6.30}$$

Here the catchment areas $\pi b_{\max}^2$ of both binary members have been added. Requiring that the maximum value of the parameter $p_{\max} = a_B$ gives the focussing factor

$$\nu = \frac{v_0^2}{u^2}. \tag{6.31}$$

The choice of the value for $p_{\max}/a_B$ depends on the problem at hand. The factor ν could be augmented by another factor, as we will do in Example 6.1. At present the second factor does not matter.

Now we may return to the constant R_a. At very large values of u^2/v_0^2 the function $(v_0^2/u^2)\ln(u^2/v_0^2) \to 0$, and it may well be approximated by a function $\propto v_0^2/u^2$. Our theory does not apply at the opposite limit $u^2/v_0^2 \to 0$, but it is reasonable to assume that R_a should approach some constant value as $u^2/v_0^2 \to 0$. At the lower end of the range of applicability of the theory, at $u^2/v_0^2 \approx 2$, we find $R_a \approx -1$ for equal binary masses, $m_a = m_b$. A function which satisfies the above requirements is

$$R_a = -\frac{6.5}{1 + 0.6 \dfrac{(m_a + m_b)^2}{m_a m_b} \dfrac{u^2}{v_0^2}}. \tag{6.32}$$

It has been found to agree well with numerical experiments (Fig. 6.7; Mikkola and Valtonen 1992, Quinlan 1996).

Besides the semi-major axis the eccentricity of the binary orbit is also of interest. Eccentricity e is related to the semi-major axis a and the angular momentum per unit mass k by Eq. (3.28):

$$k^2 = a\mu(1 - e^2) = \mu^2(1 - e^2)/2|h|.$$

Therefore

$$e^2 = 1 - \frac{2|h|k^2}{\mu^2}$$

and

$$\frac{\mathrm{d}(e^2)}{\mathrm{d}|h|} = -\frac{2k^2}{\mu^2} - \frac{4|h|\boldsymbol{k} \cdot \mathrm{d}\boldsymbol{k}}{\mu^2\,\mathrm{d}|h|}$$

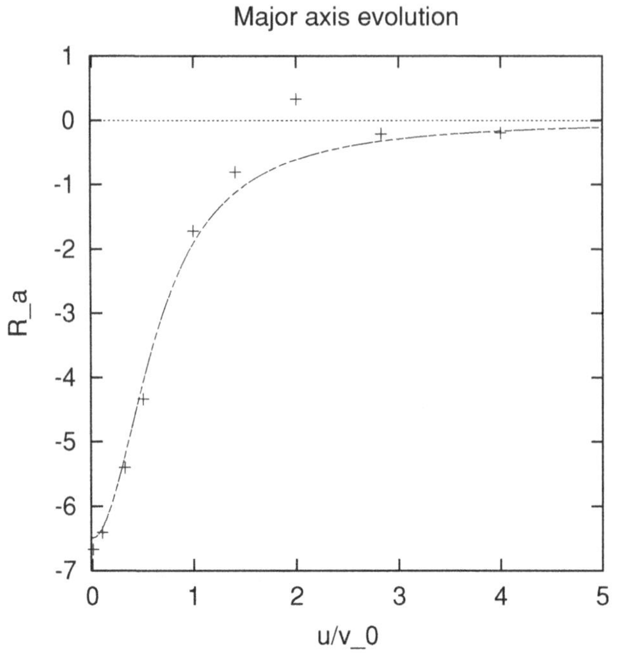

Figure 6.7 A comparison of Eq. (6.32) (curve) with numerical experiments (+, Mikkola and Valtonen 1992). Experiments use a range of binary eccentricities from 0.1 to 0.99, except for the last three points on the right which have $e = 0.5$ only.

from which it follows

$$\frac{d(e^2)}{d(\ln |h|)} = -(1-e^2) - n\frac{dk_\zeta}{d|h|}(1-e^2)^{1/2}. \qquad (6.33)$$

Here dk_ζ is the ζ-component of the change in the angular momentum, i.e. the component perpendicular to the binary plane (Example 5.2), and the mean motion n is

$$n = \frac{(2|h|)^{3/2}}{\mu}. \qquad (6.34)$$

In the restricted three-body problem the absolute value of the change in orbital energy of the third body $d|h|$ is related to its change in angular momentum dk_ζ (per unit mass) by (Example 5.2)

$$d|h| = -n\, dk_\zeta. \qquad (6.35)$$

In the present case we do not quite meet one of the central criteria of the restricted problem, i.e. $m_s = 0$. However, since $m_s \ll m_a + m_b$ we may assume that the above relation holds at least in some limiting sense. Since the energy and the

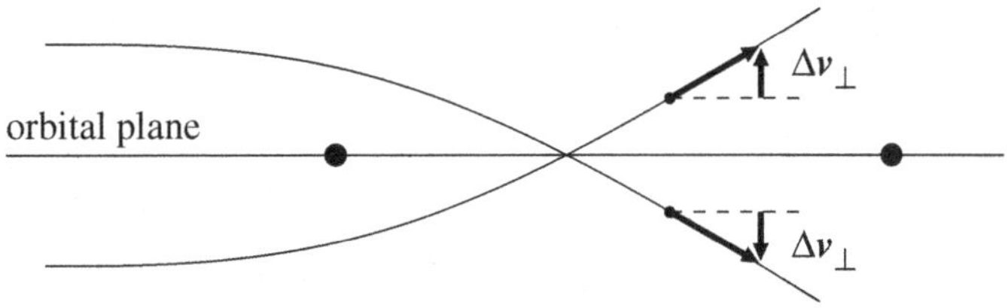

Figure 6.8 The third body passes the binary on two sides of one of the components. The corresponding changes in angular momenta d$\boldsymbol{k}$ are equal in magnitude but opposite in sign. These components of angular momenta cancel each other pairwise.

angular momentum are conserved, we expect Eq. (6.35) to be true also for the binary: the energy gain by the third body is a loss to the binary, and the same is also true for the angular momenta. Therefore for a circular binary

$$\frac{\mathrm{d}k_\zeta}{\mathrm{d}|h|} = -\frac{1}{n} \tag{6.36}$$

and consequently

$$\frac{\mathrm{d}(e^2)}{\mathrm{d}(\ln|h|)} = -(1-e^2) + (1-e^2)^{1/2}. \tag{6.37}$$

Putting $e = 0$ we get our first result, true for every individual orbit and therefore also for the orbit average: *a circular binary remains circular.*

For eccentric binaries Eq. (6.36) does not hold exactly. However, it may still be true on average if the binary orbits fast. Fast rotating binaries with all different major axis orientations form a collection of orbits which looks somewhat like a single circular binary from the point of view of a slowly approaching third body. Thus at the limit of $u/v_0 \to 0$, Eq. (6.37) should be correct.

At the opposite limit, $u/v_0 \to \infty$, the scattering is almost like a pure two-body scattering from a stationary binary component. Then we expect that for every orbit passing the binary component from one side there is a corresponding orbit passing it from the other side, with equal but opposite contributions to the change in the angular momentum (Fig. 6.8). Therefore, at this limit we expect $\mathrm{d}(e^2)/\mathrm{d}(\ln|h|) \to 0$. A function which satisfies these requirements and also gives the correct value at $e = 0$ is

$$\frac{\mathrm{d}(e^2)}{\mathrm{d}(\ln a)} = (1-e^2) - (1-e^2)^{1/2+f(u/v_0)}, \tag{6.38}$$

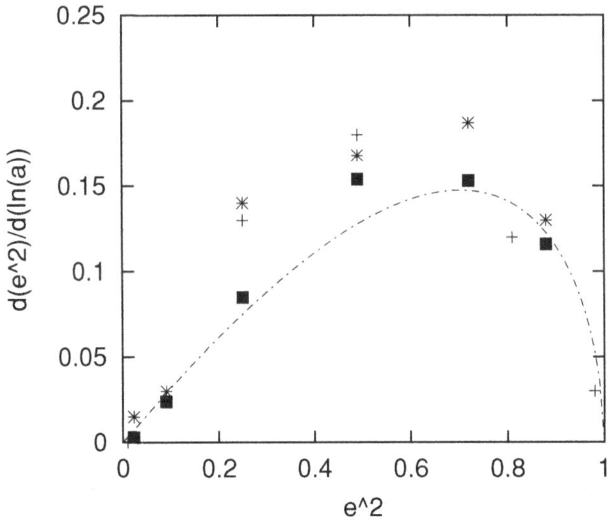

Figure 6.9 A comparison of Eq. (6.38) with numerical experiments (+ Mikkola and Valtonen 1992; ■ Quinlan 1996, equal masses; $*$ Quinlan 1996, unequal masses). The relative approach speed is $u/v_0 = 0.1$.

where $f(u/v_0) \to 0$ when $u/v_0 \to 0$ and $f(u/v_0) \to \frac{1}{2}$ when $u/v_0 \to \infty$. Here we have used the identity $\mathrm{d}(\ln|h|) = -\mathrm{d}(\ln a)$ to make the derivation with respect to $\ln a$.

On the basis of numerical experiments different forms of $f(u/v_0)$ have been suggested (Mikkola and Valtonen 1992, Pietilä 1999). A simple function which may be used is

$$f(u/v_0) = \begin{cases} u/v_0 & \text{if } u/v_0 < 0.5 \\ 0.5 & \text{if } u/v_0 \geq 0.5. \end{cases} \tag{6.39}$$

Figure 6.9 shows how this function compares with the numerical experiments when $u/v_0 = 0.1$.

6.3 Capture of small bodies by a circular binary

When a small body passes by a circular binary it may lose enough energy to become bound to the binary and thus to form a temporary triple system with the binary components. Let us consider this capture problem in the approximation of the previous section, i.e. let us calculate the velocity change Δv_s of the small body of mass m_s when it is scattered by a binary component of mass m_a (the small body mass $m_s \ll m_a$). At a large distance from the binary the small body has a speed $\boldsymbol{u}_s$ relative to the binary while the relative velocity of the binary components is $\boldsymbol{v}_0$.

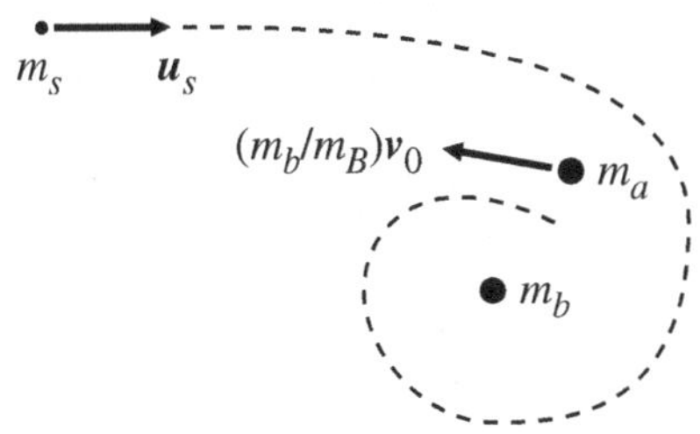

Figure 6.10 The capture of a particle of mass m_s by a binary. The particle comes from a large distance with speed $\boldsymbol{u}_s$ but remains close to the binary after scattering.

Therefore at a large distance from the binary, the velocity of the small body relative to the binary component is $\boldsymbol{u} = \boldsymbol{u}_s - (m_b/m_B)\boldsymbol{v}_0$, where m_b is the mass of the other component and $m_B = m_a + m_b$ (Fig. 6.10). If the scattering angle is θ, the change of speed is (Eq. (6.8))

$$\Delta v_s = 2u \sin\frac{\theta}{2}. \tag{6.40}$$

The differential cross-section $d\sigma$ for scattering into the interval $\theta, \theta + d\theta$ is (Eqs. (3.58) and (3.59))

$$\begin{aligned} d\sigma &= 2\pi \frac{a^2}{4} \frac{\sin\theta \, d\theta}{\sin^4(\theta/2)} \\ &\approx \pi a_B^2 \left(\frac{m_a}{m_B}\right)^2 \frac{v_0^4}{u^4} \frac{\cos(\theta/2)}{\sin^3(\theta/2)} \frac{d\theta}{d(\Delta v_s)} d(\Delta v_s). \end{aligned} \tag{6.41}$$

We have made use of $u^2 = Gm_a/a$ and $v_0^2 = Gm_B/a_B$ where a and a_B are the semi-major axes of the small body orbit relative to the binary component and of the binary, respectively. The differential $d\theta$ has been written as a product

$$\frac{d\theta}{d(\Delta v_s)} d(\Delta v_s) = \frac{d(\Delta v_s)}{u\cos(\theta/2)}.$$

At the limit of $v_s \ll v_0$ roughly half of the scatterings lead to a positive Δv_s and the other half to negative Δv_s. In case of capture we are interested only in the latter. Thus for captures the differential cross-section is, after substituting $\Delta v_s/2u$ for $\sin(\theta/2)$,

$$d\sigma_{\text{cap}} \approx 4\pi a_B^2 \left(\frac{m_a}{m_B}\right)^2 \frac{v_0^4}{u^2} \frac{d(\Delta v_s)}{(\Delta v_s)^3}. \tag{6.42}$$

The total capture cross-section σ_{cap} is obtained by integrating $d\sigma_{\text{cap}}$ over Δv_s. The lower limit of integration is u_s which corresponds to the total loss of the initial kinetic energy of the small body. The upper limit corresponds to $\sin\theta/2 = 1$, i.e.

$\Delta v_s = 2u$. Therefore

$$\begin{aligned}\sigma_{\text{cap}} &\approx 4\pi a_B^2 \left(\frac{m_a}{m_B}\right)^2 \frac{v_0^4}{u^2} \int_{u_s}^{2u} \frac{\mathrm{d}(\Delta v_s)}{(\Delta v_s)^3} \\ &\approx 2\pi a_B^2 \left(\frac{m_a}{m_B}\right)^2 \frac{v_0^2}{u_s^2}.\end{aligned} \tag{6.43}$$

The last near equality is based on the assumption that $u_s \ll v_0$ and $u \approx v_0$.

In equal mass binary systems both components contribute equally to the capture cross-section. When the two contributions are added together

$$\sigma_{\text{cap}} \left(m_a = \frac{1}{2} m_B\right) \approx \pi a_B^2 \frac{v_0^2}{u_s^2}. \tag{6.44}$$

We will find later (Eq. (8.23)) essentially the same result in the general three-body problem (for zero eccentricity binary; for high eccentricities there is a correction factor of about 1.8). Numerical experiments (Hills 1989, 1992) demonstrate that the capture probability is indeed insensitive to the mass m_s of the incoming body.

6.4 Orbital changes in encounters with planets

The Sun, a planet and a small body is a three-body system which is an application of the restricted three-body problem. The orbits of planets are generally close to circular, and the mass of the small body (asteroid, comet) is negligible in comparison with the mass of a planet. Since the mass of the planet is also much smaller than the mass of the Sun, we may view the small body moving in an elliptical or hyperbolic orbit around the Sun, until by chance a planet happens to come close to its path and a two-body scattering from the planet takes place. After the scattering the small body finds itself in a new orbit about the Sun. We will now study some of the properties of the orbital changes resulting from the planet encounters (see Fig. 6.11). Here we follow the discussion by Everhart (1968, 1969).

In Section 6.1 we found that when a small body (mass m_s) passes a planet (mass m_a), the latter receives a velocity change (Eq. (6.8))

$$\Delta \boldsymbol{v} = \frac{2m_s}{m_a} u \sin\frac{\theta}{2} \hat{\boldsymbol{e}}. \tag{6.45}$$

Here we have cast the equation in vectorial form by using the unit vector $\hat{\boldsymbol{e}}$ which points from the planet towards the pericentre of the orbit (Fig. 6.11 (a)). As long as the scattering angle θ is small, $\hat{\boldsymbol{e}}$ is nearly perpendicular to the incoming velocity $\boldsymbol{u}$

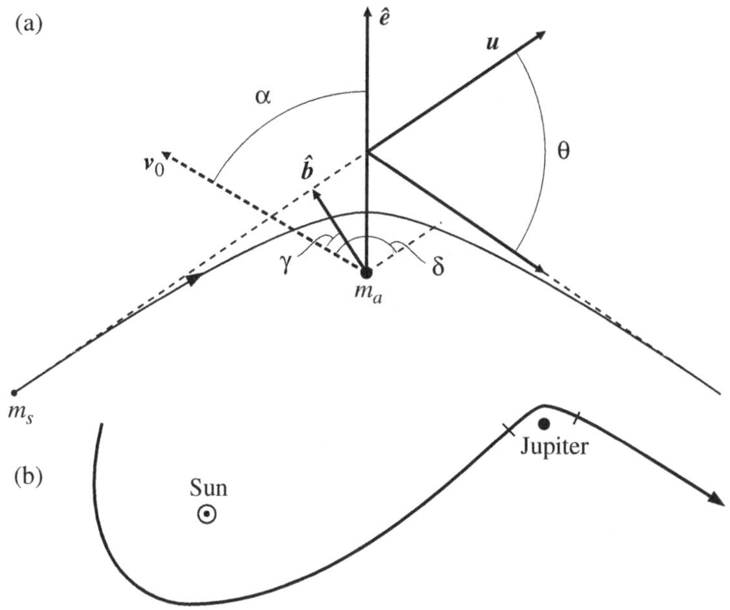

Figure 6.11 Hyperbolic encounter of a comet with Jupiter as seen from the jovicentric frame (a) and from the heliocentric frame (b). The heliocentric velocity of Jupiter $\boldsymbol{v}_0$ is generally off the plane of the two-body encounter shown in (a). Similarly, the different portions of the orbit in (b) may be in different planes.

(at large distance from the planet). Under the same assumption

$$\sin\frac{\theta}{2} \approx \tan\frac{\theta}{2} = \frac{Gm_a}{u^2 b} \tag{6.46}$$

where b is the impact parameter.

The velocity change of the planet is generally so small that we can ignore the $(\Delta v)^2$ term in Eq. (6.9). Thus the energy change of the planet is

$$\Delta E_a = m_a \boldsymbol{v}_0 \cdot \Delta\boldsymbol{v} = 2m_s u \sin\frac{\theta}{2}\boldsymbol{v}_0 \cdot \hat{\boldsymbol{e}}. \tag{6.47}$$

The change in the orbital energy of the small body is equal to this but opposite in sign. The orbital energy of the small body is

$$E_s = -\frac{Gm_s m_b}{2a} \tag{6.48}$$

where m_b is the mass of the Sun and a is the semi-major axis of the elliptic orbit of the small body relative to the Sun. The change in its orbital energy is

$$\Delta E_s = -\frac{Gm_s m_b}{2}\Delta\left(\frac{1}{a}\right) = -2m_s u \sin\frac{\theta}{2}\,\boldsymbol{v}_0 \cdot \hat{\boldsymbol{e}}. \tag{6.49}$$

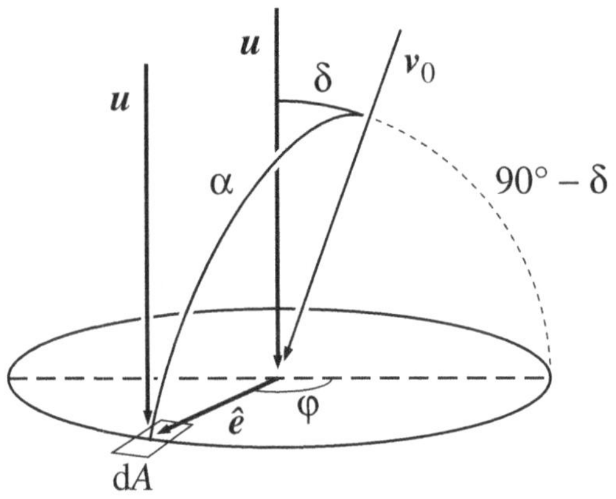

Figure 6.12 Collision plane.

Thus

$$\Delta\left(\frac{1}{a}\right) = 4\left(\frac{m_a}{m_b}\right)\frac{1}{ub}\boldsymbol{v}_0\cdot\hat{\boldsymbol{e}}. \tag{6.50}$$

Let us define the dimensionless energy change U by

$$U = \frac{m_b}{m_a}\Delta\left(\frac{a_B}{a}\right) = 4\frac{v_0 a_B}{ub}\cos\alpha, \tag{6.51}$$

where α is the angle between the vectors $\boldsymbol{v}_0$ and $\hat{\boldsymbol{e}}$ (Fig. 6.11 (a)). We now ask how U is distributed in encounters between small bodies and a planet when the encounters happen in a random way. We have to consider the distributions of b and $\cos\alpha$ in the encounters.

In Fig. 6.12 we have drawn the 'collision plane', the plane perpendicular to $\boldsymbol{u}$ which passes through the planet. We take a surface element $\mathrm{d}A = b\,\mathrm{d}b\,\mathrm{d}\varphi$ of this plane, at distance b from the planet. Angle φ measures the angular distance from the projection of $\boldsymbol{v}_0$ onto this plane to the crossing point. The relative orientation of $\boldsymbol{v}_0$ is further specified by δ, the angle between $\boldsymbol{v}_0$ and $\boldsymbol{u}$.

From the spherical triangle with sides φ, α and $90^\circ - \delta$ we obtain $\cos\alpha = \cos\varphi\sin\delta$. We substitute this in Eq. (6.51) above and solve for b:

$$b = \frac{4v_0 a_B}{uU}\cos\varphi\sin\delta$$

from which

$$\mathrm{d}A = b\,\mathrm{d}b\,\mathrm{d}\varphi = \left(\frac{4v_0 a_B\sin\delta}{u}\right)^2\frac{\mathrm{d}U}{U^3}\cos^2\varphi\,\mathrm{d}\varphi. \tag{6.52}$$

The cross-section for scattering such that the dimensionless energy change U lies in the interval $[U, U + \mathrm{d}U]$ is

$$\mathrm{d}\sigma(U)\,\mathrm{d}U = \left(\frac{4v_0 \sin\delta}{u}\right)^2 a_B^2 \frac{\mathrm{d}U}{U^3} \int_{-\pi/2}^{+\pi/2} \cos^2\varphi\,\mathrm{d}\varphi. \tag{6.53}$$

The integration is extended only through one half of the φ range since one half of the phase angle range produces positive U and the other half produces negative U. Writing $|U|^3$ in place of U^3 covers both situations. Therefore the cross-section from either positive or negative U is given by

$$\begin{aligned} \mathrm{d}\sigma(U)\,\mathrm{d}U &= \left(\frac{4v_0 \sin\delta}{u}\right)^2 a_B^2 \frac{\mathrm{d}U}{|U|^3} \bigg|_{-\pi/2}^{\pi/2} \frac{1}{2}\left(\varphi + \frac{\sin 2\varphi}{2}\right) \\ &= \frac{1}{2} F a_B^2 \frac{\mathrm{d}U}{|U|^3}, \end{aligned} \tag{6.54}$$

where

$$F = 16\pi (v_0/u)^2 \sin^2\delta. \tag{6.55}$$

The value of F depends on the parameters of the small body orbits. The angle δ is the polar angle for which we may assume that $\cos\delta$ is a uniform random variable (say R) between -1 and $+1$. Thus

$$\langle \sin^2\delta \rangle = 1 - \langle \cos^2\delta \rangle = 1 - \langle R^2 \rangle = \frac{2}{3}. \tag{6.56}$$

The relative speed u is generally greater than v_0; we may estimate that typically u averages to $\sqrt{2}v_0$, the parabolic pericentre speed at the distance a_B when the total mass is m_b. Here we are thinking of bodies like comets on nearly parabolic trajectories. For randomly distributed orbits we therefore expect $F = (16/3)\pi$; experimentally $\langle F \rangle \approx 4.5\pi$ (Everhart 1969). An extension of the experiments to large values of $|U|$ shows that in fact the power-law is steeper than $|U|^{-3}$, more like $|U|^{-3.5}$. At small values of $|U|$, $|U| \lesssim 6$, the power-law flattens to $|U|^{-1}$ (Valtonen and Innanen 1982; Fig. 6.13).

6.5 Inclination and perihelion distance

The parameters which are usually used to describe the small body orbit are its inclination ι_0 and pericentre distance q_0 relative to the binary. The initial pericentre distance is normalised to the binary semi-major axis: $Q = q_0/a_B$. If the remaining orbital elements of the small body orbit (ω and Ω) are not well known or can be considered random in some sense, then it is not immediately clear how specific values of ι_0 and Q should enter into $\sin\delta$ which is needed in Eq. (6.55) above.

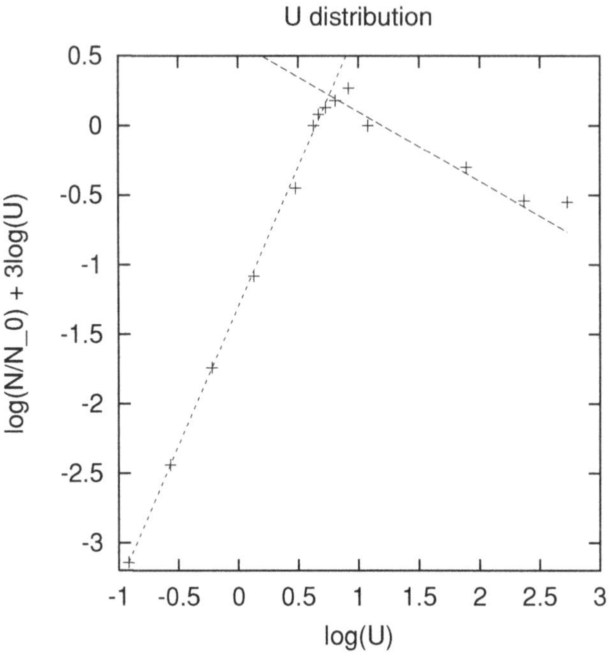

Figure 6.13 A comparison of the theoretical expression (Eq. (6.54)) for the cross-section $d\sigma(U)$ with numerical experiments from Valtonen and Innanen (1982). On the vertical axis we plot $|U|^3 d\sigma(U)$ and therefore a horizontal line would represent Eq. (6.54). The data points follow more closely two lines of slopes +2 (i.e. $\sigma(U)$ proportional to $|U|^{-1}$) and -0.5 (i.e. $\sigma(U)$ proportional to $|U|^{-3.5}$), the dividing line being at about $|U| = 6$.

It is best to use numerical orbit calculations to derive the (ι_0, Q) dependence of F.

Everhart (1968) produced extensive tables of F at specific values of ι_0 from 3° to 177° and at the values of Q ranging from 0.005 to 2. We cannot expect to find a simple analytic formula which fits all the data. However, it is possible to represent the data with fair accuracy by using two helpful features. First, the dependence on ι_0 is typically a function of $1 + \cos\iota_0$. We will learn more about this in Section 10.4. Second, the distribution of F peaks close to $\iota_0 = 0$ and $\iota_0 = 180°$; at both ends of the inclination spectrum these peaks can be fairly represented by a $(\sin\iota_0)^{-1}$ factor, as long as $\iota_0 > 1°$ and $\iota_0 < 179°$ (Valtonen *et al.* 1992).

The use of the $(\sin\iota_0)^{-1}$ factor is also handy since in randomly distributed orbits the inclination distribution is

$$f(\iota_0)\, d\iota_0 = \frac{1}{2}\sin\iota_0\, d\iota_0. \tag{6.57}$$

When we take the product $Ff(\iota_0)\,d\iota_0$, the $\sin\iota_0$ factors cancel each other, and we get a distribution which is relatively smooth from $\iota_0 = 1°$ to $\iota_0 = 179°$. This kind of distribution is expected, for example, for the comets captured from the Oort Cloud. The original inclination of the Oort Cloud comets follows Eq. (6.57) above quite well; therefore the numbers of comets captured from the Cloud at different inclinations ι_0, which are proportional to $Ff(\iota_0) \propto F\sin\iota_0$, vary rather smoothly with ι_0. In the following we will thus search for an expression for $F\sin\iota_0$ as a function of some power of $1 + \cos\iota_0$.

A fit to the data by Everhart (1968) shows that at small values of Q, $0.002 \le Q \le 0.1$, the function is linear:

$$F\sin\iota_0/\pi \approx 0.8 + 4(0.1 - Q) + [0.7 - 5(0.1 - Q)]\,(1 + \cos\iota_0). \qquad (6.58)$$

The expression may be extended even down to $Q = 0.005$ by putting $Q = -0.01$ in the above equation. With increasing Q the distribution becomes more strongly biased towards $\cos\iota_0 = 1$.

For $Q \ge 0.1$ this bias becomes so strong that a linear function of $1 + \cos\iota_0$ is not satisfactory but a higher power of $1 + \cos\iota_0$ is needed. The actual value of this power increases with Q. At the same time the linear coefficients of Q have to be replaced by power-laws. Altogether a fitting function

$$F\sin\iota_0/\pi = (1 - Q)\left[1 + (1 - Q)^{-1.75} Q^{0.3} (1 + \cos\iota_0)^{1.75 - 0.5(1-Q)}\right] \qquad (6.59)$$

is obtained. It is derived for the range $0.1 \le Q \le 0.9$, but its validity may be extended to the range $0.9 \le Q \le 1.25$ by the following trick: in place of Q, put

$$Q' = 0.938\exp\left(-|Q - 0.97|^{1.2}\right) \qquad (6.60)$$

in Eq. (6.59). Then the data from Everhart (1968) are well represented, except at $Q > 1$ close to $\iota_0 = 0°$ where $F\sin\iota_0$ is actually greater than the predicted value. In other words, at $Q \ge 1$ the distribution of F is more strongly peaked towards $\iota_0 = 0$ than the simple $(\sin\iota_0)^{-1}$ factor would indicate. Figures 6.14 and 6.15 show how well the fitting functions agree with numerical data.

The strong inclination dependence is easy to see qualitatively from the form of F. In direct orbits the small body and the planet move in the same sense and the relative speed u is small and F is large. On the contrary, in retrograde orbits the two bodies move against each other and u is relatively large and F small. When the orbital plane of the small body is close to the orbital plane of the planet, there is less free room for the bodies to miss each other and scatterings are more likely; thus the two peaks in the function $f(\iota)$ at the two ends of the spectrum.

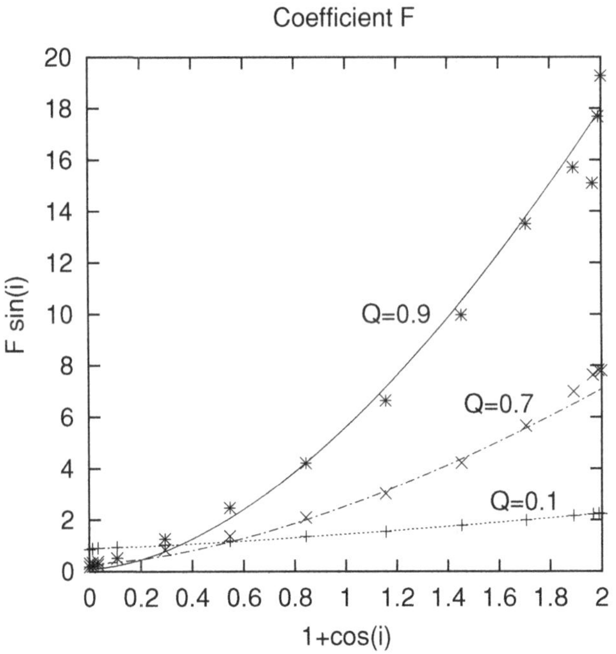

Figure 6.14 The coefficient F in Eq. (6.59) according to numerical experiments by Everhart (1968) and the corresponding fitting functions. The data are for $Q = 0.1$ (+), $Q = 0.7$ (×) and $Q = 0.9$ (∗).

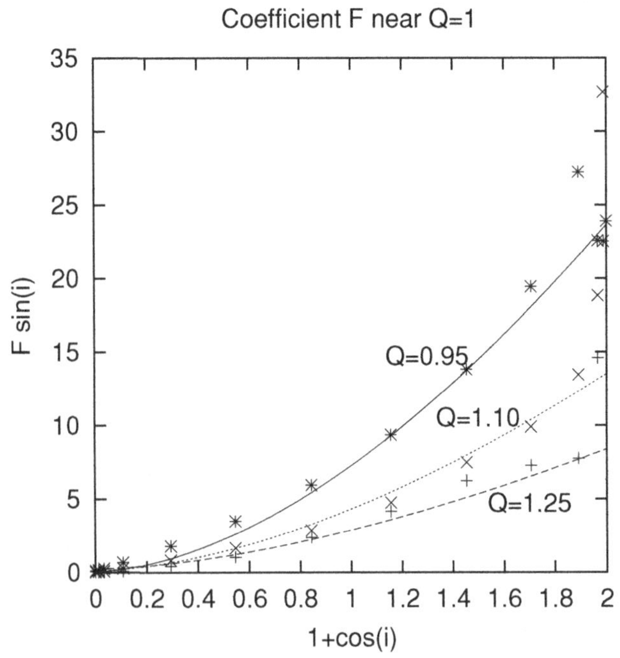

Figure 6.15 The same as the previous Figure, except that the Q values are now $Q = 0.95$ (∗), $Q = 1.10$ (×) and $Q = 1.25$ (+). In the fitting function, the modified parameter Q' is used instead of Q.

Example 6.1 Capture of interstellar bodies by the Solar System.

If small bodies come in hyperbolic orbits from outside the Solar System we have to consider the focussing of the stream of bodies by the Sun. It is this focussed, more dense stream of bodies which crosses the collision plane. Let us say that the speed of the stream far away from the Sun is u_s and that a_3 is the semi-major axis of the small body orbit relative to the Sun. The corresponding impact parameter is b_3, the parameter of the orbit is p_3 and the pericentre distance is q_3.

The focussing factor is

$$\begin{aligned}\nu &= \frac{\pi b_3^2}{\pi a_B^2} = \frac{a_3 p_3}{a_B^2} = \frac{a_3(1+e_3)q_3}{a_B^2} \\ &\approx (1+e_3)\frac{a_3}{a_B} = (1+e_3)\frac{v_0^2}{u_s^2}\end{aligned} \tag{6.61}$$

since $q_3 \approx a_B$ in order that the small body stream passes the Sun at the distance of the planet (see Fig. 6.3 and discussion there). Here e_3 is the eccentricity of the small body orbit, $e_3 > 1$. Its value depends on the velocity u_s. At values of u_s/v_0 appropriate for Galactic bodies we may choose $e_3 = 9$. The right hand side of Eq. (6.54) is to be multiplied by this augmented focussing factor.

Considering that the actual power-law slope is $|U|^{-3.5}$ rather than $|U|^{-3}$, we may write the probability distribution as (Fig. 6.13)

$$f(|U|) = 1.3 \times 10^{-3} \left(\frac{10}{|U|}\right)^{3.5}.$$

If we require a capture of the incoming body, then the energy change has to be at least

$$GM_\odot \Delta \left(\frac{1}{a_3}\right) = u_s^2$$

or

$$u_s^2 = v_0^2\, 10^{-3} |U|.$$

Transform to the distribution of u_s:

$$f(u_s) = f(|U|)\frac{\mathrm{d}|U|}{\mathrm{d}u_s} = 2.6 \times 10^{-7} v_0^{-1} (v_0/u_s)^6.$$

This is multiplied by the focussing factor ν and integrated from u_s to infinity to obtain the capture probability:

$$P_{\text{capture}} = 2.6 \times 10^{-6} v_0^{-1} \int_{u_s}^{\infty} \left(\frac{v_0}{u_s}\right)^8 \mathrm{d}u_s = 3.7 \times 10^{-7} \left(\frac{v_0}{u_s}\right)^7.$$

Since the orbital speed of Jupiter is $v_0 = 13.1$ km/s,

$$P_{\text{capture}} \approx 25 \left(\frac{\text{km/s}}{u_s} \right)^7$$

and the capture cross-section is

$$\sigma_{\text{capture}} \approx 25\pi a_J^2 \left(\frac{\text{km/s}}{u_s} \right)^7,$$

where $a_J = 5.2$ AU is the radius of Jupiter's orbit (Valtonen and Innanen 1982).

6.6 Large angle scattering

In the previous derivation it was assumed that the scattering angle θ is small. If it is not small, more exact expressions have to be used:

$$\sin\frac{\theta}{2} = \frac{\tan(\theta/2)}{\left(1 + \tan^2(\theta/2)\right)^{1/2}} = \frac{K}{(1 + K^2)^{1/2}} \tag{6.62}$$

where

$$K = \tan\frac{\theta}{2} = \frac{Gm_a}{u^2 b}. \tag{6.63}$$

The unit vector $\hat{\boldsymbol{e}}$ is not exactly perpendicular to $\boldsymbol{u}$. Let us say that the unit vector perpendicular to $\boldsymbol{u}$ is $\hat{\boldsymbol{b}}$ and that it lies along the line of the impact distance b. Then the unit vector $\hat{\boldsymbol{e}}$ is obtained from $\hat{\boldsymbol{b}}$ by rotating the latter through an angle $\theta/2$ in the plane common to $\boldsymbol{u}$ and $\hat{\boldsymbol{b}}$. Therefore

$$\hat{\boldsymbol{e}} = \hat{\boldsymbol{b}} \cos\frac{\theta}{2} + \frac{\boldsymbol{u}}{u} \sin\frac{\theta}{2} = \left(\hat{\boldsymbol{b}} + K\frac{\boldsymbol{u}}{u} \right) \Big/ (1 + K^2)^{1/2}. \tag{6.64}$$

Then

$$\begin{aligned} \Delta\left(\frac{1}{a}\right) &= \frac{4u \sin(\theta/2)}{Gm_b} \boldsymbol{v}_0 \cdot \hat{\boldsymbol{e}} \\ &= 4\left(\frac{m_a}{m_b}\right) \frac{1}{b} \frac{v_0}{u} \frac{1}{(1 + K^2)} (\cos\gamma + K \cos\delta), \end{aligned} \tag{6.65}$$

where γ is the angle between $\boldsymbol{v}_0$ and $\hat{\boldsymbol{b}}$ (Fig. 6.11 (a)). When $K \to 0$ we recover our earlier result (Eq. (6.50)).

Let us now consider the special case of coplanar orbits and the pericentre distance of the incoming orbit $q \approx a_B$, i.e. a 'grazing' encounter with the planet (see Fig. 6.16). Then $\gamma = \pi/2 \pm \delta$ and $\cos\gamma \approx \pm\delta$ as long as δ is a small (positive) angle. For direct orbits $\cos\delta \approx 1$, for retrograde orbits $\cos\delta \approx -1$. Since

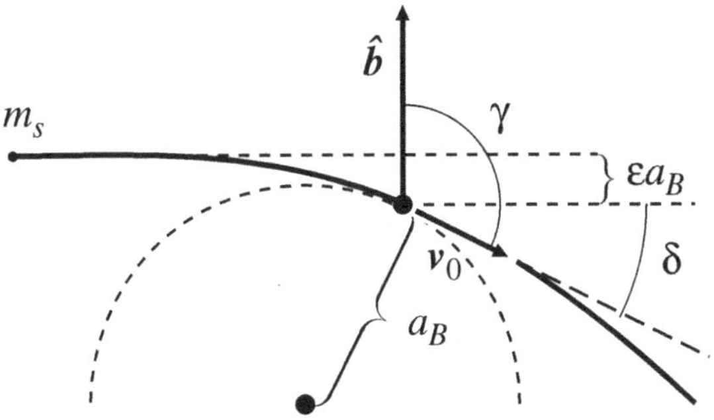

Figure 6.16 A grazing encounter between a small body of mass m_s and a planet. The orbital speed of the planet is $\boldsymbol{v}_0$, and it is shown at the point of contact of the two orbits. The normal in the orbital plane to the approach direction of the small body, at the time of its entry to the influence of the planet, is vector $\hat{\boldsymbol{b}}$, and γ is the angle between $\boldsymbol{v}_0$ and $\hat{\boldsymbol{b}}$. It differs from $\pi/2$ by the small angle δ.

$K = \tan\theta/2$, $(1+K^2)^{-1} = \cos^2(\theta/2)$ and we have for direct orbits,

$$\Delta\left(\frac{1}{a}\right)_{\text{direct}} \approx 4\left(\frac{m_a}{m_b}\right)\frac{1}{b}\frac{v_0}{u}\cos^2\frac{\theta}{2}\left(\tan\frac{\theta}{2} \pm \delta\right). \tag{6.66}$$

As long as $|\delta| \ll \tan(\theta/2)$,

$$\Delta\left(\frac{1}{a}\right) \approx 2\left(\frac{m_a}{m_b}\right)\frac{v_0}{u}\frac{\sin\theta}{b}. \tag{6.67}$$

For retrograde orbits the sign is reversed.

We should note that for direct orbits $\Delta(1/a)$ is always positive (θ is positive by definition), i.e. a decreases in the encounter. In retrograde counters $\Delta(1/a)$ is always negative. This is in contrast to our earlier result of small angle scattering that $\Delta(1/a)$ is equally likely to be positive or negative, and that the net overall change is zero. Figure 6.17 illustrates the ratio of positive to negative changes $\Delta(1/a)$ as a function of the normalized pericentre distance $Q = q/a_B$ and initial inclination ι_0 of the small body orbit. The corner $Q \approx 1$, $\iota_0 \approx 0°$ is dominated by positive $\Delta(1/a)$ while the corner $Q \approx 1$, $\iota \approx 180°$ is dominated by negative $\Delta(1/a)$. But the effect is not restricted entirely to these special points and the influence is seen also in the surrounding regions of the (Q, ι_0) initial value space.

As before, we calculate the probability for an energy change $U = (m_b/m_a)a_B\Delta(1/a)$. The calculation is simplified if we approximate $\tan(\theta/2) \ll 1$, or $\theta \ll \pi/2$, and thus (using Eq. (6.63) in the last step)

$$\begin{aligned}\sin\theta &= 2(\tan(\theta/2))/(1+\tan^2(\theta/2))\\ &\approx 2\tan(\theta/2) \approx 2Gm_a/\left[\left(\sqrt{2}-1\right)^2 v_0^2 b\right]\end{aligned}$$

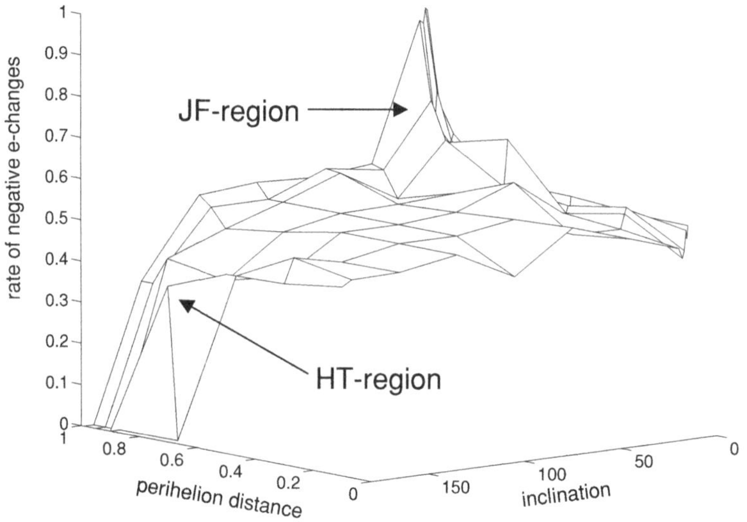

Figure 6.17 The probability (cross-section) of a parabolic comet obtaining the semi-major axis $a \leq 40$ AU after an encounter with Jupiter or obtaining the corresponding positive energy change. The diagram illustrates the fraction which the negative energy changes form of the total probability (negative + positive). The results are plotted as a function of perihelion distance ($a_J = 1$) and inclination of the initial orbit. In the diagram the arrows point to Everhart's capture region (JF) and to the retrograde excess region of Halley type comets (HT).

since $u \approx u_s - v_0 \approx \sqrt{2}v_0 - v_0 = \left(\sqrt{2} - 1\right) v_0$ for a direct parabolic incoming orbit. This leads to

$$U \approx \frac{4}{\left(\sqrt{2}-1\right)^3} \frac{a_B^2}{b^2}, \qquad\qquad (6.68)$$
$$dU \approx -\frac{2U}{b}\, db.$$

The probability for an impact within the interval $[b, b + db]$ is

$$d\sigma \approx -\frac{db}{\epsilon a_B} = \frac{1}{2\epsilon}\frac{b}{a_B}\frac{dU}{U} = \frac{1}{\left(\sqrt{2}-1\right)^{3/2}\epsilon}\frac{dU}{U^{3/2}}, \qquad (6.69)$$

where ϵa_B is a typical value of b which can be considered to give an effective 'grazing' encounter. Clearly $\epsilon < 1$, since the distance of dominance of Jupiter's gravitational influence is $(\rho_2)_d = 0.048$; numerical experiments suggest that $\epsilon = 0.02$ is a good value to use. With this choice Eq. (6.69) is found to agree well with experiments (Fig. 6.18; Valtonen *et al.* 1998). For retrograde orbits $\sqrt{2} - 1$ is replaced by $\sqrt{2} + 1$ which predicts lower probability than in direct orbits. Since ϵ

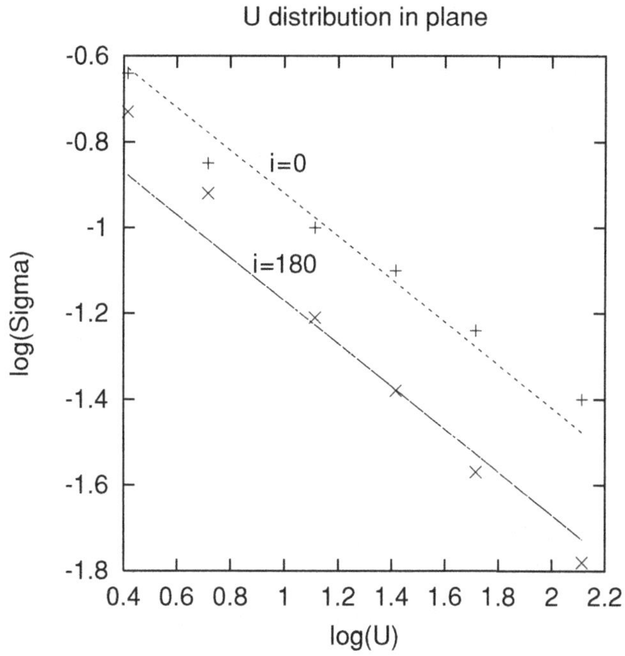

Figure 6.18 A comparison of the distribution of U in theory (Eq. (6.69)) and in experiments (Valtonen *et al.* 1998). A planar system with $\iota = 0°$ (+) and $\iota = 180°$ (×) is considered, with $Q = 0.99$. For direct orbits $U < 0$, for retrograde orbits $U > 0$. The line for the retrograde orbits is drawn a factor of 1.8 below the line for direct orbits.

is also smaller for retrograde orbits than for direct orbits, we cannot predict how much smaller the retrograde probability is in comparison with the probability for direct orbits. Experimentally the ratio is about 1.8 (Fig. 6.18). Note that there are also positive energy changes in direct planar orbits; their probability is lower than the probability of negative energy changes by a factor of about 30 at all levels of $|U|$. Similarly, the probability for negative energy changes in retrograde planar orbits is lower than the probability for positive energy changes by a factor of about 14 (Valtonen *et al.* 1998).

6.7 Changes in the orbital elements

When the orbital energy changes by U there is also a change in other orbital elements. The change is controlled by the Tisserand parameter

$$T = \frac{1}{a} + 2\sqrt{a(1-e^2)}\cos\iota \tag{6.70}$$

which must remain constant (Problem 5.2). Consider for example, a parabolic initial orbit for which $a = \infty$ and $a(1-e^2) = 2q_0$, q_0 being the original pericentre

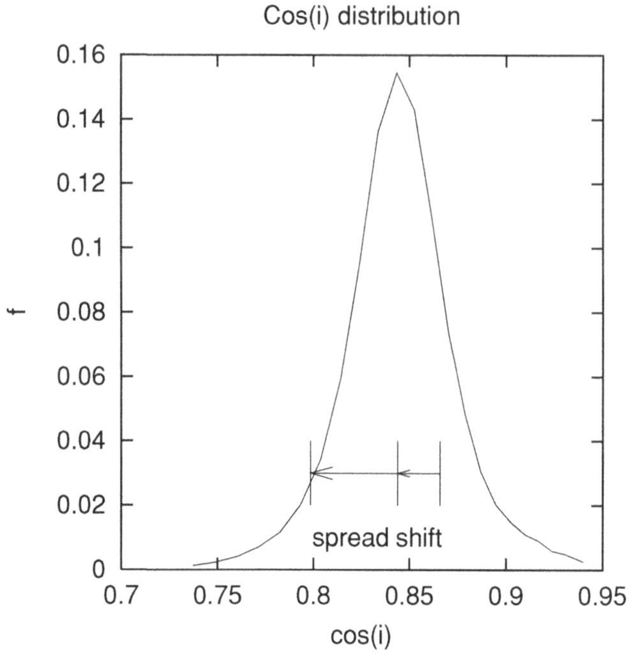

Figure 6.19 The distribution of the cosine of the new inclination after scattering from a parabolic orbit to a new orbit of $a = 5.77a_B$. The cosine of the initial inclination (30°) is shown by a vertical line while an arrow indicates where the centre of the distribution is shifted. Another arrow tells the amount of spread of the distribution, defined so as to leave 10% of the distribution outside in the tails. The curve refers to $Q = 0.5$.

distance. Then the initial inclination ι_0 and the final inclination ι are related by

$$\cos\iota = \sqrt{\frac{2q_0}{q(1+e)}}\cos\iota_0 - \frac{1}{2a\sqrt{q(1+e)}} = A\cos\iota_0 - \frac{B}{a} \tag{6.71}$$

where A and B are functions of $q(1+e)$, a is the semi-major axis of the final elliptical orbit and q is the corresponding pericentre distance. It is obvious that even for a fixed a and q_0 there is no single value of ι which would result from the scattering since $q(1+e)$ may adopt different values. To find out how ι is distributed we use numerical experiments (Valtonen *et al.* 1992, Zheng 1994).

The inclination cannot change if $\iota_0 = 0°$ (except into $\iota = 180°$). At this limit (when $\sin\iota_0 = 0$) $A \to 1$ and $B \to 0$. This suggests that we use a fitting function of the form

$$\cos\iota_c = \cos\iota_0 - 0.38\sin^2\iota_0 Q^{-1/2}(a_B/a) \tag{6.72}$$

to describe the central value ι_c of the distribution of ι (see Fig. 6.19). From Eq. (6.71) we see that $B \propto q^{-1/2}$; it is then quite natural that $B \propto q_0^{-1/2}$, and

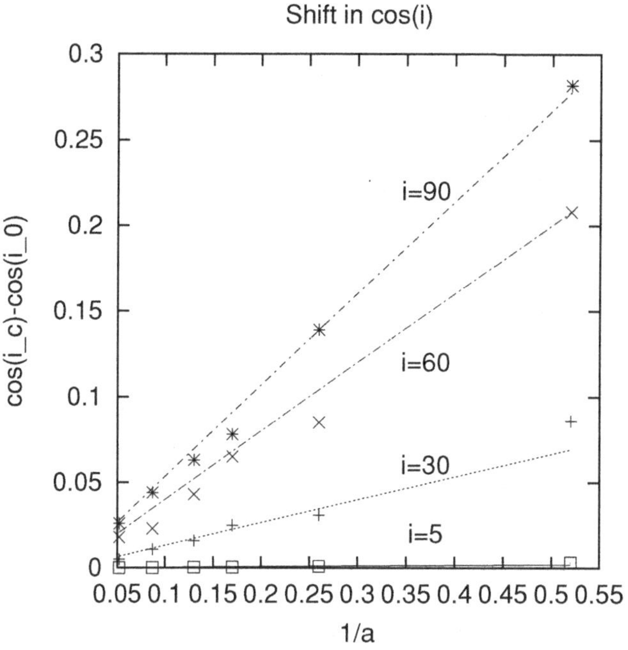

Figure 6.20 The shift in the centre of the inclination distribution as a function of the inverse of the new semi-major axis a (in units of a_B). Data points are from Zheng (1994) for four inclinations: $\iota = 5^\circ$ ($\square$), $\iota = 30^\circ$ ($+$), $\iota = 60^\circ$ ($\times$) and $\iota = 90^\circ$ ($*$). The lines refer to Eq. (6.72), and $Q = 0.5$.

since $Q = q_0/a_B$, $B \propto Q^{-1/2}$. Thus the shift in $\cos \iota$ is greatest at small values of Q. Equation (6.72) should not be used below $Q = 0.1$, except by putting $Q = 0.1$ independent of the actual value of Q; otherwise we get a singularity when $Q \to 0$.

The details of Eq. (6.72) come from numerical experiments. They show that the peak of the distribution of new $\cos \iota$ is shifted from $\cos \iota_0$ and that the distribution is spread by $\Delta \cos \iota$. Figure 6.19 shows the distribution of $\cos \iota$ for $Q = 0.5$, $\iota_0 = 30^\circ$ and $a_B/a = 0.173$. The centre of the distribution $\cos \iota_c$ is shifted from the original value $\cos \iota_0$ by an amount which is derived from Eq. (6.72). The shift varies with Q, a and ι_0. Figures 6.20 and 6.21 show that Eq. (6.72) agrees with experiments at different ι_0, Q and a.

In terms of U, Eq. (6.72) becomes

$$\cos \iota_c = \cos \iota_0 + 0.38 \times 10^{-3} \sin^2 \iota_0 Q^{-1/2} U. \qquad (6.73)$$

If U is negative, $\cos \iota_c < \cos \iota_0$ and $\iota_c > \iota_0$. Thus the inclination tends to increase when a body is captured. The opposite happens when a body is expelled to a more distant orbit.

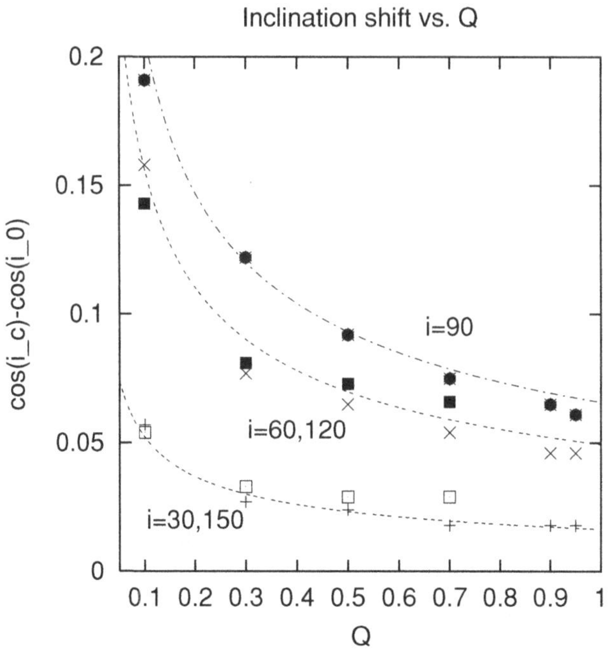

Figure 6.21 The shift in cosine of inclination as a function of Q in scattering with $U = -173$. The data points from Zheng (1994) are for $\iota = 30^\circ$ (+), 60° ($\times$), 90° ($\ast$), 120° (■) and 150° (□). The lines follow Eq. (6.72). Note that at high Q and retrograde orbits the Tisserand parameter seriously restricts the $\cos\iota$ distribution, and therefore the data points for $\iota = 120^\circ$ and 150° are missed at the high Q end.

When these expressions are used one must always check that ι_c is consistent with the Tisserand parameter. The limitation set by the Tisserand parameter truncates the allowable distributions of ι to such an extent that in extreme cases even the predicted central value ι_c is outside the range of possible values.

The spread of $\cos\iota$ values around the central value $\cos\iota_c$ may be described similarly. Numerical data are well fit by

$$\begin{aligned}\Delta\cos\iota &= 0.68\sin\iota_0\left(0.15+0.85\cos^2\iota_0\right)Q^{-1/2}(a_B/a)\\ &= 0.68\times 10^{-3}\sin\iota_0\left(0.15+0.85\cos^2\iota_0\right)Q^{-1/2}U.\end{aligned} \tag{6.74}$$

The definition of the spread $\Delta\cos\iota$ is such that 90% of scatterings end up in the interval (see Fig. 6.19)

$$\cos\iota = \cos\iota_c \pm \Delta\cos\iota. \tag{6.75}$$

6.8 Changes in the relative orbital energy of the binary

Finally, we will return briefly to the changes in the binary orbital energy. The changes of orbital energy are relatively minor in the Sun–planet binary due to small-body interactions. However, a very similar problem arises in connection with three stars interacting with each other. This is really the topic of the next chapters, but here we prepare the ground a little by using the above approximations in the general three-body problem.

The relative change in the binding energy of the binary is approximately (Eqs. (6.46), (6.47) and (6.51))

$$\Delta \approx \frac{\Delta E_a}{E_B} \approx \frac{2m_s u G m_a/(u^2 b)}{G m_a m_b/(2a_B)} \mathbf{v}_0 \cdot \hat{\mathbf{e}} = \frac{m_s}{m_b} U. \tag{6.76}$$

Including the focussing factor $\nu \approx (v_0/u_s)^2$, the differential cross-section for changes in the relative binary energy Δ is (Eqs. (6.54) and (6.55), after change of variable and multiplication by ν)

$$\frac{d\sigma(\Delta)}{d\Delta} d\Delta \approx 8\pi \left(\frac{v_0}{u}\right)^2 \sin^2 \delta a_B^2 \left(\frac{m_s}{m_b}\right)^3 \left(\frac{v_0}{u_s}\right)^2 \frac{d\Delta}{|\Delta|^3}. \tag{6.77}$$

In place of v_0^2/u_s^2 we may use a new variable v^2 which will be justified in Section 8.1 (see Eq. (8.11)):

$$\frac{v_0^2}{u_s^2} = \frac{m_s m_B^2}{M m_a m_b} \frac{1}{v^2}. \tag{6.78}$$

In the stellar case the two components of the binary may act equally well as scattering centres. Therefore the factor $(m_s/m_b)^3$ in Eq. (6.77) could be $(m_s/m_a)^3$ equally well. To obtain the required symmetry relative to m_a and m_b, we arrange the mass values in two factors:

$$\frac{m_s^3}{M m_a m_b} \frac{m_s m_B^2}{m_b^3}.$$

We replace the second, asymmetric factor by 4, its value in the equal mass case. If $\langle (v_0^2/u^2) \sin^2 \delta \rangle = 1/3$ as before (Eq. (6.56) and $u = \sqrt{2} v_0$), we have

$$\frac{d\sigma(\Delta)}{d\Delta} d\Delta \approx \frac{32}{9} \frac{3m_s^3}{M m_a m_b} \left(\pi a_B^2/v^2\right) \frac{d\Delta}{|\Delta|^3}. \tag{6.79}$$

This expression has been derived more rigorously by Heggie (1975) who has shown that it applies to incoming speeds $v \gg |\Delta|^{-1}$. Also numerical experiments have shown that the expression is good for high incoming velocities only. A more detailed discussion is deferred to Section 10.7 where Figs. 10.16 and 10.17 compare the theory with experiments. The limitation to high velocities is quite reasonable since our central assumption, scattering from one binary component while the other 'stands still', requires quick action and thus a high incoming speed. The experiments also show the excess of $\sigma(\Delta < 0)$ over $\sigma(\Delta > 0)$. This is due to the $\boldsymbol{v}_0 \cdot \Delta\boldsymbol{v}_\perp$ term (Eq. (6.9) and the discussion following it in Section 6.1) which causes a net loss of the binary energy. So far we have neglected this term.

Problems

Problem 6.1 A comet comes from the Oort Cloud and has a semi-major axis $a = 3 \times 10^4$ AU, pericentre distance relative to Jupiter $Q = 0.2$ and inclination $\iota_0 = 90°$. What is the probability that it will encounter Jupiter and that its new semi-major axis will lie between 29 AU and 31 AU? What is the most likely value of its new inclination, assuming that the new semi-major axis is 30 AU? What is the corresponding new value of Q? The semi-major axis of Jupiter is 5.2 AU.

Problem 6.2 The nearest stellar system to us (besides the Sun) is the binary star α Centauri. Its orbit has a semi-major axis $a = 26$ AU. One of the stars is a little more massive ($1.19M_\odot$) and the other a little less massive ($0.9M_\odot$) than the Sun. Close to these two stars is a third one (Proxima Centauri), a low mass star. Find the cross-section for capture of Proxima Centauri by the α Centauri binary system for the Galactic field where the typical speed is $u \approx 20$ km/s. The number density of stars in the Galactic field is about $0.1/\text{pc}^3$. How long would it take on average to have one star captured?

Problem 6.3 Let us assume that the α Centauri binary was born in a star cluster with the mass density of stars $\rho = (10^5 M_\odot/\text{pc}^3)t^{-3/2}$ and with the typical relative velocity between the stars $u = 4.3$ km/s $t^{-1/2}$, where t is the age of the cluster in millions of years ($t \geq 2$). Calculate the original value of the semi-major axis of the binary at $t = 2$ if the α Centauri binary stayed in the cluster for a period of 10^8 years. Hint: use an average value for R_a, calculated at $t = 50$.

Problem 6.4 Calculate the change of eccentricity of the α Centauri binary using the assumptions of the previous problem. The correct eccentricity of the binary is taken to be 0.96.

Problem 6.5 Calculate the value of the focussing factor for the α Centauri binary when the speed of the incoming stream is (a) 1 km/s, (b) 10 km/s, (c) 100 km/s.

7

Escape in the general three-body problem

7.1 Escapes in a bound three-body system

When three self-gravitating bodies are placed inside a small volume, the three-body system becomes unstable. Sooner or later one of the bodies leaves the volume and the two other bodies form a binary system. By recoil, the binary also leaves the original volume and escapes in the opposite direction from the single body. This instability is not at all obvious and the breakup of the bound three-body system was established as a general evolutionary path only after extensive computer simulations in the late 1960s and early 1970s. As mentioned in Chapter 1, there are exceptions to this but generally they do not represent much of the initial value space.

The breakup may be permanent in which case we say that the third body has escaped from the binary. However, sometimes the third-body motion is slowed down sufficiently that the third body returns and a vigorous three-body interaction resumes again. Then the breakup stage is called an ejection. We start by studying escape orbits, and will come to ejections in Section 8.3.

These orbit calculations and later ones have shown that the orbit behaviour of a three-body system is essentially chaotic. The chaoticity can be shown, for example, as follows. Take a given three-body configuration with position vectors $\boldsymbol{r}_1, \boldsymbol{r}_2$ and $\boldsymbol{r}_3$ and velocity vectors $\dot{\boldsymbol{r}}_1, \dot{\boldsymbol{r}}_2$ and $\dot{\boldsymbol{r}}_3$ for the three bodies labelled 1, 2 and 3. Let their masses be m_1, m_2 and m_3. From this initial configuration we calculate the future orbits of the three bodies using a computer. We may plot the orbits as a function of time or, for purposes of simplicity, choose some quantity which describes the motion and plot this quantity as a function of time. In Fig. 7.1 we show the time evolution of the distance d from the centre of mass of the three bodies. Also shown are the corresponding plots of orbits which start from slightly different positions $\boldsymbol{r}_1 + \mathrm{d}\boldsymbol{r}_1, \boldsymbol{r}_2 + \mathrm{d}\boldsymbol{r}_2, \boldsymbol{r}_3 + \mathrm{d}\boldsymbol{r}_3$. For a while, for about two larger periods of the system, the two sets of orbits deviate from each other only slightly, but during the third cycle the orbits start to diverge and after that no similarity remains.

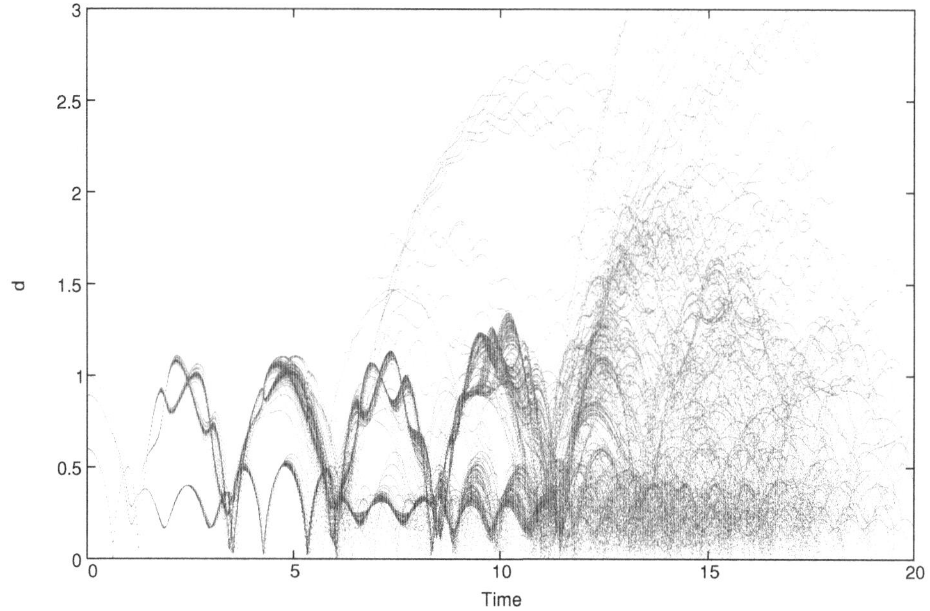

Figure 7.1 The time evolution of a three-body system with small variations is shown. The y axis displays the distances of the three bodies from the centre of mass while the x axis is time in units of crossing time. In addition, a slight variation is introduced to the original system by moving the initial position of one of the bodies by at most 10^{-3} distance units. In this way 100 almost identical three-body systems are studied, and the time evolutions of all of the systems are superimposed on the same plot. We see that the time evolution tracks start to diverge strongly after about six crossing times, and beyond about 12 crossing times chaos sets in (Heinämäki *et al.* 1999).

When slightly different initial paths lead to quite different end results, the system is chaotic. Typically the breakup of a three-body system happens after several cycles of the system and therefore the breakup happens in a chaotic way. Thus it may seem at first sight that a physical description of the breakup process is difficult. Indeed, the details of breakup of any given bound three-body system are unpredictable prior to the actual orbit calculation. However, the chaoticity allows one to describe the breakup in a statistical sense. In a large number of systems which have similar (but not identical) initial states, the statistical distributions of the orbital properties are stable and predictable. Often this is enough, just as in the theory of gases, we do not normally want to know the detailed orbital history of every molecule but rather it is enough to know the bulk properties such as temperature and pressure. In the same way, it is often enough to know the statistical distributions of the three-body breakup and the corresponding bulk numbers which describe these distributions.

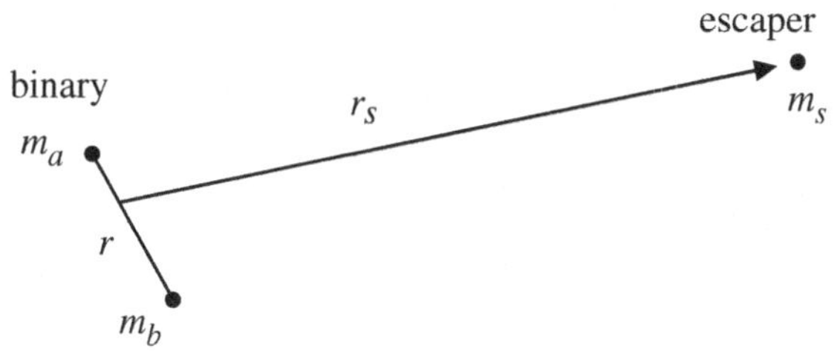

Figure 7.2 The basic configuration of three bodies after the body with mass m_s has escaped from the binary with component masses m_a and m_b. The current speed of the escaper relative to the binary centre of mass is v_s and the current distance r_s. The separation of the binary components is r at this time.

In the statistical theory of the disintegration of three-body systems (Monaghan 1976a, b) one assumes that the probability of a given escape configuration is proportional to the volume in phase space available to this configuration. This is the case if the escaping body makes use of all channels available to it, without any preferences. That this should be the case is not immediately obvious, but results from other theoretical approaches as well as from numerical orbit calculations support this assumption.

Let us divide the three-body system in two parts: a binary of components m_a and m_b, total mass $m_B = m_a + m_b$, and a third body (escaper) of mass m_s. The total mass of the system is then $M = m_B + m_s$. Let $\boldsymbol{r}$ be the separation vector between the two binary components and $\boldsymbol{r}_s$ the position vector of the third body relative to the centre of mass of the binary (see Fig. 7.2). The kinetic energy of the binary motion is $0.5(m_a m_b/m_B)\dot{\boldsymbol{r}}^2$. As before, we write $\mathcal{M} \equiv m_a m_b/m_B$ and $m \equiv m_B m_s/M$, which are the reduced masses of the relative motions of the binary and the third body, respectively. When the third body is far from the binary, its potential energy is $V_s = -Gm_s m_B/r_s$, while the potential energy of the binary is $-Gm_a m_b/r$. Thus the total energy of the three-body system is

$$E_0 = \frac{1}{2}m\dot{\boldsymbol{r}}_s^2 + \frac{1}{2}\mathcal{M}\dot{\boldsymbol{r}}^2 - G\frac{m_a m_b}{r} - G\frac{m_s m_B}{r_s} \qquad (7.1)$$

or

$$E_0 = E_s + E_B \qquad (7.2)$$

where the energy is divided between the escaper energy E_s and the binary energy E_B.

The density of escape configurations per unit energy σ is obtained by integrating over the phase space volume, with the phase space coordinates $\boldsymbol{r}, \boldsymbol{r}_s, \boldsymbol{p}, \boldsymbol{p}_s$ where $\boldsymbol{p}$ and $\boldsymbol{p}_s$ are the momenta of the relative motions corresponding to $\boldsymbol{r}$ and $\boldsymbol{r}_s$. In the integral Eq. (7.2) has to be satisfied. Mathematically, this is achieved by placing the delta function $\delta(E_s + E_B - E_0)$ in the integrand. Then there is a contribution to the integral only if (7.2) is satisfied, i.e. when $\delta = \infty$ and the integration over δ is unity. Otherwise $\delta = 0$. Thus

$$\sigma = \int \cdots \int \delta\left(\frac{p_s^2}{2m} + V_s + E_B - E_0\right) \mathrm{d}\boldsymbol{r}_s \, \mathrm{d}\boldsymbol{p}_s \, \mathrm{d}\boldsymbol{r} \, \mathrm{d}\boldsymbol{p} \tag{7.3}$$

where we have put $E_s = p_s^2/2m + V_s$.

Besides the total energy E_0, the second conserved quantity is $\boldsymbol{L}_0$, the total angular momentum of the system. It is made up of two vector components, the angular momentum of the binary $\boldsymbol{L}_B = \mathcal{M}(\boldsymbol{r} \times \dot{\boldsymbol{r}})$ and the angular momentum of the escaper relative to the binary $\boldsymbol{L}_s = m(\boldsymbol{r}_s \times \dot{\boldsymbol{r}}_s)$. We may require that $\boldsymbol{L}_0 = \boldsymbol{L}_B + \boldsymbol{L}_s$ is a constant for all possible escape orbits in which case the integrand of the phase space integral also contains a factor $\delta(\boldsymbol{L}_B + \boldsymbol{L}_s - \boldsymbol{L}_0)$. This forces the possible choices of $\boldsymbol{L}_B$ and $\boldsymbol{L}_s$ such that $\boldsymbol{L}_0 =$ constant. The discussion is started by neglecting the second δ-function which means that $\boldsymbol{L}_0$ can vary from case to case. However, there are some limitations on the angular momentum of the escaper $\boldsymbol{L}_s$ which must be justified.

A useful approach to take is the so-called *loss cone* method. Let us assume that the escaper at a distance r_s from the binary centre has come to this point along a straight line orbit. It must have come from the neighbourhood of the binary; otherwise it would not have acquired the escape velocity. The neighbourhood of the binary may be defined for our purposes as a circular area, perpendicular to the vector $\boldsymbol{r}_s$, with the radius of some simple multiple na, where a is the semi-major axis of the binary. The latter is related to the binary energy E_B by

$$a = -\frac{Gm_am_b}{2E_B}. \tag{7.4}$$

The multiple n in the definition of the neighbourhood of the binary comes from experience with orbit calculations, and is $n \approx 7$. It is further discussed in Chapter 8. The choice of the value seven which at first sight appears rather arbitrary leads to a good agreement between analytic theory and numerical experiments.

The straight lines drawn from the point $\boldsymbol{r}_s$ through the circle of radius $7a$ define a cone; this is called the loss cone because particles travelling in reverse direction from the apex of the cone to the binary will generally be scattered away from the cone, and thus these orbits are 'lost'. In the current problem we can say that only the orbits within the loss cone are true escape orbits since they have been strongly influenced by the binary in the past. The loss cone directions contain approximately the fraction of $\pi(7a)^2/4\pi r_s^2 = 12.25(a/r_s)^2$ of the whole sphere of radius r_s surrounding the escaper (see Fig. 7.3).

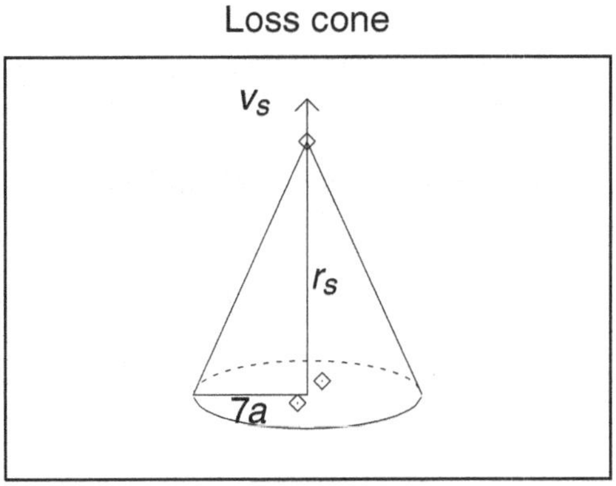

Figure 7.3 The loss cone. The escaping body, at the apex of the cone, must have come from inside the cone in order for there to have occurred a strong interaction with the binary in the past. In reverse, bodies falling into the cone will interact efficiently with the binary and will be perturbed to new orbits. Thus these orbits are 'lost'.

Let us first carry out the integration over the momentum space $\boldsymbol{p}_s$, with a uniform distribution of directions over the whole sphere. Then

$$\begin{aligned}
&\iiint \delta\left(\frac{p_s^2}{2m} + V_s + E_B - E_0\right) \mathrm{d}\boldsymbol{p}_s \\
&= 4\pi \int_0^\infty \delta\left(\frac{p_s^2}{2m} + V_s + E_B - E_0\right) p_s^2\, \mathrm{d}p_s \\
&= 4\pi m \int_0^\infty \delta\left(\frac{p_s^2}{2m} + V_s + E_B - E_0\right) p_s\, \mathrm{d}\left(\frac{p_s^2}{2m}\right) \\
&= 4\pi m \int_0^\infty \delta\left[x - (E_0 - V_s - E_B)\right] \sqrt{2mx}\, \mathrm{d}x \\
&= 4\pi m \sqrt{2m(E_0 - V_s - E_B)}
\end{aligned} \tag{7.5}$$

where we have substituted $x = p_s^2/2m$. The last equality follows from the property of the delta function:

$$\int_0^\infty f(x)\delta(x-a)\,\mathrm{d}x = f(a),$$

where $f(x)$ is any function.

The right hand side of Eq. (7.5) has to be multiplied by the loss cone factor $(49/4)(a/r_s)^2$. Then we proceed to carry out the integration over $\boldsymbol{r}_s$. Since the integrand depends only on the radial coordinate r_s, $\mathrm{d}\boldsymbol{r}_s = 4\pi r_s^2 \mathrm{d}r_s$, and the integral becomes

$$4 \times 49\pi^2\sqrt{2}m^{3/2}a^2 \int_0^R \sqrt{E_0 - E_B + Gm_s m_B/r_s}\,\mathrm{d}r_s, \tag{7.6}$$

where the upper limit R of the r_s range is considered a free parameter. Let us denote $x^2 = r_s$, i.e. $2x\,\mathrm{d}x = \mathrm{d}r_s$ and

$$y = \frac{E_0 - E_B}{Gm_s m_B/R}.$$

Then the integration is easily carried out:

$$\begin{aligned}
&\int_0^{\sqrt{R}} 2\sqrt{Gm_s m_B}\sqrt{\frac{y}{R}x^2 + 1}\,\mathrm{d}x \\
&= \Big|_0^{\sqrt{R}} 2\sqrt{Gm_s m_B} \\
&\quad \times \left[\frac{x\sqrt{\frac{y}{R}x^2 + 1}}{2} + \frac{1}{2\sqrt{y/R}}\ln\left(x\sqrt{\frac{y}{R}} + \sqrt{\frac{y}{R}x^2 + 1}\right)\right] \\
&= \sqrt{Gm_s m_B R}\left[\sqrt{y+1} + \frac{1}{\sqrt{y}}\ln\left(\sqrt{y} + \sqrt{y+1}\right)\right].
\end{aligned}$$

The function in the square brackets has the value of 2.3 when $y = 1$ and it approaches 2 when $y \to 0$. If R is relatively small, say $R = 3a_0$, this is more or less the range of interest of y. As a slowly varying function of y the square bracket may therefore be put equal to 2. Here a_0 is the initial semi-major axis of the binary before the three-body interaction. Using $m_s m_B = mM$ and Eq. (7.4) we get finally

$$\sigma = 98\sqrt{2}\pi^2(GMR)^{1/2}m^2(Gm_a m_b)^2 \int \cdots \int \frac{\mathrm{d}\boldsymbol{r}\,\mathrm{d}\boldsymbol{p}}{|E_B|^2}. \tag{7.7}$$

In order to see the significance of the remaining integrals, spherical polar coordinates (r, θ, ϕ) are useful. Then (Eq. (4.88), $\theta \to \pi/2 - \theta$; we now measure the θ-coordinate from the pole rather than from the equator, contrary to Chapter 4)

$$\begin{aligned}
E_B &= \frac{1}{2}\frac{p^2}{\mathcal{M}} - \frac{Gm_a m_b}{r} \\
&= \frac{1}{2\mathcal{M}}\left(p_r^2 + \frac{1}{r^2}p_\theta^2 + \frac{1}{r^2\sin^2\theta}p_\phi^2\right) - \frac{Gm_a m_b}{r}
\end{aligned}$$

from which

$$\begin{aligned}
\frac{\mathrm{d}E_B}{\mathrm{d}p_r} &= \frac{1}{\mathcal{M}}p_r, \\
\mathrm{d}p_r &= \frac{\mathcal{M}\,\mathrm{d}E_B}{p_r}
\end{aligned}$$

and

$$p_r = \left[2\mathcal{M}E_B + Gm_am_b2\mathcal{M}/x - k^2/x^2 - p_\phi^2/x^2\sin^2\theta\right]^{1/2} \tag{7.8}$$

where we have used $x \equiv r$ and $k \equiv p_\theta$.

The square of the total angular momentum vector is (see Eq. (4.94), noting that $\theta \to \pi/2 - \theta$)

$$L^2 = k^2 + \frac{p_\phi^2}{\sin^2\theta},$$

$$p_\phi = \sin\theta\sqrt{L^2 - k^2},$$

$$\mathrm{d}p_\phi = \sin\theta\, L\, \mathrm{d}L/\sqrt{L^2 - k^2},$$

from which it follows that

$$p_r = \left[2\mathcal{M}E_B + Gm_am_b2\mathcal{M}/x - L^2/x^2\right]^{1/2}. \tag{7.9}$$

The integral in Eq. (7.7) is then written as

$$\int\int \frac{\mathrm{d}E_B}{E_B^2} \frac{\mathcal{M}\mathrm{d}k\left(\sin\theta/\sqrt{L^2 - k^2}\right) L\,\mathrm{d}L\,\mathrm{d}x\,\mathrm{d}\theta\,\mathrm{d}\phi}{\left[2\mathcal{M}E_B + Gm_am_b2\mathcal{M}/x - L^2/x^2\right]^{1/2}}. \tag{7.10}$$

The integrals over θ and ϕ give $\int_0^\pi \int_0^{2\pi} \sin\theta\,\mathrm{d}\theta\,\mathrm{d}\phi = 4\pi$. The integral of k is

$$\int_{-L}^{L} \frac{\mathrm{d}k}{\sqrt{L^2 - k^2}} = \Big|_{-L}^{L} \arcsin\left(\frac{k}{L}\right) = \pi.$$

The integral over x is

$$\int_{x_1}^{x_2} \frac{x\,\mathrm{d}x}{\left[-L^2 + 2Gm_am_b\mathcal{M}x - 2\mathcal{M}|E_B|x^2\right]^{1/2}},$$

which is of the form

$$\int \frac{x\,\mathrm{d}x}{\sqrt{C + Bx - Ax^2}} = -\frac{1}{A}\sqrt{C + Bx - Ax^2} + \frac{B}{2A^{3/2}} \arcsin\frac{2Ax - B}{\sqrt{B^2 + 4AC}}.$$

The limits of the integration are the zero points of $Ax^2 - Bx - C = 0$ or

$$x_{1,2} = \frac{B \pm \sqrt{B^2 + 4AC}}{2A}.$$

At the integration limits

$$\frac{2Ax - B}{\sqrt{B^2 + 4AC}} = \pm 1$$

and the second term gives

$$\frac{\pi B}{2A^{3/2}} = \frac{\pi}{2} \frac{2Gm_a m_b \mathcal{M}}{(2\mathcal{M}|E_B|)^{3/2}} = \frac{\pi}{2\sqrt{2}} \frac{G(m_a m_b)^{1/2} m_B^{1/2}}{|E_B|^{3/2}}.$$

Since the first term goes to zero at both limits, this is also the total contribution of the integral over x. Combining the integrations over θ, ϕ, k and x together with Eq. (7.7), not forgetting the $\mathcal{M}$-factor in Eq. (7.10), we find

$$\sigma = 2 \times 98\pi^5 (Gm_a m_b)^{7/2} R^{1/2} m_B^{3/2} M^{-3/2} m_s^2 \int\int \frac{\mathrm{d}E_B}{|E_B|^{7/2}} L\,\mathrm{d}L.$$

Since

$$\begin{aligned} L^2 &= \mathcal{M} \frac{(Gm_a m_b)^2}{2|E_B|}(1 - e^2), \\ L\,\mathrm{d}L &= -\mathcal{M} \frac{(Gm_a m_b)^2}{2|E_B|} e\,\mathrm{d}e, \end{aligned} \tag{7.11}$$

the final form is

$$\sigma = 98\pi^5 (Gm_a m_b)^{11/2} R^{1/2} m_B^{3/2} M^{-3/2} m_s^2 \mathcal{M} \int\int \frac{\mathrm{d}E_B}{|E_B|^{9/2}} e\,\mathrm{d}e. \tag{7.12}$$

For the moment, let us consider only the quantities which follow the integral signs and forget the coefficients in front of them. These quantities represent the distributions over which one has to integrate in order to obtain the total phase space volume. These are the fundamental distributions in which we are interested. Thus the distribution of $|E_B|$, normalized to unity, is

$$f(|E_B|)\,\mathrm{d}|E_B| = 3.5|E_0|^{7/2}|E_B|^{-9/2}\,\mathrm{d}|E_B|. \tag{7.13}$$

This result has also been derived starting from a different theoretical concept (Heggie 1975). Assume that the binaries are in a field of single bodies which constantly interact with each other. Then the concept of detailed balancing requires that the formation and disruption of binaries are reverse processes and that they occur at such rates that an equilibrium is established. Equation (7.13) represents such an equilibrium distribution. We will discuss this point further in Section 8.2. Figure 7.4 compares the distribution with numerical data. The case of $V_s = 0$ is left as an exercise (Problem 7.1); it leads to a slightly different distribution of binding energies (Monaghan 1976a). In some older sources (Jeans 1928) one finds a

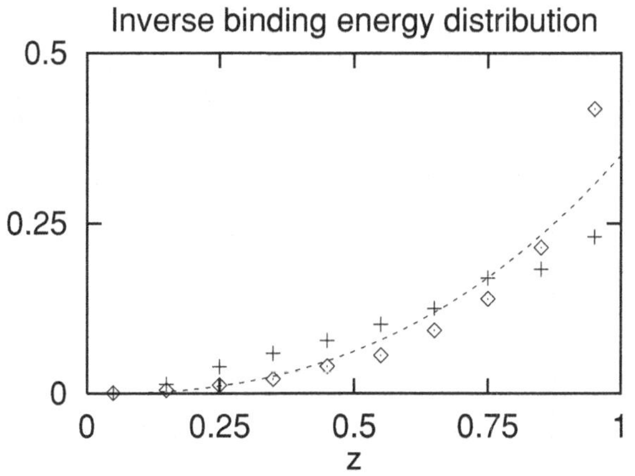

Figure 7.4 A comparison of Eq. (7.13) with experimental data (Mikkola 1994). In the numerical experiments all masses are equal. In the first set of experiments (+) the third body falls straight into a binary; in the second set (◇) the third body has an additional transverse velocity component uniformly distributed between given limits. The quantity plotted is $z = |E_0|/|E_B|$. Due to change of variables, the expected distribution is of the form $f(z) = 3.5z^{2.5}$, shown by the dashed line. The experimental points come from two data sets of 10 000 experiments each, and thus have such small error bars that they are not shown. What is more important is the value of the total angular momentum: the first set has $L = 0.15$ and the second one has a range from $L = 0.15$ to $L = 0.92$, with the median around $L = 0.48$ (see Section 7.5 for the definition of the units). We notice that Eq. (7.13) gives a fair description of the data in some average sense, but that the value of the total angular momentum plays an important role in changing the shape of the distribution.

power-law index $-5/2$ instead of $-9/2$ in Eq. (7.13) due to a different derivation. This is well outside the experimental range as we will see later (Section 7.5).

The corresponding distribution of eccentricities is

$$f(e)\,\mathrm{d}e = 2e\,\mathrm{d}e. \tag{7.14}$$

See Fig. 7.5 for a comparison with numerical experiments.

7.2 A planar case

When the motions are restricted to a plane, Eq. (7.5) gives $2\pi m$ which is to be multiplied by

$$2\pi \int_0^R \frac{14a}{2\pi r_s} r_s\,\mathrm{d}r_s = 14aR = 7\frac{Gm_a m_b}{|E_B|}R.$$

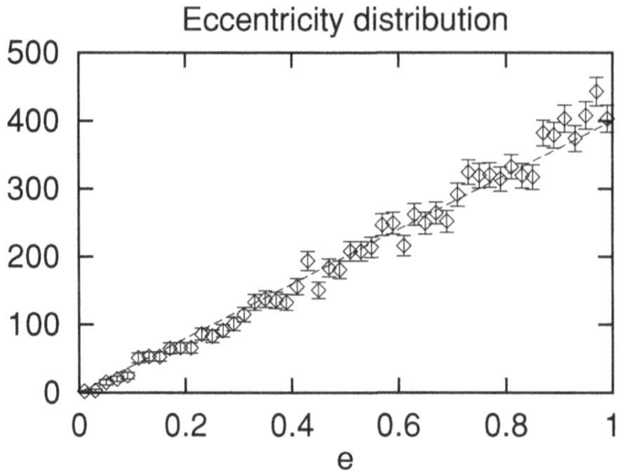

Figure 7.5 A test of the eccentricity distribution of Eq. (7.14) (straight line) using 10 000 computer orbit solutions from Mikkola (1994), with angular momentum $L = 0.15$.

This leads to

$$\sigma = 14\pi m G m_a m_b R \int \frac{1}{|E_B|} \mathrm{d}\boldsymbol{r}\, \mathrm{d}\boldsymbol{p},$$

and after we put $\theta = \pi/2$, $p_\phi = 0$, Eq. (7.12) becomes

$$\sigma = 7\pi^3 m (G m_a m_b)^3 \mathcal{M} R \int\int \frac{\mathrm{d}E_B}{|E_B|^3} \frac{e\,\mathrm{d}e}{\sqrt{1-e^2}}. \tag{7.15}$$

Thus the planar case gives

$$f(|E_B|)\,\mathrm{d}|E_B| = 2|E_0|^2 |E_B|^{-3}\,\mathrm{d}|E_B|, \tag{7.16}$$

and

$$f(e)\,\mathrm{d}e = e(1-e^2)^{-1/2}\,\mathrm{d}e \tag{7.17}$$

which is also found in numerical orbit calculations (see Fig. 7.6). The details of the derivation of Eq. (7.15) are left as an exercise (Problem 7.2).

7.3 Escape velocity

The escape energy $E_s = |E_B| - |E_0|$ is obtained from Eq. (7.2) when the binary energy E_B is known. We substitute this into Eq. (7.12) and integrate over e, $\int_0^1 e\,\mathrm{d}e = 1/2$, after which there remains

$$\sigma = 49\pi^5 (G m_a m_b)^{11/2} R^{1/2} m_B^{3/2} M^{-3/2} m_s^2 \mathcal{M} \int \frac{\mathrm{d}E_s}{(|E_0| + E_s)^{9/2}}.$$

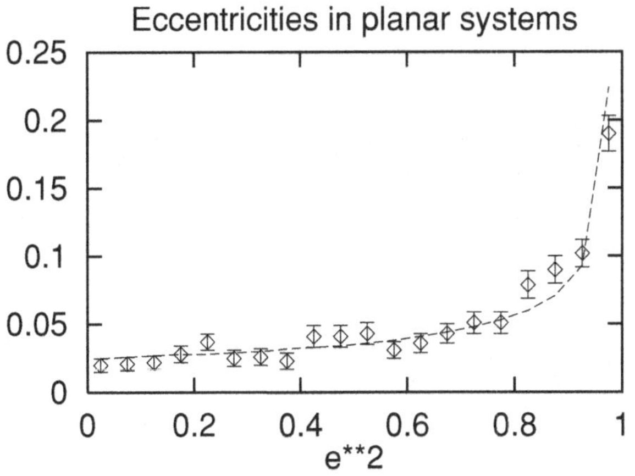

Figure 7.6 The distribution of eccentricities in the breakup of planar three-body systems. Equation (7.17) is compared with numerical experiments from Saslaw *et al.* (1974), excluding very high angular momentum systems. Note that the horizontal axis is e^2 which makes the distribution of Eq. (7.14) flat. The rise of the distribution towards $e \to 1$ is very prominent due to the $(1 - e^2)^{-1/2}$ factor in Eq. (7.17).

Let us denote the velocity of the escaper in the centre of mass coordinate system by v_s. Then

$$\begin{aligned} v_s &= \frac{m_B}{M}|\dot{\boldsymbol{r}}_s|, \\ E_s &= \frac{1}{2}m|\dot{\boldsymbol{r}}_s|^2 = \frac{1}{2}\frac{m_s m_B}{M}\frac{M^2}{m_B^2}v_s^2 = \frac{1}{2}\frac{m_s M}{m_B}v_s^2, \\ \mathrm{d}E_s &= \frac{m_s M}{m_B}v_s\,\mathrm{d}v_s, \end{aligned}$$

and

$$\begin{aligned} \sigma &= 49\pi^5(Gm_a m_b)^{11/2}R^{1/2}m_B^{1/2}M^{-1/2}m_s^3\mathcal{M} \\ &\quad\times\int\frac{v_s\,\mathrm{d}v_s}{\left(|E_0| + \frac{1}{2}(m_s M/m_B)v_s^2\right)^{9/2}}. \end{aligned} \tag{7.18}$$

The escape velocity distribution is therefore

$$f(v_s)\,\mathrm{d}v_s = \frac{\left(3.5|E_0|^{7/2}m_s M/m_B\right)v_s\,\mathrm{d}v_s}{\left(|E_0| + \frac{1}{2}(m_s M/m_B)v_s^2\right)^{9/2}}. \tag{7.19}$$

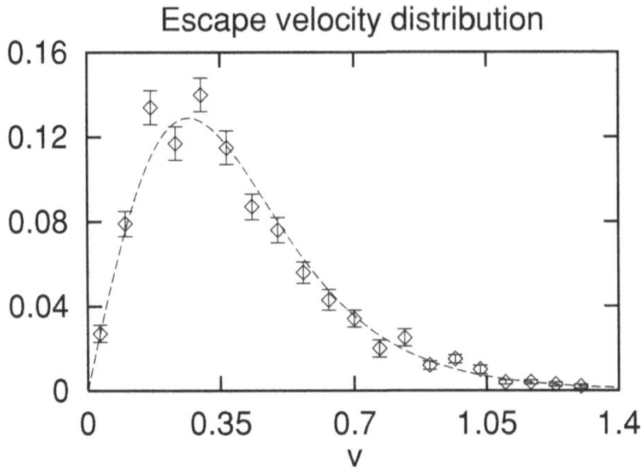

Figure 7.7 The escape velocity (v_s) distribution of Eq. (7.19) is compared with numerical experiments from Saslaw *et al.* 1974. The appropriate mean escaper mass value is $m_s \approx 0.2M$. The escape speed v_s is in units of the original binary speed v_0.

The peak of this distribution is obtained by putting $\mathrm{d}f(v_s)/\mathrm{d}v_s = 0$:

$$\left(|E_0| + \frac{1}{2}\frac{m_s M}{m_B}v_s^2\right)^{-11/2}\left(-4\frac{m_s M}{m_B}v_s^2 + |E_0|\right) = 0,$$

$$(v_s)_{\text{peak}} = \frac{1}{2}\sqrt{\frac{(M - m_s)}{m_s M}}\sqrt{|E_0|}.$$

The corresponding calculation in the two-dimensional case gives (Problem 7.4)

$$\sigma = 7\pi^3(Gm_a m_b)^3 \mathcal{M} R m_s^2 \int \frac{v_s\,\mathrm{d}v_s}{\left(|E_0| + \frac{1}{2}(m_s M/m_B)v_s^2\right)^3},$$

$$(v_s)_{\text{peak}} = \sqrt{\frac{2(M - m_s)}{5m_s M}}\sqrt{|E_0|}. \qquad (7.20)$$

See Fig. 7.7 for a comparison with numerical data. In this and in the following displays of numerical data we use the unit system of Saslaw *et al.* (1974): The gravitational constant $G = 1$, the initial semi-major axis of the binary $a_0 = 1$, the binary mass $m_B = 1$, and the mean orbital speed $v_0 = 1$. The idea is that the bound three-body system can be divided into a close binary and a more distant third body at the initial moment, and the semi-major axis of the close binary serves as a yardstick for measuring distances.

For velocities, a good standard is the original binary speed v_0. It is defined by Eq. (6.24) and by assuming that initially the binary has practically all the energy of the system:

$$|E_0| = \frac{1}{2}\mathcal{M}v_0^2.$$

Then the normalised escape velocity $v = v_s/v_0$ is distributed as

$$f(v)\,\mathrm{d}v = \left(7\frac{m_s M}{m_a m_b} v\,\mathrm{d}v\right)\left(1 + \frac{m_s M}{m_a m_b} v^2\right)^{-9/2}.$$

At the limit of small escaper mass ($m_s \lesssim 0.17$), and at the limit of large escaper mass ($m_s = m_B$) the corresponding accumulated distributions are (putting $m_a = m_b$)

$$\begin{aligned} F(v) &= (1 + v^2)^{-3.5}, \\ F(v) &= (1 + 8v^2)^{-3.5}, \end{aligned} \tag{7.21}$$

respectively. These two limits will be discussed in Chapter 11. The peak velocity is

$$v_{\text{peak}} = \frac{1}{2\sqrt{2}}\sqrt{\frac{m_a m_b}{m_s M}},$$

which gives $v_{\text{peak}} = 0.35$ and $v_{\text{peak}} = 0.125$ at the two limits. Note that for escaper masses smaller than $m_s = 0.17$ this limiting value is used in the above formulae. At small m_s the velocities do not go to inifinity as the formulae would predict, but approach the values given by $m_s = 0.17$ (Valtonen 1976a).

7.4 Escaper mass

Now we ask what is the probability that the body of mass m_s escapes, rather than one of the other two (m_a or m_b). In order to complete the integration of Eq. (7.18) to obtain the phase space volume per unit energy, one needs to calculate the integral

$$\begin{aligned} \int_0^\infty & \frac{v_s\,\mathrm{d}v_s}{\left(|E_0| + \frac{1}{2}(m_s M/m_B)v_s^2\right)^{9/2}} \\ &= -\left|_0^\infty \frac{2}{7}\frac{m_B}{m_s M}\frac{1}{\left(|E_0| + \frac{1}{2}(m_s M/m_B)v_s^2\right)^{7/2}}\right. \\ &= \frac{2}{7}\frac{m_B}{m_s M}\frac{1}{|E_0|^{7/2}}. \end{aligned}$$

Therefore

$$\begin{aligned} \sigma &= 56\sqrt{2}\pi^5 G^3 R^{1/2} a_0^{5/2}\left(\frac{m_B}{M}\right)^{0.5} M^{-1} m_s^2 (m_a m_b)^4 |E_0|^{-1} \\ &\propto (m_a m_b)^4 m_s^2 \end{aligned} \tag{7.22}$$

for a given energy $|E_0|$, for a given volume $a_0^{2.5}R^{0.5}$ and for a given total mass M. Note that use has been made of Eq. (7.4) to connect $|E_0|$ and a_0, assuming that the initial binary binding energy is $\approx |E_0|$, and that it corresponds to the initial binary semi-major axis a_0 before the strong three-body interaction. This procedure brings out the volume factor $a_0^{2.5}R^{0.5}$ and the energy factor $|E_0|^{-1}$ explicitly. There is an additional weak dependence on m_s in the m_B/M factor since $m_B = M - m_s$, but this is a small correction which can be generally ignored.

The probability that m_s is the escaper is thus

$$\sigma_s \propto (m_a m_b)^4 m_s^2$$

while the corresponding probabilities for the masses m_a and m_b to escape are

$$\sigma_a \propto (m_s m_b)^4 m_a^2,$$
$$\sigma_b \propto (m_s m_a)^4 m_b^2.$$

These have to be normalized to unit probability, i.e. to $\sigma_a + \sigma_b + \sigma_s$. The normalized probability for mass m_s to escape is therefore

$$P_s = \frac{(m_a m_b)^4 m_s^2}{(m_a m_b)^4 m_s^2 + (m_s m_b)^4 m_a^2 + (m_s m_a)^4 m_b^2}$$

which, after division by $(m_a m_b m_s)^4$ becomes:

$$P_s = \frac{m_s^{-2}}{m_s^{-2} + m_a^{-2} + m_b^{-2}} \qquad (7.23)$$

(Problem 7.5).

The corresponding results for the planar case (Problem 7.6) are

$$\sigma = \frac{7}{2}\pi^3 G^3 (m_a m_b)^4 R(m_s/M)|E_0|^{-1}$$

and

$$P_s = \frac{m_s^{-3}}{m_s^{-3} + m_a^{-3} + m_b^{-3}} \qquad (7.24)$$

which is compared with numerical experiments in Fig. 7.8.

7.5 Angular momentum

In numerical orbit calculations it has been found that the distributions of binary energy $|E_B|$ as well as the probability that the body with mass m_s escapes are functions of the total angular momentum L_0. The first indications of this trend have been seen already: the distribution of $|E_B|$ was found to be different for three-dimensional and planar systems. Since $L_0 = 0$ implies that $\boldsymbol{L}_B$ and $\boldsymbol{L}_s$ are equal

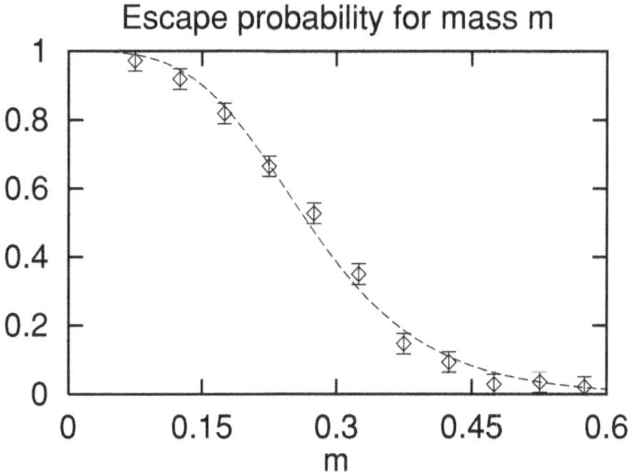

Figure 7.8 The probability of escape of mass m following Eq. (7.23) (dashed line) and in numerical experiments in Saslaw *et al.* (1974). The latter are a roughly even mixture of planar cases and three-dimensional cases. The mass ratio m_a/m_b influences the result somewhat. Here it was assumed to be 3/2, a typical value in the experiments.

and opposite, the motion must be restricted to a plane (Problem 7.7). Therefore it is not surprising that the distributions derived for the planar motion are generally good for all $L_0 = 0$ systems.

At the end of large L_0, close to its maximum possible value $L_{\max}$, we may argue that the escaper's angular momentum $\boldsymbol{L}_s$ dominates. Then $\boldsymbol{L}_B$ has to be small in order that the vector sum $\boldsymbol{L}_s + \boldsymbol{L}_B$ is large independent of the relative orientations of the two vectors. Equation (7.11) tells us that this is the case when $|E_B|$ is large. Equation (7.13) indicates that the probability for the binding energy to be greater than $|E_B|$ is proportional to $|E_B|^{-7/2}$. When Eq. (7.12) is multiplied by this factor, we deduce that for very high angular momentum

$$\sigma \propto \int\int \frac{\mathrm{d}|E_B|}{|E_B|^8} e\, \mathrm{d}e. \tag{7.25}$$

More generally, from numerical calculations the distribution of $|E_B|$ becomes

$$f(|E_B|)\,\mathrm{d}|E_B| = (n-1)|E_0|^{n-1}|E_B|^{-n}\,\mathrm{d}|E_B| \tag{7.26}$$

where the power-law index varies smoothly from $n = 3$ at $L_0 = 0$ to $n = 14.5$ at $L_0 = 0.8L_{\max}$ (see Figs. 7.9 and 7.10). The value of the maximum angular momentum is (Mikkola 1994; see Sections 2.12 and 5.7)

$$L_{\max} = 2.5G\left(m_0^5/|E_0|\right)^{1/2} \tag{7.27}$$

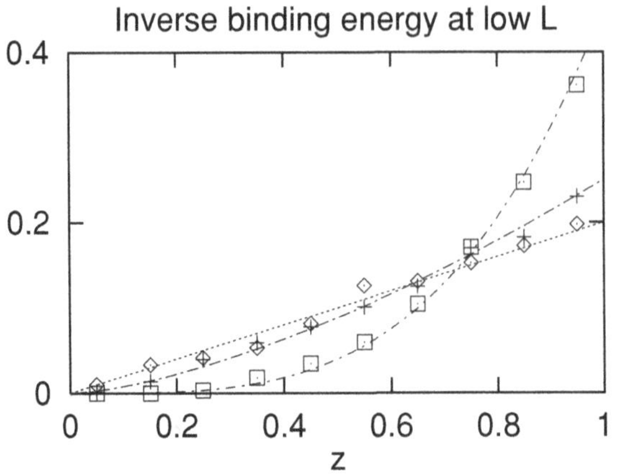

Figure 7.9 The distribution of $f(z) = (n-1)z^{n-2}$ where n is the power-law index of Eq. (7.32). The three lines are for $n = 3$ (dotted line), $n = 3.5$ (dash-dot line) and $n = 5.5$ (strongly curving dash-dot line). The data points are from experimental sets with $L = 0$, $L = 0.4$ (Mikkola and Valtonen 1986, 600 experiments in each set) and $L = 0.15$ (Mikkola 1994, 10 000 experiments). The close association between the power law index n and the angular momentum L is seen.

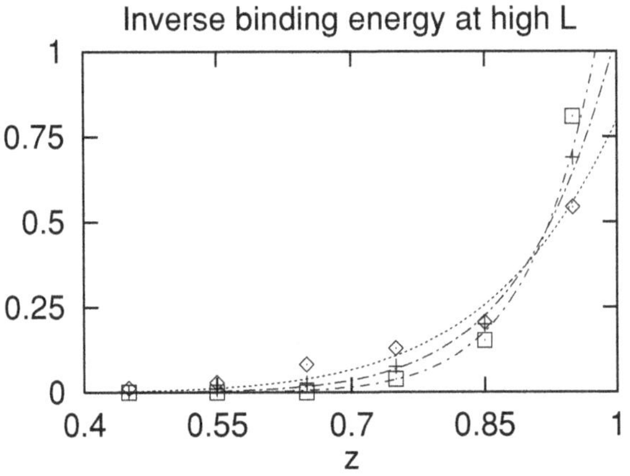

Figure 7.10 As Fig. 7.9, but for higher angular momenta. The comparison curves are for the power-law indices $n = 9$, 11.5 and 14.5, and the corresponding data come from Mikkola and Valtonen (1986) (600 experiments per set). The functional dependence between the power-law index n and the angular momentum L is $n = 18L^2 + 3$ on the basis of numerical experiments.

where m_0 is the average mass of the three bodies. This average is defined by (Marchal *et al.* 1984)

$$m_0 = \sqrt{\frac{m_a m_b + m_a m_s + m_b m_s}{3}}. \tag{7.28}$$

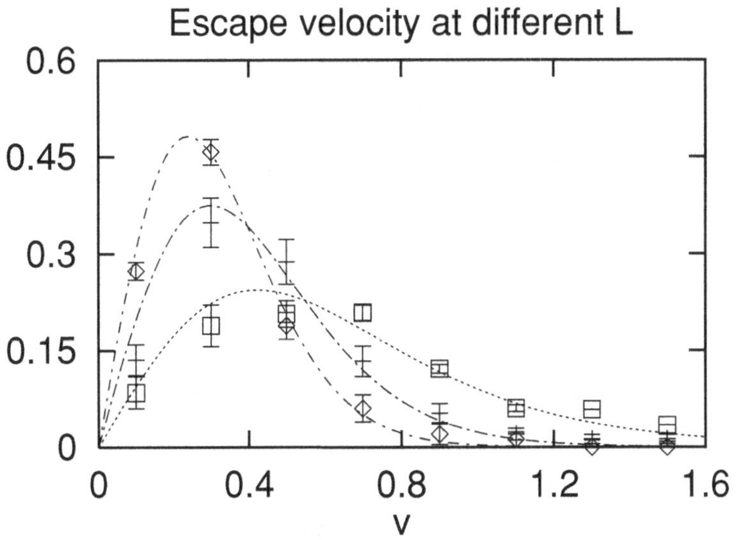

Figure 7.11 The distribution of the relative escape velocity v_∞ (in units of original binary speed v_0) between the binary and the escaping body at three different values of L. The experimental data are from Valtonen (1974) and the theoretical curves use the power-law index $n = 18L^2 + 3$. The curves are drawn for $m_s = 0.16M$ which is a typical value in the experimental sample.

Generally $L_{\max} > 3.5L_B$. It is convenient to normalise the angular momentum L_0 with $L_{\max}$. We label the normalised quantity simply by L in the following: $L = L_0/L_{\max}$.

The peak value of the escape velocity varies together with the changing $|E_B|$ distribution. If we write

$$(v_s)_{\text{peak}} = \alpha\sqrt{\frac{M - m_s}{m_s M}|E_0|} \tag{7.29}$$

then $\alpha = \sqrt{2/5} = 0.63$ when $L_0 = 0$, $\alpha = \sqrt{1/7} = 0.38$ when $L_0 = 0.5L_{\max}$, and $\alpha = 0.5$ at the typical intermediate value of L_0 (see Fig. 7.11).

The probability that a body of mass m_s escapes has been found to be

$$P_s = \frac{m_s^{-q}}{m_s^{-q} + m_a^{-q} + m_b^{-q}} \tag{7.30}$$

where we previously derived $q = 2$. Also q is a function of L_0: $q = 3$ at $L_0 = 0$ and it decreases to $q = 1.5$ at $L_0 = 0.8L_{\max}$. A typical intermediate value has been found to be $q = 2.5$ (Monaghan 1977, Mikkola and Valtonen 1986, Mikkola 1994).

At zero angular momentum the eccentricity should be distributed as Eq. (7.17), at large angular momentum the power law index of $(1 - e^2)$ should be more like $+1/2$ instead of $-1/2$ (Standish 1972, Monaghan 1976b). Both cases are included in the general expression

$$f(e) = 2(p+1)e(1-e^2)^p, \tag{7.31}$$

where the index p is found from numerical experiments (Valtonen *et al.* 2003):

$$2p = L - 0.5.$$

The angular momentum dependence of (7.26) is well described by

$$n - 3 = 18\,L^2 \tag{7.32}$$

(Mikkola and Valtonen 1986, Valtonen 1976a, 1988, Valtonen and Mikkola 1991) where L is the total angular momentum normalized to $L_{\max}$. The corresponding formula for q has the form:

$$q = 3/(1+2L^2). \tag{7.33}$$

7.6 Escape angle

In this section the direction of escape in the decay of a bound three-body system is discussed. Do the escapers go in every direction with equal probability, or are there preferred escape directions? Numerical experiments have shown that the escape directions are strongly concentrated in the plane perpendicular to the total angular momentum (Saslaw *et al.* 1974, Valtonen 1974, Anosova and Orlov 1986).

In order to understand this concentration of escape directions, let us consider what are the available breakup channels for the three-body system. Figure 7.12 illustrates the angular momentum vectors $\boldsymbol{L}_0$ and $\boldsymbol{L}_s$, separated by an angle ϕ, and the momentum vector of the escaper $\boldsymbol{p}_s$. This momentum vector lies in the plane perpendicular to $\boldsymbol{L}_s$, and its direction is described by two angles, ψ (position angle in the plane) and θ (its angular distance from $\boldsymbol{L}_0$).

First, consider the simple case $\psi = \pi/2$, which, by the definition of the angle ψ in Fig. 7.12, means that the above mentioned three vectors lie in a single plane. In this case a simple relation holds between ϕ and θ: $\theta = \pi/2 - \phi$. The more general expression, easily obtained from Fig. 7.12 by use of spherical trigonometry (Problem 7.8), reads

$$\cos\theta = \sin\phi \sin\psi. \tag{7.34}$$

Figure 7.13 shows the triangle formed by the angular momentum vectors $\boldsymbol{L}_0$, $\boldsymbol{L}_s$ and $\boldsymbol{L}_B$. From the triangle,

$$L_B^2 = L_0^2 + L_s^2 - 2L_0L_s\cos\phi \tag{7.35}$$

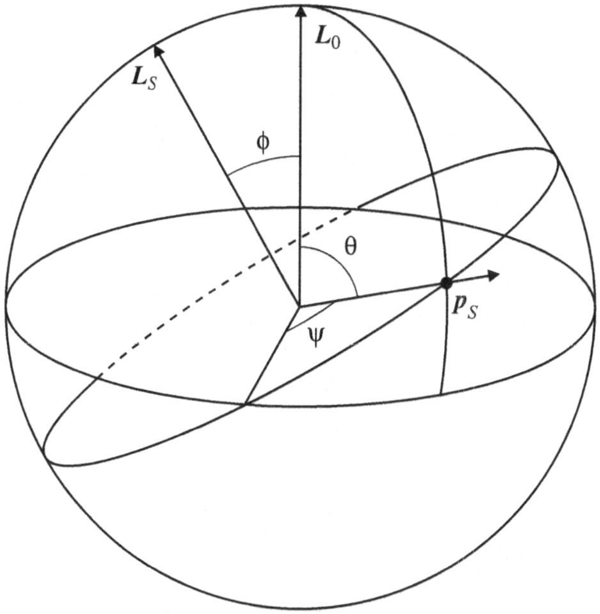

Figure 7.12 The geometry of the vectors $\boldsymbol{L}_0$, $\boldsymbol{p}_s$ and the angles between them. The momentum vector $\boldsymbol{p}_s$ lies in a plane perpendicular to $\boldsymbol{L}_s$ while this plane is at an angle ϕ relative to the fundamental plane perpendicular to $\boldsymbol{L}_0$. The direction of $\boldsymbol{p}_s$ is specified by the two angles θ and ψ.

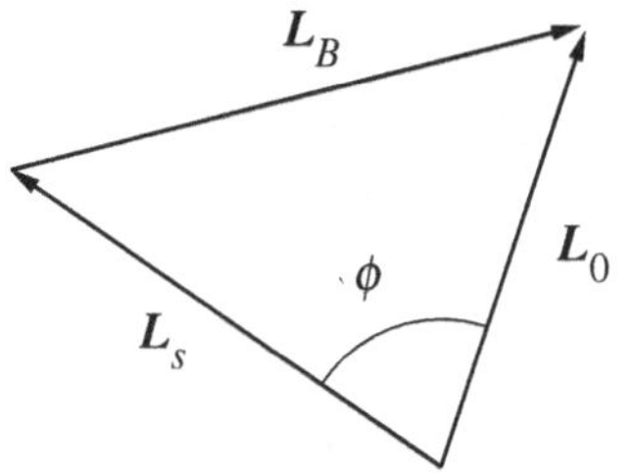

Figure 7.13 The relation between vectors $\boldsymbol{L}_0$, $\boldsymbol{L}_s$ and $\boldsymbol{L}_B$. $\boldsymbol{L}_0$ is the vector sum of $\boldsymbol{L}_s$ and $\boldsymbol{L}_B$, and the angle between $\boldsymbol{L}_s$ and $\boldsymbol{L}_0$ is ϕ.

from which we solve L_0:

$$L_0 = L_s \cos\phi \left[1 + \sqrt{1 + \left(L_B^2 - L_s^2\right) / L_s^2 \cos^2\phi} \right]. \tag{7.36}$$

The use of the squares of angular momenta rather than the angular momenta themselves is preferred because it was previously found (Section 7.1) that the squares of angular momenta tend to be uniformly distributed in phase space. For example, Eq. (7.11) tells us that L^2 is proportional to $1 - e^2$, and $1 - e^2$ is uniformly distributed between 0 and 1 since e^2 is uniformly distributed over the same range, by Eq. (7.14). This applies most definitely to the binary but is not a bad first assumption for the third-body orbit either. In that particular case, by Eq. (7.11), $L_s^2 \propto a_3(e_3^2 - 1) = p_3$, where $a_3 = Gm_s m_B/2E_3$ and e_3 is the eccentricity of the escape orbit. The quantity p_3 is the parameter of the orbit which would be uniformly distributed for randomly incoming orbits. It does not need to be so for outgoing orbits, but $L_s^2 \propto p_3$ is a natural phase space coordinate for describing escape orbits. The natural phase space coordinate associated with ϕ is $\zeta = \cos\theta$, since for random relative orientations $\cos\theta$ is uniformly distributed between -1 and $+1$ (see Problem 7.9). In the following, assume that the range of ζ is from 0 to 1. Experimentally, it has been found (Anosova and Orlov 1986) that the distribution of ζ is symmetric with respect to $\zeta = 0$, and thus here it is not necessary to deal with the negative ζ values.

Thus the new 'natural' coordinates are $x = L_s^2$, $y = L_B^2$ and $1 - \zeta^2 = \cos^2\phi$ in Eq. (7.36), so that

$$L_0 = \sqrt{x}\sqrt{1-\zeta^2}\left(1 + \sqrt{1 + (y-x)/(x(1-\zeta^2))}\right). \tag{7.37}$$

The volume of the available phase space, on the condition that L_0 is a constant, is proportional to

$$\begin{aligned}\sigma = \int\!\!\int\!\!\int \delta\Big[&\sqrt{x}\sqrt{1-\zeta^2} \\ &\times \left(1 + \sqrt{1 + (y-x)/(x(1-\zeta^2))}\right) - L_0\Big]\,\mathrm{d}x\,\mathrm{d}y\,\mathrm{d}\zeta.\end{aligned} \tag{7.38}$$

The limits of integration are x from 0 to ∞, ζ from 0 to 1, and y from 0 to the finite maximum value $y_{\max}$:

$$y_{\max} = \mathcal{M}(Gm_a m_b)^2/2|E_B|. \tag{7.39}$$

We note that the quantity inside the square root in Eq. (7.37) cannot be negative, i.e.

$$1 + (y-x)/(x(1-\zeta^2)) \geq 0 \tag{7.40}$$

or

$$\zeta^2 \leq y/x. \tag{7.41}$$

This restriction excludes considerable portions of phase space volume which would be available otherwise. For all values of $x \gg y$, $|\zeta|$ is confined to the neighbourhood of $|\zeta| \approx 0$, i.e. θ has to be close to 90°. This is an important restriction since x can range, in principle, from 0 up to ∞ while the y range extends only up to y_{max}. Therefore the escapers should be concentrated to the invariable plane, i.e. to the plane perpendicular to the total angular momentum vector (Saari 1974).

To be able to handle the restriction imposed by Eq. (7.41) let us define a new variable k in place of y

$$y/x = (k^2 + 1)\zeta^2 \tag{7.42}$$

where k is a real number. Replace y/x in the square root of Eq. (7.38) by $(k^2 + 1)\zeta^2$ to obtain a new equation

$$\sigma = \int\int\int \delta\left[\sqrt{x}\left(\sqrt{1-\zeta^2} + k\zeta\right) - L_0\right] 2xk\zeta^2 \, \mathrm{d}x \, \mathrm{d}k \, \mathrm{d}\zeta. \tag{7.43}$$

In order to carry out the integration of the δ-function we introduce another new variable w in place of x

$$w = \sqrt{x}\left(\sqrt{1-\zeta^2} + k\zeta\right) \tag{7.44}$$

after which

$$\begin{aligned} \sigma &= \int\int\int \delta(w - L_0) 4k\zeta^2 \left(\sqrt{1-\zeta^2} + k\zeta\right)^{-4} w^3 \, \mathrm{d}w \, \mathrm{d}k \, \mathrm{d}\zeta \\ &= \int\int 4k\zeta^2 L_0^3 \left(\sqrt{1-\zeta^2} + k\zeta\right)^{-4} \mathrm{d}k \, \mathrm{d}\zeta. \end{aligned} \tag{7.45}$$

The integration over k is carried out with the help of

$$\int \frac{x\,\mathrm{d}x}{(Ax+B)^4} = -\frac{1}{2A^2(Ax+B)^2} + \frac{B}{3A^2(Ax+B)^3} \tag{7.46}$$

and the limits of integration are from k_0 to ∞. The result is

$$\sigma = \int F(\zeta)\,\mathrm{d}\zeta \tag{7.47}$$

where, using only the first term on the right-hand side of Eq. (7.46) which dominates when $Ax + B \gg 1$, i.e. $l_0\zeta \gg 1$:

$$F(\zeta) \approx \frac{2l_0^3}{\left(\sqrt{1-\zeta^2} + k_0\zeta\right)^3}. \tag{7.48}$$

The preceding analysis has dealt with the special case $\theta = \pi/2 - \phi$ or $\zeta = \zeta_0 = \sin\phi$. Recall that, in general, $\zeta = \zeta_0 \sin\psi$ (Fig. 7.12), so that the distribution of Eq. (7.48) is really for ζ_0, not for ζ.

How does one then extract the distribution $f(\zeta)$ from Eq. (7.48)? Note that for each value of ζ_0, there is a whole range of escape orbits corresponding to the different values of ψ. These orbits have the property $0 \le \zeta \le \zeta_0$. There also exist escape orbits in the same range of ζ which are related to larger ζ_0 values but because of the steep fall-off of $F(\zeta_0)$ with ζ_0, these orbits are relatively uncommon. Thus, in the first approximation, $F(\zeta_0)$ gives the relative weight of all escape orbits up to $\zeta \leq \zeta_0$. In other words,

$$\frac{F(\zeta_0) - F(0)}{F(1) - F(0)} \approx \int_0^{\zeta_0} f(\zeta)\,\mathrm{d}\zeta. \tag{7.49}$$

In this approximation, $F(\zeta)$ is identified as being related to the accumulated distribution of $f(\zeta)$.

It is possible to simplify Eq. (7.48) by noting that when $\zeta \to 0$, $\zeta^2 \approx 0$. After normalizing to make it a proper accumulated distribution, we adopt

$$F(\zeta) = \frac{(1+k_0)^2}{k_0(k_0+2)}\left[1 - \frac{1}{(1+k_0\zeta)^2}\right], \tag{7.50}$$

which has the desired property that $F(0) = 0$ and $F(1) = 1$ when $k_0 \gg 1$. It follows that

$$f(\zeta) = \frac{(1+k_0)^2}{k_0+2}\frac{2}{(1+k_0\zeta)^3}. \tag{7.51}$$

Equation (7.42) suggests that the proper value for k_0 is zero. However, numerical experiments (Saslaw *et al.* 1974, Valtonen 1974, Anosova and Orlov 1986, Valtonen *et al.* 2004) show that at high values of L, the effective range of k starts from k_0 such that

$$k_0 = 9L^{1.25}. \tag{7.52}$$

When $L \to 0$ the distribution of $f(\zeta)$ is flat, indicating that there is no preferred escape direction relative to $\boldsymbol{L}_0$, as there cannot be when $\boldsymbol{L}_0 = 0$. A qualitatively similar dependence was found by Nash and Monaghan (1978) using analytical theory. Figure 7.14 illustrates the corresponding distributions of θ from the previous equations at two values of L, together with experimental data:

$$f(\theta) = \frac{(1+k_0)^2}{k_0+2}\frac{2\sin\theta}{(1+k_0\cos\theta)^3}. \tag{7.53}$$

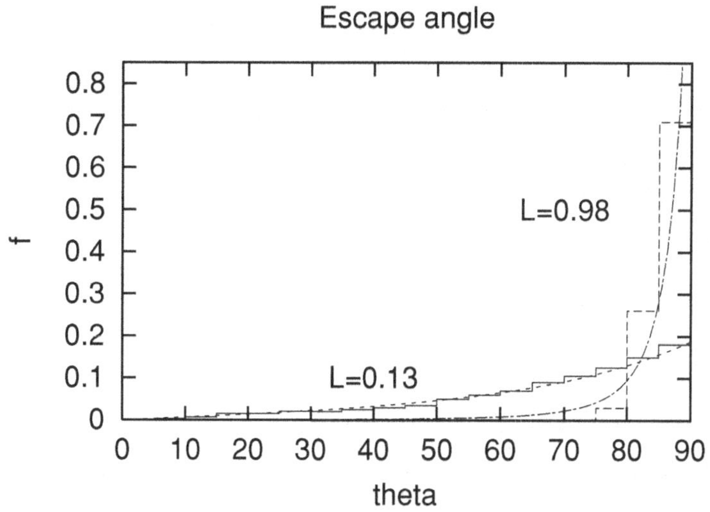

Figure 7.14 The distribution of the escape angle θ in experiments (Valtonen *et al.* 2004) and by Eq. (7.53). The theoretical curves are for $k_0 = 2.2$ and 17 while the experiments have values $L = 0.13$, and 0.98. Experimental results are shown by histograms, the theory by continuous lines.

Example 7.1 When a three-body system breaks up, the escape happens close to the fundamental plane (Fig. 7.14). Then we may ask what is the projected angle between the escape direction and the normal to the plane, if the system is viewed from a random direction. Let us say for simplicity that the escape happens exactly in the plane ($\theta = 90°$).

The answer is easiest to obtain by computer simulations. The result is different depending on whether we consider an angle between the total angular momentum vector and the escape velocity vector, or whether the angle is the acute angle between the line of escape and the normal to the plane. In the former case the projection onto the plane of the sky produces a distribution ranging from 0° to 180°, while the latter distribution is obtained from the former by folding the distribution from above 90° to below 90°. The two distributions are shown by dashed lines in Figs. 7.15 and 7.16.

Together with the theoretical curves we plot data points from two different sets of observational data. In Fig. 7.15 we have data for the angle between small scale jets and jets in extended radio sources (Valtonen 1996 and references therein). In general the data points agree well with the theoretical line, except at difference angle $\approx 0°$. There the observations show an excess. At present it is not clear whether this excess is due to observational selection, or whether it signifies two separate classes of extended radio sources. One class is definitely related to bodies escaping close to the fundamental plane while the second may signify the escape of large

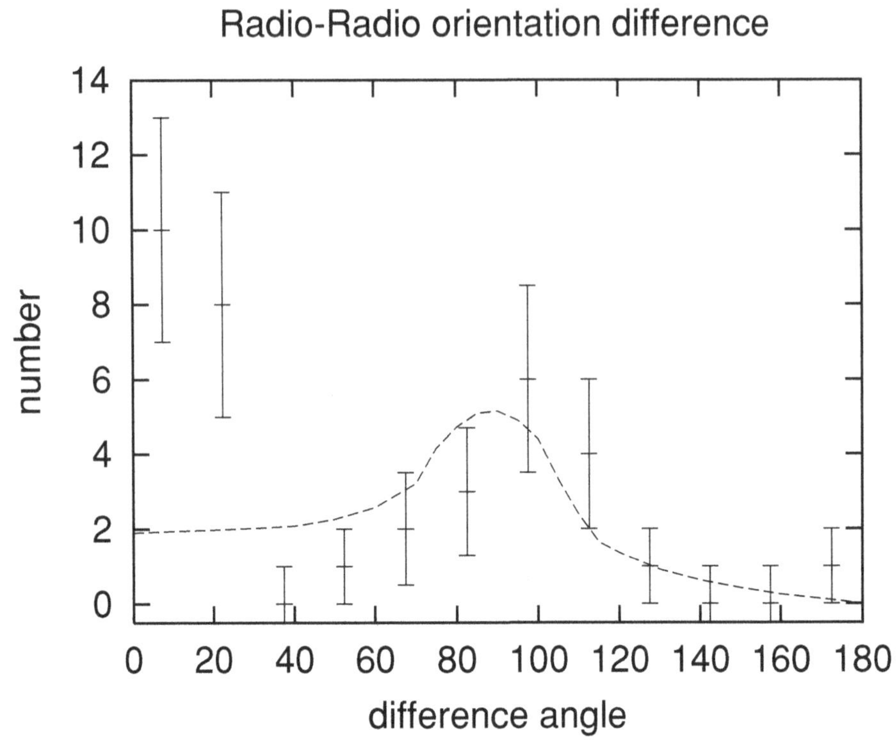

Figure 7.15 The angle between a vector in a plane and a vector perpendicular to the plane, as viewed from different random directions (dashed line). The points with error bars refer to the angle between a small scale jet and a large scale jet in a sample of observed extended radio sources.

scale jets perpendicular to the fundamental plane. In both cases the small scale jet is assumed to be perpendicular to this plane.

In Fig. 7.16 we show observational points for the difference angle between the radio source axis and the minor axis of the light distribution in a radio galaxy. The minor axis of the image tells which way the circular disk of the central plane of the galaxy is tilted. It shows the projection of the symmetry axis of the disk in the plane of the sky. Imaging has been done in the spectral line of ionized gas as well as in red light coming from stars (McCarthy *et al.* 1995). The distributions from the two imaging techniques are shown separately and they agree quite well with each other as well as with the theoretical line. This agreement shows that extended radio emission may be associated with bodies escaping along the fundamental plane which is also the plane of concentration of interstellar gas and the plane of symmetry of an oblate stellar distribution. Both the oblate stellar distribution and the plane of gas concentration may be regarded as resulting from a merger of two galaxies whose orbital angular momentum determines largely the fundamental plane for

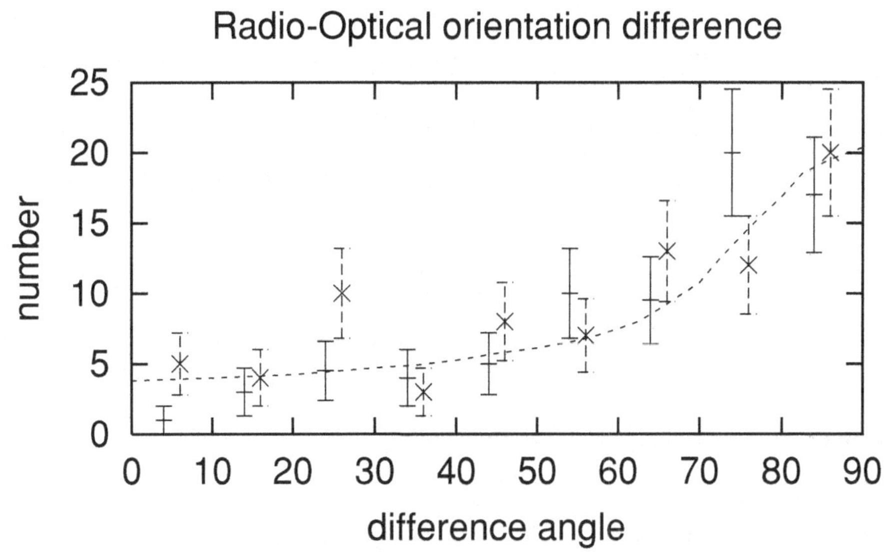

Figure 7.16 The angle between a line in a plane and a line perpendicular to the plane, as viewed from different random directions (dotted line). The points with error bars refer to the angle between the minor axis of gas distribution (+), or the minor axis of stellar distribution (×), and the axis of extended radio emission in a sample of radio galaxies.

gas, stars and the system of supermassive black holes in the centre of the merged galaxy.

Problems

Problem 7.1 Show that the normalised distribution of binding energies is

$$f(|E_B|)\,\mathrm{d}|E_B| = \frac{105}{16}|E_0|^3 \frac{\sqrt{|E_B| - |E_0|}}{|E_B|^{9/2}}\,\mathrm{d}|E_B| \tag{7.54}$$

if one uses

$$\sigma \propto \int \cdots \int \frac{\sqrt{E_0 - E_B}}{E_B^2}\,\mathrm{d}\boldsymbol{r}\,\mathrm{d}\boldsymbol{p}$$

instead of Eq. (7.7). Show that this is the result of putting $V_s = 0$ in Eq. (7.5). Why is this distribution not acceptable? Hint:

$$\int \frac{(x - z)^{1/2}\,\mathrm{d}x}{x^{9/2}} = \frac{2}{7}\frac{(x - z)^{3/2}}{zx^{7/2}}\left(1 + \frac{4}{5}\frac{x}{z} + \frac{8}{15}\frac{x^2}{z^2}\right).$$

Problem 7.2 Derive Eq. (7.15) starting from Eq. (7.3) in the case that the motion is limited to a plane. Derive further the distributions of binding energy, Eq. (7.16), and eccentricity, Eq. (7.17).

Problem 7.3 Using the binding energy distribution of Problem 7.1 instead of Eq. (7.13), show that the peak of the escape velocity distribution is

$$(v_s)_{\text{peak}} = \sqrt{\frac{4}{7}\frac{(M - m_s)}{m_s M}|E_0|}. \tag{7.55}$$

Problem 7.4 Starting from Eq. (7.15), derive Eq. (7.20) and the corresponding peak velocity $(v_s)_{\text{peak}}$.

Problem 7.5 Starting from the binding energy distribution of Problem 7.1 instead of Eq. (7.13), show that P_s remains as in Eq. (7.23).

Problem 7.6 Starting from Eq. (7.20), show that in the case of planar motion, the escape probability P_s is given by Eq. (7.24).

Problem 7.7 Show that if the angular momentum of the three-body system $L_0 = 0$, the motion is restricted to a plane.

Problem 7.8 Using spherical trigonometry and the definitions of the angles θ, ϕ and ψ as shown in Fig. 7.12, prove Eq. (7.34).

Problem 7.9 Show that if ϕ is an angle between two vectors which are randomly oriented relative to each other, $\cos\phi$ is uniformly distributed between -1 and $+1$.

8

Scattering and capture in the general problem

8.1 Three-body scattering

Three-body scattering is a process where a third body comes from a large distance in a hyperbolic orbit and interacts with a binary. The interaction may result in a capture of the third body into the vicinity of the binary. Then we say that a *resonance* (a long lasting state, as in atomic physics) has formed. We expect that the resonance will finally end with an escape of one of the bodies. The other alternatives are an *exchange* where the interaction leads to an immediate expulsion of one of the binary members, or a *flyby* when the third body immediately leaves the scene of the close interaction with the binary. These processes will be discussed in turn, in the following sections. Here the basic theoretical groundwork is formulated, using the results from Section 7.1. At very high energies the three bodies may fly apart separately; then the process is called *ionisation*.

Sometimes a different definition of exchange is used: whenever one of the original binary members is ejected, the process is called exchange. It may happen immediately (prompt exchange) or after an intermediate resonance (resonance exchange). We do not follow this wider definition but define exchange as prompt exchange.

The calculation of the scattering process is performed in two steps: (1) the probability that the third body meets the binary is calculated, and (2) the probability that the binary gains or loses a given amount of energy in the interaction with the third body is estimated.

The scattering probability $d\sigma$ thus has two parts: the geometrical cross-section Σ for the two systems to meet each other and the probability $f(\Delta)$ for the energy change to be in the interval $(\Delta, \Delta + d\Delta)$. Define the relative energy change

$$\Delta = \frac{|E_B| - |E_B|_0}{|E_B|_0} \tag{8.1}$$

where $|E_B|_0$ is the original value of the binary binding energy and $|E_B|$ is its the final value. By this definition

$$|E_B| = (1 + \Delta)|E_B|_0. \tag{8.2}$$

The geometrical cross-section is a matter of definition. The weaker the interaction to be considered, the larger is the interaction radius around the binary. In order to get a well defined interaction radius and the corresponding geometrical cross-section, we use an analogy with escape orbits. For escape orbits the geometrical cross-section is the base area of the loss cone (see Fig. 7.3), i.e. πb_{max}^2 where b_{max} is the maximum impact distance which gives a significant three-body interaction. Before we adopted $b_{max} = 7a_0$. The quantity b is the semi-minor axis of the orbit of the third body relative to the binary centre of mass and it gives the asymptotic distance between the binary centre of mass and the projected line of motion of the third body coming from far away and without gravitational focussing. In a hyperbolic orbit, the semi-major axis a_3 and the maximum semi-minor axis b_{max} are related by $b_{max}^2 = a_3 p_{max}$. A typical value of the semi-latus rectum p_{max} in three-body breakup is $p_{max} = 4.7a_0$ which corresponds to $b_{max} = 7a_0$ (Problem 8.1).

The interaction radius for three-body scattering is now defined to be the same as for escape. It is not immediately obvious why this should be a good definition. However, considering the time reversability of orbits, it is known that for every escape orbit there is a corresponding scattering orbit, the two being identical except for the sense of time. Therefore for the whole ensemble of escape orbits with the maximum interaction radius b_{max} there is a corresponding ensemble of scattering orbits of the same strength of three-body interaction and of the same interaction radius b_{max}. Note that in this ensemble the escape velocities $v_s \lesssim 0.5v_0$ (see Fig. 7.7), and thus our definition should be valid if the ensemble of the approach speeds in the scattering also satisfies $v_3 \lesssim 0.5v_0$. It turns out that this is a good way to define the geometrical cross-section for most purposes, but the cross-section must be increased if very weak encounters are to be included. Therefore the geometrical cross-section for the incoming body to meet the binary is

$$\Sigma = \pi b_{max}^2 = 4.7\pi a_3 a_0. \tag{8.3}$$

The semi-major axis is related to the asymptotic velocity at very large distance v_∞ by

$$a_3 = \frac{GM}{v_\infty^2} \tag{8.4}$$

while the corresponding velocity v_3 in the centre-of-mass system is

$$v_3 = (m_B/M)v_\infty. \tag{8.5}$$

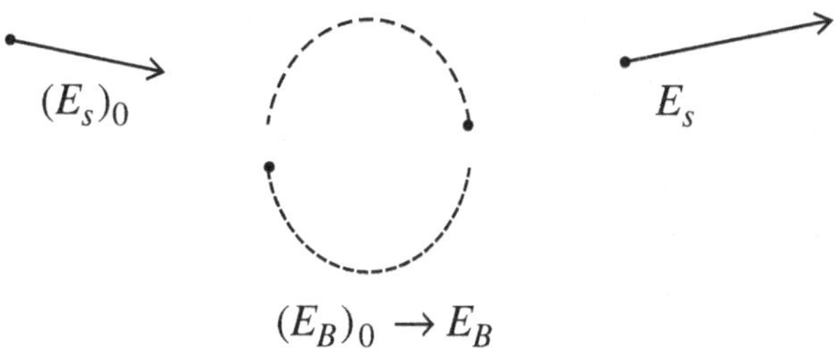

Figure 8.1 Scattering of a single body off a binary. The single body comes with energy $(E_s)_0$ and interacts with the binary of energy $(E_B)_0$. As a result, one of the bodies (not necessarily the incoming body) escapes with energy E_s and leaves behind a binary of energy E_B.

Therefore

$$\Sigma = 4.7\pi \frac{Gm_B^2}{v_3^2 M} a_0 = 4.7 \frac{m_B}{M} \frac{Gm_B}{a_0} \frac{1}{v_3^2} \pi a_0^2 \equiv \nu \pi a_0^2 \tag{8.6}$$

where ν is the focussing factor (really an augmented focussing factor, cf. Chapter 6):

$$\nu = 4.7 \frac{m_B}{M} \left(\frac{v_0}{v_3} \right)^2. \tag{8.7}$$

Here the asymptotic approach speed v_3 is normalised to the average original binary speed v_0:

$$v_0^2 = \frac{Gm_B}{a_0}. \tag{8.8}$$

The factor ν tells how much bigger the effective interaction area is than the binary area πa_0^2, including the effect of gravitational focussing. Note that this factor refers to bodies arriving from a very large distance.

To obtain the second factor $f(\Delta)$, let us consider a binary of binding energy $|E_B|_0$ meeting a third body of energy $(E_s)_0$. After the three-body interaction, there remains a binary of binding energy $|E_B|$ and a third body escapes with energy E_s. The total energy of the system is $E_0 = E_B + E_s = (E_B)_0 + (E_s)_0$. As a limiting case, when $E_s \leq 0$ and nothing escapes, we have a capture (see Fig. 8.1).

There are basically two different situations to consider. (1) A resonance forms and the three bodies are in a bound system for some time until one of them breaks loose. It is justifiable to assume that during the resonance stage, the information from the initial approach conditions has been lost, except for the total energy, total angular momentum and the mass values. Then the theory of Section 7.1 can be used to predict the breakup results. (2) One of the three bodies escapes immediately after the close three-body encounter. Now the applicability of the preceding theory is not so obvious, and experience with numerical experiments has shown that the theory is valid only with certain limits.

Let us then assume that the three-body interaction provides an efficient shuffle in phase space and that the theory of Section 7.1 is applicable (Mikkola 1986, Heggie and Sweatman 1991). Then from Eq. (7.13),

$$f(|E_B|) = 3.5|E_0|^{7/2}|E_B|^{-9/2}, \tag{8.9}$$

or, using Δ,

$$f(\Delta) = 3.5(|E_0|/|E_B|_0)^{7/2}(1+\Delta)^{-9/2}. \tag{8.10}$$

The ratio $|E_0|/|E_B|_0$ is easily calculated using Eqs. (7.4), (8.4) and (8.8) and its value is (noting that $|E_0| = |E_B|_0 - |E_s|_0$ and $|E_s|_0 = Gm_s m_B/2a_3$)

$$1 - \frac{m_s m_B^2 v_\infty^2}{M m_a m_b v_0^2} = 1 - v^2.$$

Here v^2 is defined by

$$v^2 = \frac{m_s m_B^2}{M m_a m_b}\frac{v_\infty^2}{v_0^2} = \frac{m_s M}{m_a m_b}\frac{v_3^2}{v_0^2}. \tag{8.11}$$

Its significance lies in the fact that $v = 1$ corresponds to zero total energy, and often it is convenient to express the incoming velocity in this scale. In that case

$$\frac{d\sigma}{d\Delta} = \Sigma f(\Delta) = 33(1-v^2)^{7/2}\frac{m_s m_B}{2m_a m_b}\frac{1}{v^2}\pi a_0^2(1+\Delta)^{-9/2}. \tag{8.12}$$

As we have pointed out above, our theory should be used only for $v_3 \lesssim 0.5v_0$ which generally means $v < 1$.

However, Eq. (8.12) is not expected to be quite exact since we have used the distribution in Eq. (7.26) with a single value of the power-law index $n = 4.5$. We know in fact that n is a function of the total angular momentum L (Eq. (7.32)). The escapers with large impact distances b typically come from systems with large L requiring $n \gg 4.5$. This is because the angular momentum of the escaper L_s dominates and mostly determines the value of L in escapes with large b. Therefore we should use a weighted average distribution, averaged over the whole range of the indices n in place of Eq. (8.10). The full calculation, however, is rather complicated (Mikkola 1986).

To get an idea how the distributions with different values of n contribute to produce the average distribution, let us consider just three such distributions. Let one of them, with $n = 4.5$, represent low and intermediate L values, while two others with $n = 8.2$ and 10.6 are made to represent the end of high L. With the

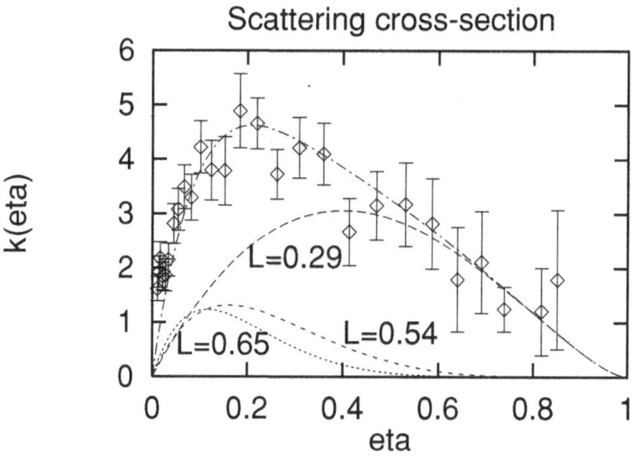

Figure 8.2 This shows how the differential cross-section is built up from contributions of different L. The low L values (corresponding to $n = 3 - 7$) are represented by an index $n = 4.5$ in Eq. (8.14) while the high values are represented by $n = 8.2$ (for the range 7–9.4, and with relative weight 60%) and $n = 10.6$ (for the range 9.4–11.8, also with 60% weight relative to the $n = 4.5$ distribution). When the three distributions are added, they reproduce the experimental data from Heggie and Hut (1993) fairly well.

general power law index n, the distributions have the form (Eqs. (7.26) and (8.10))

$$f(\Delta) = (n-1)(1-v^2)^{n-1}(1+\Delta)^{-n}. \tag{8.13}$$

The experimental testing is conveniently carried out after transforming to a new variable $\eta = \Delta/(1+\Delta)$. For positive Δ this has a range from 0 to 1. The corresponding transformation of the differentials is $d\eta = d\Delta/(1+\Delta)^2$. On the y axis we plot the quantity $4.7\Delta f(\Delta) = k(\eta)$. Not only does the multiplication of $f(\Delta)$ by Δ make the plots more compact, but also the quantity $k(\eta)$ has direct physical significance, as we will learn in Section 8.5. The factor 4.7 is our usual '4.7', indicating how distant are the encounters (i.e. $p_{\max}$) being considered relative to the binary semi-major axis. After the coordinate transformation, the function to be confronted by experimental data is

$$k(\eta) = 4.7(n-1)(1-v^2)^{n-1}\eta(1-\eta)^{n-3}. \tag{8.14}$$

In Fig. 8.2 experimental data points from Heggie and Hut (1993) are plotted and compared with Eq. (8.14), using, in part, the distribution for $n = 4.5$, in part, for $n = 8.2$ and 10.6. Even with the simple representation of the whole n range by only three discrete values, a fair match with the data (except at the very small values of η, $\eta \lesssim 0.05$) is achieved. The proper integral over all n obviously improves the agreement. Also it is found that Eq. (8.14) works well up to $v \leq 0.5$ (Mikkola 1986). The comparison in Fig. 8.2 has been carried out at $v \approx 0.1$.

It is obvious that the scattering at large distance from the binary, and thus with large L, makes a significant modification to Eq. (8.12) at small values of Δ. It would be useful to describe this modification analytically; this is done by introducing a correction factor $A(\Delta)$ by which the right hand side of Eq. (8.12) is to be multiplied. Again, a full calculation is too complicated to follow here, but the correct answer arises by considering the weighted mean values of the distributions of Eq. (8.13) at different n. Let us assume that rather than producing a distribution of Δ, the scattering at angular momentum L always results in a fixed value of the energy change $1+\Delta$, which is equal to the weighted mean of the distribution:

$$\langle 1+\Delta \rangle = \frac{\int_1^\infty (1+\Delta)(1+\Delta)^{-n}\, d(1+\Delta)}{\int_1^\infty (1+\Delta)^{-n}\, d(1+\Delta)}.$$

This mean value is a function of n which in turn is a function of L, by Eq. (7.32).

The probability that the extrapolated orbit passes within the annulus of unit width is proportional to the area of the annulus, i.e. to $2\pi b$. Since at large impact distances $L \propto b$, the probability of a scattering event with a value L is proportional to L. We leave it as an exercise (Problem 8.2) to show that for positive Δ and with the above mentioned assumptions the probability for an energy change Δ is proportional to $\sqrt{\Delta^{-1}}$ when $\Delta \ll 1$. The latter is true at large b interactions.

We have thus concluded that a correction factor of the form

$$A(\Delta) = \frac{1}{2}\sqrt{|\Delta|^{-1}} \tag{8.15}$$

should be applied to the right hand side of Eq. (8.12). Here we have used $|\Delta|$ to allow for negative values, and the factor $1/2$ has been added.

As before, the comparison in the $k(\eta)$–η plot is carried out except that now $k(\eta)$ includes also the $A(\Delta)$ factor:

$$k(\eta) = \Delta \frac{d\sigma}{d\Delta} \frac{2m_a m_b}{m_s m_B} v^2 \frac{1}{\pi a_0^2} \frac{1}{2}\sqrt{|\Delta|^{-1}} = 16(1-v^2)^{7/2}\sqrt{\eta}(1-\eta)^2. \tag{8.16}$$

This distribution is compared with experimental data (Heggie and Hut 1993) in Fig. 8.3. Further justification for the correction factor is provided in Section 10.8 where we compare the energy transfer rate arising from Eq. (8.12) with experimental data.

We now see more clearly the significance of the choice $7a$ as the maximum range of interaction between a binary of semi-major axis a and a third body (Section 7.1). It is good choice for the 'common' range of angular momenta, the range which is usually met in three-body breakup situations. However, the scattering events are strongly biased towards large L since it is more likely that the encounter is distant and has large L rather than small L. Therefore the interaction range going up to $7a$ is adequate only for 'strong' scattering, i.e. for $\Delta \geq 0.25$. For weaker scattering

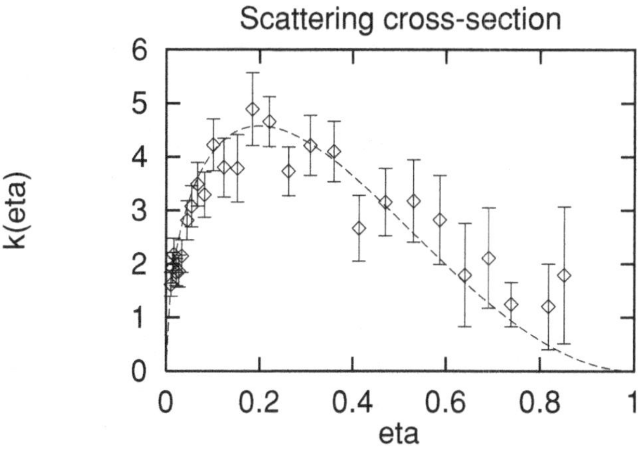

Figure 8.3 A comparison of experimental scattering cross-section data from Heggie and Hut (1993) with Eq. (8.16), including the correction factor $A(\Delta)$ arising from scattering with large angular momentum L.

the interaction range has to be increased, the result of which may be represented by the correction factor $A(\Delta)$.

Fortunately, there is a practical upper limit as to how far one needs to go in extending the interaction range. Figure 8.2 demonstrates that a little more than doubling the interaction area, by adding other distributions to the 'standard' range distribution, gives a fair representation of the numerical experiments in the $k(\eta)$–η plot. The reason why very distant encounters do not significantly contribute to this plot is that we are looking at the quantity $k(\eta)$ which is the product of the differential cross-section $\mathrm{d}\sigma/\mathrm{d}\Delta$ and Δ. While the former includes a $\sqrt{1/|\Delta|}$ factor which would take the differential cross-section to ∞ when $\Delta \rightarrow 0$, the latter factor counteracts to bring $k(\eta)$ to zero at the same limit. It is the energy transfer rate involved in the scattering event, represented by $k(\eta)$, which is usually more significant than the probability of the event itself.

In three-body scattering the range of the strong interaction also depends on the mass of the incoming body as we will learn in the next chapter. In Section 10.7 we will find out that the interaction range (called stability boundary) scales as $(m_s/m_B)^{1/3}$, and the interaction area scales as $(m_s/m_B)^{2/3}$. When $m_s/m_B \geq 1$, the right hand side of Eq. (8.12) should be multiplied by this factor (Hills 1992).

8.2 Capture

In case of capture, a third body coming with initial velocity v loses much of its kinetic energy $(E_s)_0$ to the binary and becomes bound to it. Correspondingly, the binary energy E_B increases upwards toward zero. Let us study this increase with the help of the distribution in (8.9). The presentation is made a little simpler

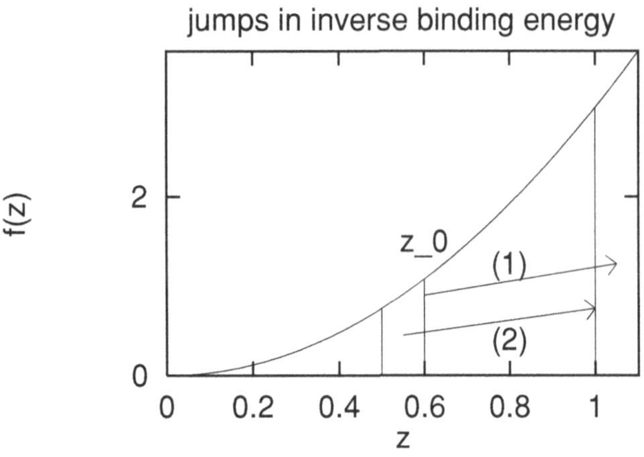

Figure 8.4 An illustration of the distribution $f(z)$ of inverse binding energies $z = |E_0|/|E_B|$ for the power-law index $n = 4$. Before the three-body interaction $z = z_0$, and the three-body encounter brings z to the capture zone $[1, 1 + \mathrm{d}z]$, as indicated by the arrow (1). An equivalent transition is shown by arrow (2) which takes z from the interval $[z_0 - \mathrm{d}z, z_0]$ to $z = 1$. The relative number of states within the latter interval is $3z_0^2\mathrm{d}z = 3(1 - v^2)^2\mathrm{d}z$ which gives the capture probability for $n = 4$.

if the variable is changed to

$$z = |E_0|/|E_B| \tag{8.17}$$

whereby the distribution using the general power-law index n becomes:

$$f(z) = (n - 1)z^{n-2}. \tag{8.18}$$

Let us study how the various states of the three-body system are represented by the new variable z.

Since

$$z = |E_0|/|E_B| = (|E_B| - E_s)/|E_B| = 1 - E_s/|E_B|, \tag{8.19}$$

taking $z > 1$ signifies that we are extending the range of E_s to negative numbers. The negative energy of the third body means that it is in a bound elliptical orbit relative to the binary, and $|E_s|$ represents its binding energy. The range of $|E_s|$ is limited by the requirement that the major axis of the third body has to be much greater than the major axis of the binary in order that we can apply the theory of Section 7.1 and Eq. (8.9). This consideration leads us to state that $|E_s|/|E_B|$ should not be greater than, say, 0.25 (Problem 8.3). Thus, the condition for a capture is that a binary with the initial value z_0 moves in energy space to the region between $z = 1$ and $z = 1 + \mathrm{d}z$, where $\mathrm{d}z$ is typically 0.25 (see Fig. 8.4).

Let us now consider what is the probability that a binary specified by the z value z_0 before the capture ends up in the interval $(1, 1 + \mathrm{d}z)$ after the capture. Experience with numerical experiments in the three-body problem has shown that what is of primary importance is the magnitude of the relative energy change $\Delta|E_B|/|E_B| \equiv \Delta z/z$, not the initial and final states. In other words, the relative jump $\Delta z/z_0 = [1 - (z_0 - \mathrm{d}z)]/z_0 = [(1 + \mathrm{d}z) - z_0]/z_0$ between the lower limits of the two equally wide z-strips in Fig. 8.4 is about equally likely as the relative jump between the upper limits of these strips $\Delta z/z_0 = [(1 + \mathrm{d}z) - z_0]/z_0$, because the two jumps are equal in magnitude. Therefore, for every jump in z from z_0 to the interval $(1, 1 + \mathrm{d}z)$ there exists an equivalent jump from the interval $(z_0 - \mathrm{d}z, z_0)$ to 1. Assuming that the distribution of Eq. (8.18) is valid for the pre-encounter binaries which are able to capture a third body, then the relative area in the strip $(z_0 - \mathrm{d}z, z_0)$ gives us the probability for this jump $P(z_0 \to 1)$:

$$P(z_0 \to 1) = (n - 1)z_0^{n-2}\,\mathrm{d}z.$$

As

$$z_0 = 1 - v^2 \tag{8.20}$$

by the definition of v, the capture probability is

$$P(z_0 \to 1) = (n - 1)(1 - v^2)^{n-2}\,\mathrm{d}z. \tag{8.21}$$

With the value $\mathrm{d}z = 0.25$

$$P(z_0 \to 1) = 0.25(n - 1)(1 - v^2)^{n-2}. \tag{8.22}$$

After multiplication by the geometrical cross-section Σ, the capture cross-section can be written as:

$$\sigma_{\mathrm{cap}} = 2.35(n - 1)(1 - v^2)^{n-2}\frac{m_s m_B}{2m_a m_b}\frac{1}{v^2}\pi a_0^2. \tag{8.23}$$

At low incoming velocities we may put $1 - v^2 = 1$. The value of n depends on the total angular momentum; in typical experiments (Hut and Bahcall 1983, Hut 1993) $L \approx 0.3$ at low values of v, but it decreases towards $L = 0$ when $v \to 1$. We can understand this behaviour of $L = L_0/L_{\mathrm{max}}$ since $L_{\mathrm{max}} \propto |E_0|^{-1}$ and $|E_0| \to 0$ when $v \to 1$. Therefore we expect $n = 4.5$ at $v \to 0$ and $n = 3$ at $v \to 1$. Equation (8.22) gives us the capture probability around 0.9 at low incoming velocity, and $0.5(1 - v^2)$ at high velocity.

A similar comparison can be made with the capture cross-sections. Then it is convenient to plot the normalised capture cross-section $4.7\sigma_{\mathrm{cap}}/\Sigma$ which is shown in Fig. 8.5. Note that in this scale the value 4.7 would mean 100% capture rate. A satisfactory agreement between theory and experiments is seen.

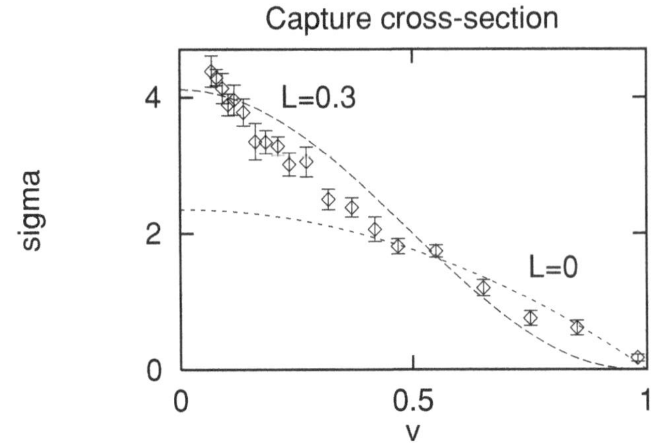

Figure 8.5 Capture cross-sections from the numerical experiments of Hut (1993) for the initial eccentricity $e = 0.7$, normalized so that σ_{cap} is divided by $\pi a_0^2 m_s m_B/(m_a m_b v^2)$. The corresponding data points for $e = 0$, which are not shown, are a factor of 1.7 below these, indicating that the 'effective' cross-sectional area of the binary is greater for higher e. In the experiments the median angular momentum varies from $L = 0$ at $v = 1$ to about $L = 0.3$ at $v = 0$. The two theoretical curves, drawn to correspond to these L values, match the data at either end of the v-range.

We should note that the capture probability depends on the geometrical cross-section Σ which we are using. If we take the interaction radius to be much greater than 7, the number of captures will not significantly increase. The three-body scattering from large impact distances is mainly of the flyby type, and therefore the increase in Σ will increase the relative proportion of flybys and decrease the probability of capture. The probability of capture is a significant number only as long as one specifies the impact distance range of interest. Numerical experiments (Saslaw *et al.* 1974, Hills 1992) show that generally the capture probability decreases with the increasing semi-latus rectum p_3 of the third-body orbit, but that it is also quite sensitive to other parameters, such as the mass of the incoming body.

In contrast, the capture cross-section σ_{cap} is a stable quantity as soon as Σ is large enough. Three-body scattering at large distances does not make further contributions to the total number of captures and thus σ_{cap} remains unchanged. For this reason, it is generally better to use the capture cross-section rather than capture probability.

In the above derivation of Eq. (8.23) it was assumed that the relative number of pre-encounter binaries with different values of z can be obtained from Eq. (8.18). This is not a trivial assumption. Why should one value of z (or one value of the semi-major axis of the binary) be better for the capture than some other value, and in such a way that Eq. (8.18) is followed? Remember that this distribution was derived for the binaries after an escape, and after all, capture is quite different from escape.

A partial resolution of this problem comes from the time-reversibility of three-body orbits. For every escape orbit, there is a corresponding time-reversed capture orbit. For every binary with the inverse energy z after an escape, there is another binary with the same inverse energy z which is capable of a capture. This suggests a *detailed balance* between emission (escape) and absorption (capture) of bodies by binaries. Using the detailed balance principle, one can understand why the after-escape distribution of z can also be used to describe the pre-capture distribution of z.

However, we should point out that the principle of detailed balance involves more than just noting the obvious time-reversibility of orbits. The time reversals prove that for a sample of escape orbits, with a definite distribution of escape velocities v_s (see Fig. 7.7), there exists a corresponding sample of capture orbits, starting from the same distribution of approach velocities v_3. Then it is quite clear that binary energies should also be distributed exactly in the same way in the two processes. However, in the derivation of Eq. (8.23) we were discussing samples with a single value of v_3. The fact that the binary energy distributions are still equivalent between escape and capture even after this breach of symmetry constitutes a further step which we take knowing that it leads to a good description of numerical experiments. The principle of detailed balance has been used as a starting point of an alternative derivation of the binary energy distributions, thus proving its generality and utility (Heggie 1975).

8.3 Ejections and lifetime

When a resonance has formed through a capture, or otherwise, we may ask how long it survives before one of the bodies escapes. We call this period the *lifetime* of the bound system. Numerical orbit calculations have shown that a three-body system evolves via numerous close triple encounters between which one of the bodies recedes to some distance from the binary and then returns.

These excursions are called ejections. Thus typically the three-body evolution consists of interplay when the three bodies orbit each other in a relatively small volume, interspersed by ejections until one of the departures happens fast enough to become an escape. The time spent in the interplay mode is typically about one half of the time spent in ejections (Anosova and Orlov 1986). The exact timeshare between the two modes depends strongly on the definition of the borderline between the two modes, i.e. on how far from the binary the third body is allowed to recede before the interplay status is lost. Here we use the definition that the third body should not go further than $r_s = 10r$ during the interplay phase where r is the binary separation (see Fig. 7.2). In the following, the time spent in ejections will be counted and in the end a 50% correction will be applied to account for the interplay time.

In some ways ejections resemble escape; during an ejection the system is clearly divided in two parts, a binary and a third body which initially recedes from the binary and then turns around and returns. The theory of Section 7.1 could be used to describe the receding phase of motion, when the distance r_s satisfies $a < r_s < R$ where R is the upper limit of integration in Eq. (7.6). Obviously the turnaround distance has to be much greater than R.

The approach taken here is to assume that the probability for obtaining a certain escape energy E_s is as derived previously, and that the corresponding binary energy distribution is given by Eq. (7.13). However, for an escape to occur, $E_s > (E_s)_{\mathrm{exc}}$, where the excess energy $(E_s)_{\mathrm{exc}}$ is required to overcome the potential 'barrier' arising from the potential energy in the gravitational field of the binary. If the excess energy is not available, the third body 'bounces back' from the 'barrier', and an ejection takes place.

How big a barrier should be considered? The orbit of the third body relative to the binary should be extensive enough to satisfy our specification of near escape. The semi-major axis of this orbit should be $a_3 \gg a$. To fix our attention to a specific number, let us say that $a_3 > 10a$.

The range $10a < a_3 < \infty$ implies an attempted escape, i.e. an escape followed immediately by a capture. Therefore, our result from Section 8.2 can be used, which tells us that the width of the capture zone (i.e. the 'barrier') is $\mathrm{d}z = 0.25$ in the distribution of Eq. (8.18). The connection between the choice $a_3 = 10a$ and $\mathrm{d}z = 0.25$ is as follows. By definition (Eq. (7.4)) $a_3/a = [(m_B m_s)/(m_a m_b)](|E_B|/|E_s|)$, and by Eq. (8.19) $\mathrm{d}z = z - 1 = |E_s|/|E_B|$. Thus $\mathrm{d}z = [(m_B m_s)/(m_a m_b)](a/a_3)$. Putting $a/a_3 = 0.1$ and equal masses gives $\mathrm{d}z = 0.2$; considering unequal but comparable masses one may choose $\mathrm{d}z = 0.25$. The probability of ejection is then the relative area of the capture zone $[1, 1 + \mathrm{d}z]$ (see Fig. 8.4):

$$P_{\mathrm{ej}} = \frac{(1 + \mathrm{d}z)^{n-1} - 1}{(1 + \mathrm{d}z)^{n-1}} = 1 - 0.8^{n-1}. \tag{8.24}$$

One observes that the ejection probability is expected to range from $P_{\mathrm{ej}} = 0.36$ when the angular momentum $L_0 = 0$ ($n = 3$), to $P_{\mathrm{ej}} \approx 0.75$ when $L_0 = 0.5L_{\max}$ ($n = 7.5$). Numerical experiments (Valtonen 1975a) carried out in the middle range of angular momenta, $L_0 \approx 0.33L_{\max}$, give the ejection rate $P_{\mathrm{ej}} = 0.5 - 0.6$, as expected. As the alternative to ejection is escape, the escape probability is $1 - P_{\mathrm{ej}}$ per one strong three-body encounter and it shows a similar variation with angular momentum.

Let us suppose that the escape takes place at the Nth attempt. The probability that the $N - 1$ previous attempts produced an ejection is $P_{\mathrm{ej}}^{(N-1)}$, and the probability that the last one was an escape has the probability $1 - P_{ej}$; thus altogether the probability that the Nth three-body encounter produced an escape is $(1 - P_{ej})P_{ej}^{N-1}$. The probability that even after N encounters there is no escape is P_{ej}^N, and this is also the probability that the lifetime of the bound three-body system exceeds $N\langle T\rangle$ where $\langle T\rangle$ is the average length of time which the system spends in one single ejection.

Let us now estimate $\langle T\rangle$ in order to calculate system lifetimes. The orbital period of an ejection orbit is

$$\begin{aligned} T_{orb} &= 2\pi\sqrt{a_3^3/GM} \\ &\geq 2\pi\, 10^{3/2} G[(m_a m_b)^3/M]^{1/2}(2|E_0|)^{-3/2}, \end{aligned} \tag{8.25}$$

where use was made of Eq. (7.4), the inequality $a_3 > 10a$, as well as the fact that in ejections, $E_B \approx E_0$. It is convenient to measure the lifetime in terms of the crossing time T_{cr} of the system. The crossing time is the time during which a body with typical speed $\langle V\rangle$ crosses through a system of dimension $\langle R\rangle$, i.e. $T_{cr} = \langle R\rangle / \langle V\rangle$. If the mass of the system is M and the energy of the system is $|E_0|$, in equilibrium we have the approximate relations $2|E_0| = M\langle V\rangle^2 = GM^2/\langle R\rangle$. From here $\langle R\rangle / \langle V\rangle = GM^2/(2|E_0|)[M/(2|E_0|)]^{1/2}$, so that

$$T_{cr} = GM^{5/2}(2|E_0|)^{-3/2}. \tag{8.26}$$

We may now normalise the orbital period T_{orb} to the crossing time:

$$T_{orb}/T_{cr} \geq 2\pi\, 10^{3/2}\left(\sqrt{m_a m_b}/M\right)^3 \tag{8.27}$$

or $T_{orb}/T_{cr} \geq 7.36$ for equal masses. Therefore the range of T_{orb} in units of T_{cr} goes from ~ 10 to ∞. If we take z from the lower part of the capture zone, e.g. $z = 1 + 0.25/3$, the corresponding value of T_{orb}/T_{cr} is 3.5 times greater than the minimum value 7.36 (Problem 8.4), and using this value one can see that typically $T_{orb} \approx 26T_{cr}$, in agreement with numerical experiments (Szebehely 1972).

The distribution of lifetimes T which follows from the process of repeated escape trials, with a given escape probability per trial, is exponential

$$f(T) = 0.69\mathrm{e}^{-0.69T/T_{1/2}}. \tag{8.28}$$

Here the half-life $T_{1/2}$ of the system is the point in time when the probability that the system remains intact is 50%. Let this happen after N orbital cycles, each of which lasts $26T_{cr}$. Then N is solved from the equation $P_{ej}^N = 0.5$, i.e. $N = 0.3/\log(1/P_{ej})$, and using this value of N, the half-life can be written $T_{1/2} = 7.8/\log(P_{ej}^{-1})T_{cr}$.

As was noted previously, the probability of escape also depends on the mass of the escaper. Systems with a large range of masses break up faster because they

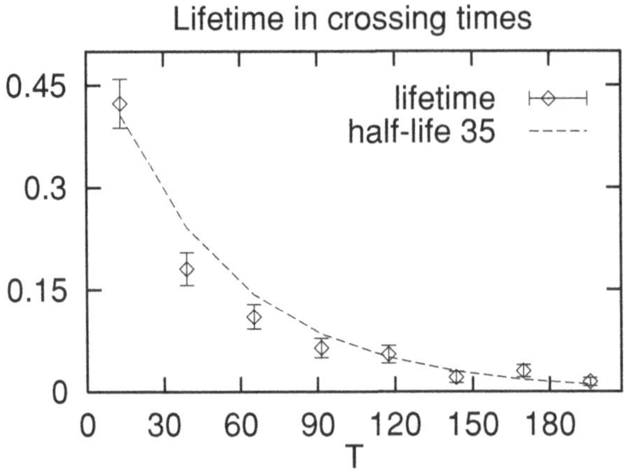

Figure 8.6 The distributions of lifetimes T, in units of the crossing time T_{cr}, for a system with equal masses and the average angular momentum $L = 0.2$ (Saslaw *et al.* 1974). A theoretical curve of the form $f(T) = 0.69e^{-0.69T/T_{1/2}}$ fits the data well when $T_{1/2} = 35$.

contain at least one body which has a low fraction of the total mass and which consequently has a high probability of escape. The half-life should be inversely proportional to P_s in Eq. (7.23):

$$T_{1/2} \propto P_s^{-1} = 1 + m_s^2(m_a^{-2} + m_b^{-2}). \tag{8.29}$$

How the mass values are divided between different bodies varies from system to system, but a fairly typical situation occurs if m_s is the smallest mass and the rest of the mass is equally divided between the other two: $m_a = m_b$. Let us call the range of masses $m = (1 - m_s)/(2m_s)$ in this case. Then $T_{1/2}$ is $\frac{1}{3}(1 + 2/m^2)$ times the half-life for equal masses. In addition, there is the factor 1.5 which adds the interplay time to the ejection time to make the total lifetime of a three-body system. Putting all the factors together,

$$T_{1/2} = 3.9(1 + 2/m^2)/\log(P_{ej}^{-1})T_{cr} \tag{8.30}$$

where P_{ej} is given by Eqs. (7.32) and (8.24). When going from low to high angular momentum, $T_{1/2}$ is expected to range from $27T_{cr}$ to over $60T_{cr}$ for equal masses, and from $11T_{cr}$ to over $24T_{cr}$ for the mass ratio $m = 3$. Numerical experiments have confirmed the exponential decay of bound three-body systems and the half-life dependence described above (see Figs. 8.6 and 8.7; Anosova 1969, Szebehely 1972, Saslaw *et al.* 1974, Valtonen and Aarseth 1977, Anosova and Polozhentsev 1978, Agekyan *et al.* 1983, Anosova and Orlov 1983, 1986, 1994, Anosova *et al.* 1984).

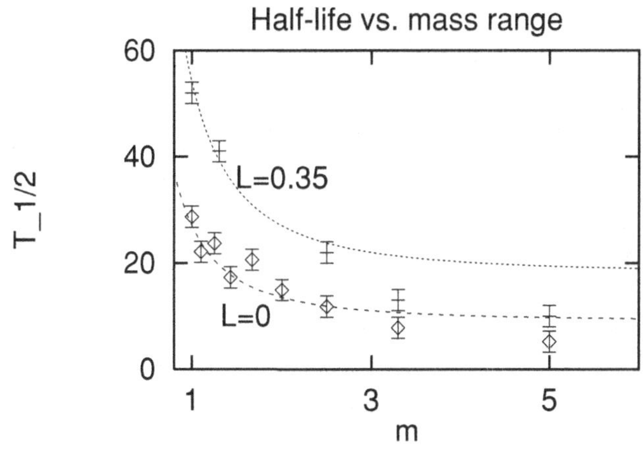

Figure 8.7 The half-life $T_{1/2}$ as a function of the maximum mass range m of the three bodies. The data points come from Valtonen (1974), Anosova and Polozhentsev (1978), Anosova and Orlov (1986), and the theoretical lines follow Eq. (8.30) with different values of $P_{\rm ej}$. The latter are derived using Eqs. (7.32) and (8.24) for $L = 0$ and $L = 0.35$. It is apparent that Eq. (8.30) is applicable only for $m < 3$.

8.4 Exchange and flyby

Often the impact of the incoming body on the binary is immediately followed by an escape. If the incoming body escapes, we call the process a flyby. The single body flies by the binary with a certain amount of change of its direction as well as speed. Correspondingly, the binary binding energy changes.

If, on the other hand, the escaped body is one of the former binary members, we call the event an exchange. Then one binary member is replaced by the incoming body, i.e. we have an exchange of bodies between the two systems. (More exactly, we are talking about prompt exchange, see Section 8.1.) Naturally also the binary energy changes in the process.

The exchange probability is a product of two factors. First, there is the probability that capture has not taken place. According to Section 8.2, this probability is about 11% at low values of v. The second factor is the probability $1 - P_s$ that the escaper is not the incoming body, of mass m_s, but rather one of the original binary members. Therefore the differential cross-section for the relative energy change is given by Eq. (8.12) multiplied by these two factors:

$$\frac{d\sigma_{\rm ex}}{d\Delta} = 3.6(1 - P_s)(1 - v^2)^{7/2}\frac{m_s m_B}{2m_a m_b}\left(\frac{1}{v}\right)^2 \pi a_0^2(1 + \Delta)^{-9/2}. \qquad (8.31)$$

For equal mass systems $P_s = 1/3$ and the coefficient $3.6(1 - P_s)$ is about 2.4 which

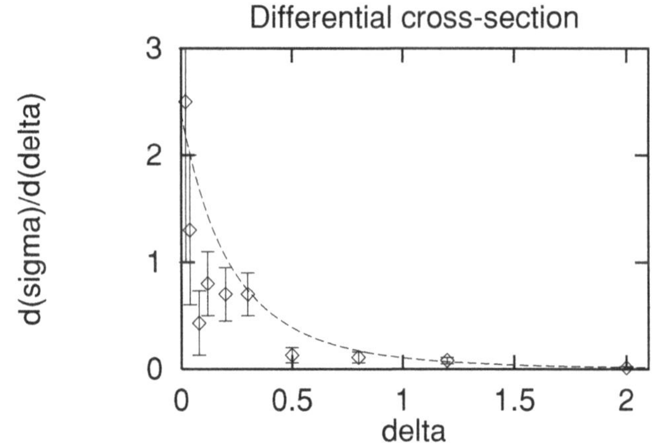

Figure 8.8 The normalized cross-section $d\sigma/d\Delta$ for exchange scattering as a function of Δ. The experimental points come from Hut (1984), and the theoretical curve follows Eq. (8.31) with $P_s = 1/3$.

is not very far from the value given by numerical experiments for $\Delta \geq 0$ (Fig. 8.8; Hut 1984, Heggie and Hut 1993).

The total cross-section for exchange σ_{ex} is obtained from Eq. (8.31) by integration over Δ. The lower limit of integration is $\Delta = -v^2$ (Problem 8.5), the upper limit is ∞. The integration gives ($P_s = 1/3$)

$$\sigma_{ex} = 0.7 \frac{m_s m_B}{2 m_a m_b} \frac{1}{v^2} \pi a_0^2. \tag{8.32}$$

This agrees very well with numerical experiments. (These are called 'direct exchange' or prompt exchange in Hut (1993); 'resonance exchange' gives a coefficient 3.5 in Eq. (8.32). The latter represent about 43% of all captures.) Note that σ_{ex} does not contain the $(1 - v^2)$ factors. The power $7/2$ of this factor in Eq. (8.31) is cancelled after inserting the lower limit to the integral. Therefore the exchange cross-section is rather independent of v for $v \leq 0.5$, except for the v^{-2} factor arising from focussing.

The cross-section for flybys is then what is left over from the total geometrical cross-section

$$\Sigma = 9.4 \frac{m_s m_B}{2 m_a m_b} \frac{1}{v^2} \pi a_0^2 \tag{8.33}$$

after the capture cross-section and the exchange cross-section are deducted from it. One must remember that the flyby cross-section is a function of p_{max}: when p_{max} is increased, the flybys dominate more and more. Thus the flyby cross-section only has meaning when the maximum interaction radius has been

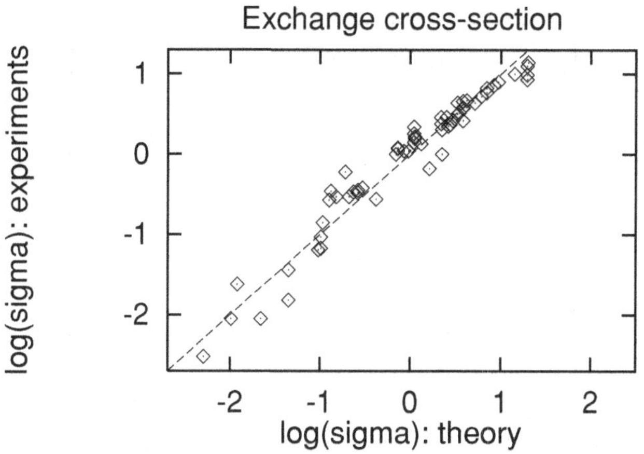

Figure 8.9 A comparison of the theoretical exchange cross-section of Eq. (8.32) (multiplied by the correction factors for unequal masses) with experimental data from Heggie *et al.* (1996). There is generally good agreement (points close to the dashed line), but also a few major disagreements. The latter are not a surprise since simplifying assumptions were made in the derivation of the correction factors.

specified. On the contrary, exchanges are obtained primarily inside the $p_{\rm max} = 4.7a_0$ radius, and thus the exchange cross-section is well defined at sufficiently large $p_{\rm max}$.

It is desirable to know which one of the two original binary components escapes in an exchange interaction. This can be found by going back to the phase space volume σ in Eq. (7.22), and by noting that $\sigma \propto m_B^{9/2}$ when we substitute $m_a m_b = \mathcal{M} m_B$. Thus the probability of escape for a body of mass m_1 should be proportional to $(m_2 + m_s)^{9/2}$ where m_2 and m_s are the masses of the new binary members after the exchange. Normalising the binary mass to the total mass M gives us the escape probability factor $3.1(1 - m_1/M)^{9/2}$. This factor is 0.5 for equal masses as it should be since then m_1 and m_2 have equal probability for escape. Thus the factor can be used as a multiplier in Eq. (8.32) where the numerical coefficient is based on numerical experiments with equal masses.

Sometimes it is convenient to normalise the cross-section so that the initial binary mass $m_a + m_b = 1$. Now M is not a constant, and we have to take note of the M dependence (normalised to the initial binary mass) $[M/(m_a + m_b)]^{1/2}$ in the phase space volume of Eq. (7.22). As a result Eq. (8.32) is then further multiplied by $0.81[1 + m_s/(m_a + m_b)]^{1/2}$. This quantity is unity for equal masses, and thus it is a suitable correction factor between equal mass and unequal mass cross-sections. Experimental data on the mass dependence of the exchange cross-sections supports the use of the above mass factors (see Fig. 8.9; Heggie *et al.* 1996).

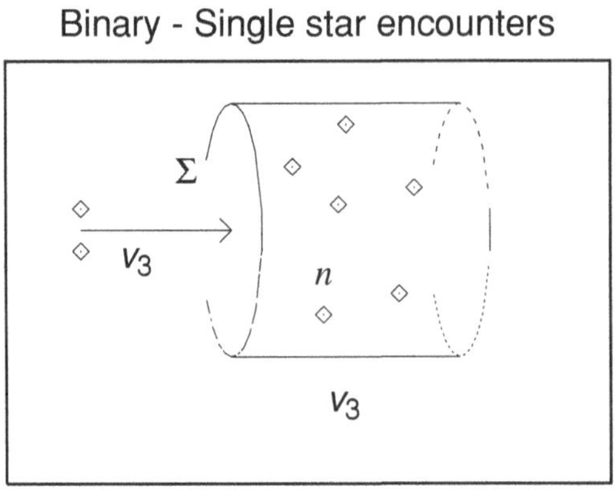

Figure 8.10 A binary passes through a field of single stars (number density n) with speed v_3. The effective interaction cross-section is Σ. The volume sampled by the binary in one time unit is Σv_3, and the number of single stars encountered is $\Sigma v_3 n$.

8.5 Rates of change of the binding energy

The cross-sections derived in previous sections are useful in many applications. They are needed especially in the study of the dynamical evolution of stellar systems, such as star clusters (Heggie 1975). Star clusters are composed both of single stars and binary stars, as well as temporary multiple star subsystems, and the interaction between binaries and single stars has been identified as a major influence on the evolution of the structure of the cluster. Clusters may either collapse or expand which means that the overall cluster potential is changing. Quite often the binaries in the cluster provide the source of energy which is required for the overall adjustment of the cluster. Therefore we will now study briefly what happens to a binary when it interacts with the single stars of the cluster.

Consider single stars of space density n (stars/pc^3) flowing past the binary at speed v_3. The number of encounters between the binary and the single stars per unit time is $n v_3 \Sigma$ (Fig. 8.10). The encounters occur at the rate

$$\begin{aligned} R &= \int n v_3 \frac{\mathrm{d}\sigma}{\mathrm{d}\Delta}\,\mathrm{d}\Delta \\ &= \int 33(1-v^2)^{3.5} \pi a_0^2 \frac{m_s m_B}{2 m_a m_b} \frac{n v_3}{v^2} (1+\Delta)^{-9/2}\,\mathrm{d}\Delta \\ &= \frac{7.4 G^2 m_B^3}{\frac{1}{2}\mathcal{M} v_0^2} \frac{\mathcal{M}}{M} \frac{n}{v_3}. \end{aligned} \tag{8.34}$$

The rate R has been obtained using Eqs. (8.8), (8.11) and (8.12). For the cross-section Σ we have used $\int \mathrm{d}\sigma$, and the integration is to be carried out over the whole range of Δ from $-v^2$ to ∞.

At each encounter, a relative amount of energy Δ is lost or gained by the binary. The rate of energy transfer is then

$$R_\Delta = \int n v_3 \Delta \frac{\mathrm{d}\sigma}{\mathrm{d}\Delta}\,\mathrm{d}\Delta = \frac{3G^2 m_B^3}{{}^1\!/_2 \mathcal{M} v_0^2} \frac{\mathcal{M}}{M} (1 - 3.5v^2) \frac{n}{v_3}. \quad (8.35)$$

Here we have made use of the integral

$$\int \frac{x\,\mathrm{d}x}{(1+x)^{4.5}} = -\frac{2}{35}\frac{2+7x}{(1+x)^{3.5}}. \quad (8.36)$$

The average amount of energy transferred per encounter is

$$\langle \Delta \rangle = R_\Delta / R = 0.4(1 - 3.5v^2). \quad (8.37)$$

Since Eq. (8.12) does not put enough weight on the small values of Δ, it is expected that the above $\langle \Delta \rangle$ is an overestimate. The correct value is about a factor of two smaller (Heggie and Hut 2003; see Problem 8.6), as one might guess by looking at Fig. 8.2.

Equation (8.37) implies that the average energy change $\langle \Delta \rangle$ is independent of the masses of the three bodies. This cannot be quite correct. Our theory was based on the reversibility between escape and capture orbits, which seems to work well for nearly equal masses. However, the scattering of a heavy third body from a light binary has no corresponding escape orbit since the probability of such an escape is practically zero (Eq. (7.23)). At the other end of the scale, the probability for the escape of a very small body is 100% while its capture cannot be so efficient. Numerical experiments (Hills 1992) show that the theory may be rectified by multiplying the right hand side of Eq. (8.37) by $(2m_3/m_B)^{1/2}$. The data points from Hills (1984) are shown in Fig. 10.14 where a more complete discussion of $\langle \Delta \rangle$ as a function of the closest approach distance q follows.

It is important to notice that $\langle \Delta \rangle$ is positive for all values $v \lesssim 0.5$. When the binary binding energy clearly exceeds the energy of the incoming body, we say that the binary is *hard* (Chapter 6). This is true when $v \lesssim 0.5$. Therefore we may state that the binding energy of the binary increases on average, i.e. *hard binaries become harder* (Aarseth and Hills 1972, Hills 1975, Heggie 1975, Gould 1991). In case of very unequal masses a more exact condition for the hardening of binaries is $v \lesssim v_0$, i.e. the approach speed of the third body at large distances should be below the average binary speed (Hills 1990). This has the important consequence that hard binaries in the cluster act as sources of energy. When the binary binding energy increases, i.e. when its energy becomes more negative, it throws out single stars with velocities higher than they came by, and subsequently the star cluster is *heated*. The heating leads to an expansion, and in case of smaller clusters, to their dispersal. A tidally limited cluster contracts, however, as mass is lost.

8.6 Collisions

During the chaotic phase of three-body evolution, two of the three bodies sometimes approach very close to each other. At this time, their speeds increase to values which are much higher than the average speed. In astrophysical systems, the bodies have finite sizes which means that in close approaches the two bodies may collide. Speed cannot exceed the speed of light in the real physical world. Much before that happens, the dynamics require modifications and Newtonian gravity no longer applies.

It is possible to estimate the frequency of close approaches by assuming that the system evolves through a succession of ejections until an escape takes place. At each ejection a temporary binary is formed which has close two-body encounters at the pericentre of the orbit. The pericentre distance is $q = a(1 - e)$. While the semi-major axis of the temporary binary does not vary greatly from one ejection to the next, the eccentricity varies a lot, over the whole range given by the distribution of Eq. (7.14).

The probability that the eccentricity is greater than e is $1 - e^2$, and since $1 - e^2 = (1 - e)(1 + e) = (q/a)(1 + e)$, at high values of e ($e \approx 1$), this probability becomes approximately equal to $2q/a$. This is the probability of close approach within distance q per ejection. During its evolution the system goes through a number of ejections, some of short duration and others of longer duration. For the present calculation, all ejections are of equal importance; even the very brief ones count for they produce temporary binaries with possibly close two-body encounters. In typical interplay phase, there is an ejection of one kind or another per crossing time and, related to it, there are on average two close approaches (Szebehely and Peters 1967). Occasional long-lasting ejections decrease the average number of ejections per crossing time somewhat but the number of close approaches is rather constant at two per crossing time. Typically, a three-body system lives through about 30 crossing times (see Section 8.3) which implies something like 60 close encounters. Assuming that Eq. (7.14) can be used to estimate the closeness of these approaches, then for very small values of q/a the probability of approach closer than q is approximately $120q/a$.

The approach distance q is normalised to the temporary semi-major axis a. Then it is necessary to estimate how large a is relative to some standard measure like a_0 which is the semi-major axis of the (temporary) binary before the strong three-body interaction begins. Typically $a \approx (1/2)a_0$, similar to the semi-major axis of the binary after an escape. Therefore the probability that the approach is closer than

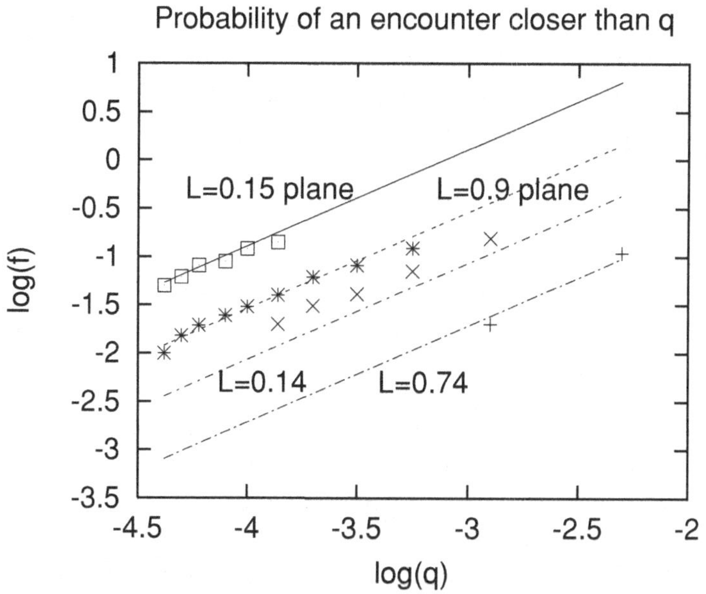

Figure 8.11 The probability $f(q)$ as a function of q in three-dimensional systems ($L = 0.74$ and 0.14) and in planar systems ($L = 0.9$ and 0.15). The experiments come from Saslaw *et al.* (1974) and the straight lines are the expected relations at low q. The latter is in units of a_0, the semi-major axis of the initial binary.

q is:

$$f(q) \approx 240\, C(L) q / a_0. \tag{8.38}$$

The angular momentum dependence has been introduced through the factor $C(L)$. This factor is needed since the system lifetime depends on L (see Fig. 8.7). The functional form $T_{1/2}(L)$ is rather complex as we see by combining Eqs. (7.32), (8.24) and (8.30). It may be described more simply by (Saslaw *et al.* 1974)

$$C(L) \approx 1 + 7.5L^2. \tag{8.39}$$

If the diameters of the bodies in question are q, then the collision probability becomes $P_{\text{coll}} = f(q)$.

When the system is two-dimensional the eccentricity distribution peaks strongly at $e \approx 1$ due to the $(1 - e^2)^{-1/2}$ factor in Eq. (7.17). For a typical value $q/a = 1/500$, $1 - e^2 = 1/250$ and $(1 - e^2)^{-1/2} \approx 15$. Thus the probability of close encounters is increased by about this factor in two-dimensional systems. Figure 8.11 illustrates how these results agree with numerical experiments.

The above theory may also be extended to scattering where we define the cross-section $\sigma(q)$ as the impact area which leads to the closest two-body encounter to be less than q. It consists of two parts: the geometrical cross-section for the third body to meet head-on the binary $\Sigma = \nu\pi a_0^2$ (Eq. (8.6)), where ν is the focussing factor, and the probability $f(q)$ that there is an approach closer than q (Eq. (8.38)). If only more or less head-on collisions are considered, we may (rather arbitrarily) put $C(L) = 1$, and $p_{\max} = \frac{1}{2}a_0$. Then Eq. (8.7) becomes

$$\nu = \frac{1}{2}\frac{m_B}{M}\left(\frac{v_0}{v_3}\right)^2 = \frac{m_B m_s}{2m_a m_b}\frac{1}{v^2}. \tag{8.40}$$

Altogether

$$\sigma(q) \approx 240\frac{q}{a_0}\frac{m_B m_s}{2m_a m_b}\frac{\pi a_0^2}{v^2}. \tag{8.41}$$

The theory and experiments agree well at small values of q/a_0, $q/a_0 \lesssim 10^{-3}$, but at large values of q/a_0 the theoretical cross-section is too large (Hut and Inagaki 1985). The large q/a_0 regime has a major contribution of flybys and exchanges which possess essentially one close encounter instead of the typical 60 close encounters following a capture. Therefore the cross-section is lowered by a factor which is somewhere between 1 and 1/60 and which decreases with increasing q/a_0.

The overall effect is to change the slope and the coefficient in Eq. (8.41). Numerical experiments (Sigurdsson and Phinney 1993) suggest

$$\begin{aligned} \sigma(q) &= 40\left(\frac{q}{a_0}\right)^{0.75}\frac{m_B m_s}{2m_a m_b}\frac{\pi a_0^2}{v^2}, \quad \frac{q}{a_0} \le 0.0137, \\ \sigma(q) &= 8.1\left(\frac{q}{a_0}\right)^{0.375}\frac{m_B m_s}{2m_a m_b}\frac{\pi a_0^2}{v^2}, \quad \frac{q}{a_0} > 0.0137. \end{aligned} \tag{8.42}$$

This applies over a wide range of mass ratios m_s/m_B and incoming velocities v (Hills 1991, Sigurdsson and Phinney 1993). In Fig. 8.12, $\sigma(q)$ is represented by the two lines, and they are seen to provide a good fit to the data.

The relative speed at the close two-body encounter is of prime interest in cases of neutron stars and black holes. A 'speed limit' $v_{\max}$ may be set which should not be exceeded in order that our Newtonian calculation is valid. Setting aside the concern about the proper value for $v_{\max}$, one can simply proceed to calculate the probability that $v_{\max}$ is not exceeded in different types of systems. This probability is $1 - P_{\text{coll}}$ when we take $v_{\max}$ to be the pericentre velocity:

$$v_{\max}{}^2 = \frac{G(m_a + m_b)}{q}(1 + e) \tag{8.43}$$

where $1 + e \approx 2$ at the limit of high eccentricities. The maximum speed may be compared with the circular speed (Eq. (8.8)):

$$\frac{v_{\max}^2}{v_0^2} = 2\frac{a_0}{q}$$

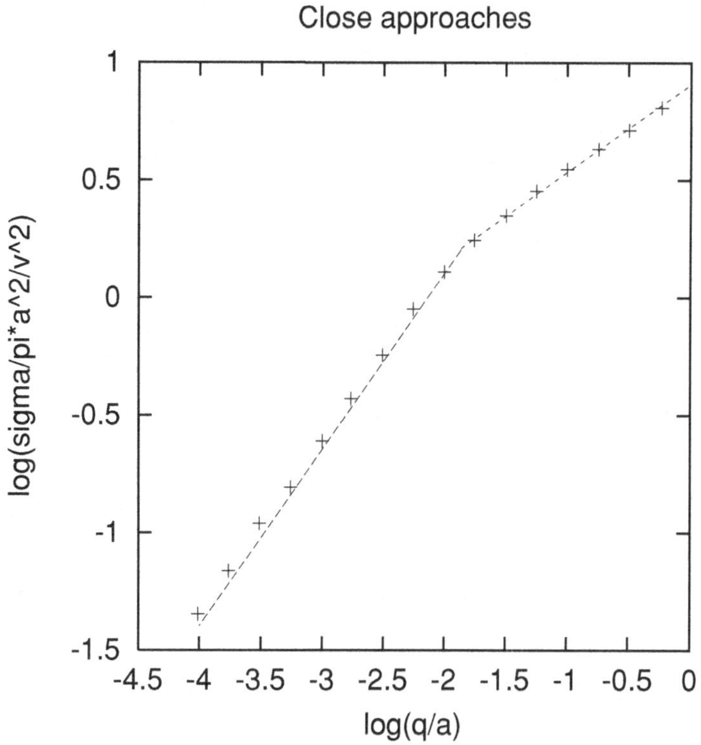

Figure 8.12 The normalised cross-section for scattering experiments (equal masses and zero eccentricity binary), which lead to a close two-body encounter below the value q. Experiments are from Sigurdsson and Phinney (1993) and the straight lines are from Eq. (8.41).

or

$$\frac{2q}{a_0} = 4\left(\frac{v_0}{v_{\max}}\right)^2. \tag{8.44}$$

Therefore the probability that the speed remains below $v_{\max}$ is

$$P_{v\max} = 1 - 480C(L)(v_0^2/v_{\max}^2). \tag{8.45}$$

Problems

Problem 8.1 Show that for an escaper with a typical escape velocity $v_s = 0.28$ and escaper mass $m_s = 0.2$ (figure 7.7), the maximum semi-latus rectum $p_{\max} \approx 4.7$ if the maximum semi-minor axis $b_{\max} = 7$. Use units where $M = G = a_0 = 1$.

Problem 8.2 Show that the probability for energy change Δ, for values $0 < \Delta \ll 1$, should include a factor of form $A(\Delta) \propto \sqrt{\Delta^{-1}}$, using the weighted average $\langle 1 + \Delta \rangle$ as $1 + \Delta$, and Eq. (7.32).

Problem 8.3 Suppose that the three bodies have equal masses, that $(E_s)_0 = 0.07|E_B|_0$, and that a capture has taken place. Let the relative energy change of the binary be $\Delta = 0.85(|E_0|/|E_B|_0) - 1$. What is the new binary binding energy $|E_B|$ and the new third-body energy E_s as a function of $|E_0|$? When $|E_s|$ is interpreted as the new orbital binding energy of the third body relative to the binary, what is the ratio of the semi-major axes of the third-body orbit and the new binary orbit, $(a_3)_{\text{new}}/a_{\text{new}}$? Consider also a more typical case of $m_s = 0.2$ and $m_a = m_b = 0.4$. How big is the jump in the z-axis?

Problem 8.4 Let us assume that an ejection orbit is in the lower part of the capture zone $[1, 1 + \mathrm{d}z]$, $z = 1 + 0.25/3$. Calculate the orbital period of this orbit in terms of crossing time T_{cr} and for equal masses.

Problem 8.5 Show that the smallest possible value of Δ in three-body scattering is $-v^2$.

Problem 8.6 Calculate $\langle\Delta\rangle$ for the distribution of Eq. (8.13) when $n = 11.5$ which corresponds to scattering with a large angular momentum. How does this compare with $\langle\Delta\rangle$ for low angular momentum ($n = 4.5$)? Hint:

$$\int \frac{x\,\mathrm{d}x}{(1+x)^{11.5}} = -\frac{1}{19}\left(\frac{4}{21} + \frac{46}{21}x + 2x^2\right)(1+x)^{-11.5}. \tag{8.46}$$

9

Perturbations in hierarchical systems

Complete analytical solutions are not available for systems with more than two bodies. However, it is possible to describe three-body orbits by approximate methods when the system is hierarchical, i.e. if there is a clearly defined binary and a third body which stays separate from the binary. These methods may be validated by comparison with numerical orbit integrations. Then we may take exact two-body orbits as a first approximation, and the effects of other bodies and other disturbances are taken into account as small forces which make the true trajectory deviate from this reference orbit.

When analysing perturbations we have to make some approximations that depend on the form of the perturbing force. Thus perturbation theory is a collection of various methods applicable in different situations rather than a single theory. In this chapter we will study a classical method that applies to the usual orbital elements. Another method will be discussed in the next chapter.

The problem which we consider by using this method is the long term evolution of a binary orbit when it is perturbed by a distant companion. This applies especially to triple stars and to the stability of planetary orbits around binary members.

9.1 Osculating elements

Consider the motion of a planet in a heliocentric xyz-frame. At the moment $t = t_0$ the planet is at (x_0, y_0, z_0). We can determine the reference ellipse E_0 by selecting the orbital elements in such a way that the ellipse goes through the point (x_0, y_0, z_0). In order to fix all six elements we also require that the velocity on the reference ellipse at (x_0, y_0, z_0) is the same as the actual velocity of the planet (Fig. 9.1). The elements of such an ellipse E_0 are called *osculating elements*. They

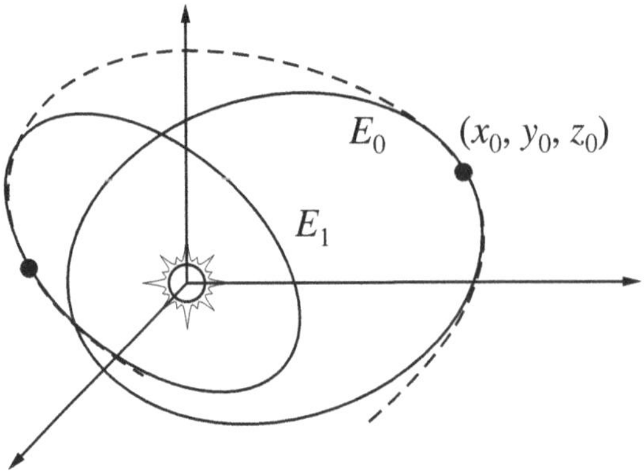

Figure 9.1 Even if the actual orbit is not elliptic, the motion at a given time can be described by the osculating ('kissing') ellipse that fits the true orbit as well as possible.

describe the orbit which the planet would follow if the perturbing force suddenly disappeared at t_0. At some later time the planet will not be in this orbit; then its motion can be described by some other ellipse E_1 with different osculating elements than E_0.

Due to perturbations the orbital elements are no longer constant but change with time. The same is true also for other integrals of the equations of motion. Our task is now to find expressions for the rate of change of the elements.

9.2 Lagrangian planetary equations

We begin by studying a set of equations that is very widely used. They can be derived from Newtonian mechanics with rather elementary methods, but the calculations are then prohibitively long. Using the more powerful Hamiltonian formalism the derivation becomes quite simple.

We include a small perturbing term R, usually called the *perturbing function*, in the Hamiltonian:

$$\mathcal{H} = -\frac{\mu^2}{2L^2} - R. \tag{9.1}$$

Then we find the time derivatives of Delaunay's elements. We assume that all elements appearing in their expressions can change with time:

$$
\begin{aligned}
\dot{L} &= \frac{\mathrm{d}}{\mathrm{d}t}\sqrt{\mu a} = \frac{1}{2}\sqrt{\frac{\mu}{a}}\dot{a}, \\
\dot{G} &= \frac{\mathrm{d}}{\mathrm{d}t}\sqrt{\mu a(1-e^2)} = \frac{1}{2}\sqrt{\frac{\mu(1-e^2)}{a}}\dot{a} - \frac{\sqrt{\mu a}e}{\sqrt{1-e^2}}\dot{e}, \\
\dot{H} &= \frac{\mathrm{d}}{\mathrm{d}t}\sqrt{\mu a(1-e^2)}\cos\iota \\
&= \frac{1}{2}\sqrt{\frac{\mu(1-e^2)}{a}}\cos\iota\,\dot{a} - \frac{\sqrt{\mu a}e}{\sqrt{1-e^2}}\cos\iota\,\dot{e} - \sqrt{\mu a(1-e^2)}\sin\iota\,\dot{\imath}, \\
\dot{l} &= \dot{M}, \\
\dot{g} &= \dot{\omega}, \\
\dot{h} &= \dot{\Omega}.
\end{aligned}
\tag{9.2}
$$

As before, we use the symbol ι (iota) for inclination , while $\dot{\imath}$ marks its time derivative. The values on the left hand side of these equations are obtained from the equations of motion:

$$
\begin{aligned}
\dot{L} &= -\frac{\partial \mathcal{H}}{\partial l} = \frac{\partial R}{\partial l}, \\
\dot{G} &= -\frac{\partial \mathcal{H}}{\partial g} = \frac{\partial R}{\partial g}, \\
\dot{H} &= -\frac{\partial \mathcal{H}}{\partial h} = \frac{\partial R}{\partial h}, \\
\dot{l} &= \frac{\partial \mathcal{H}}{\partial L} = \frac{\mu^2}{L^3} - \frac{\partial R}{\partial L}, \\
\dot{g} &= \frac{\partial \mathcal{H}}{\partial G} = -\frac{\partial R}{\partial G}, \\
\dot{h} &= \frac{\partial \mathcal{H}}{\partial \mathcal{H}} = -\frac{\partial R}{\partial H}.
\end{aligned}
\tag{9.3}
$$

The derivatives of the perturbing function must be expressed in terms of the ordinary elements. The variables l, g and h are trivial, since they are the same as M, ω and Ω. To find the three others we express a, e and ι as functions of Delaunay's elements:

$$
\begin{aligned}
a &= L^2/\mu, \\
e &= \sqrt{1 - G^2/L^2}, \\
\cos\iota &= H/G.
\end{aligned}
\tag{9.4}
$$

Using these we can evaluate the required derivatives. Finally, we express Delaunay's elements in terms of the ordinary elements:

$$
\begin{aligned}
\frac{\partial R}{\partial L} &= \frac{\partial R}{\partial a}\frac{\mathrm{d}a}{\mathrm{d}L} + \frac{\partial R}{\partial e}\frac{\mathrm{d}e}{\mathrm{d}L} = \frac{\partial R}{\partial a}\frac{2L}{\mu} + \frac{\partial R}{\partial e}\frac{G^2/L^3}{\sqrt{1-G^2/L^2}} \\
&= \frac{\partial R}{\partial a}2\sqrt{\frac{a}{\mu}} + \frac{\partial R}{\partial e}\frac{1-e^2}{e\sqrt{\mu a}}, \\
\frac{\partial R}{\partial G} &= \frac{\partial R}{\partial e}\frac{\mathrm{d}e}{\mathrm{d}G} + \frac{\partial R}{\partial \iota}\frac{\mathrm{d}\iota}{\mathrm{d}G} = \frac{\partial R}{\partial e}\frac{-G/L^2}{\sqrt{1-G^2/L^2}} + \frac{\partial R}{\partial \iota}\frac{H}{\sin\iota G^2} \\
&= \frac{\partial R}{\partial e}\left(-\frac{\sqrt{1-e^2}}{e\sqrt{\mu a}}\right) + \frac{\partial R}{\partial \iota}\frac{\cos\iota}{\sqrt{\mu a(1-e^2)}\sin\iota}, \\
\frac{\partial R}{\partial H} &= \frac{\partial R}{\partial \iota}\frac{\mathrm{d}\iota}{\mathrm{d}H} = \frac{\partial R}{\partial \iota}\frac{-1}{G\sin\iota} \\
&= \frac{\partial R}{\partial \iota}\frac{-1}{\sqrt{\mu a(1-e^2)}\sin\iota}.
\end{aligned}
\tag{9.5}
$$

Substituting these into Eq. (9.2) we get equations involving only ordinary elements:

$$
\begin{aligned}
&\frac{1}{2}\sqrt{\frac{\mu}{a}}\dot{a} = \frac{\partial R}{\partial M}, \\
&\frac{1}{2}\sqrt{\frac{\mu(1-e^2)}{a}}\dot{a} - e\sqrt{\frac{\mu a}{1-e^2}}\dot{e} = \frac{\partial R}{\partial \omega}, \\
&\frac{1}{2}\sqrt{\frac{\mu(1-e^2)}{a}}\cos\iota\,\dot{a} - e\sqrt{\frac{\mu a}{1-e^2}}\cos\iota\,\dot{e} - \sqrt{\mu a(1-e^2)}\sin\iota\,\dot{i} = \frac{\partial R}{\partial \Omega}, \\
&\dot{M} = \frac{\mu}{a\sqrt{\mu a}} - 2\sqrt{\frac{a}{\mu}}\frac{\partial R}{\partial a} - \frac{1-e^2}{e\sqrt{\mu a}}\frac{\partial R}{\partial e}, \\
&\dot{\omega} = \frac{\sqrt{1-e^2}}{e\sqrt{\mu a}}\frac{\partial R}{\partial e} - \frac{\cos\iota}{\sqrt{\mu a(1-e^2)}\sin\iota}\frac{\partial R}{\partial \iota}, \\
&\dot{\Omega} = \frac{1}{\sqrt{\mu a(1-e^2)}\sin\iota}\frac{\partial R}{\partial \iota}.
\end{aligned}
\tag{9.6}
$$

From the first three equations we get

$$\dot{a} = 2\sqrt{\frac{a}{\mu}}\frac{\partial R}{\partial M},$$
$$\dot{e} = -\sqrt{\frac{1-e^2}{\mu a}}\frac{1}{e}\frac{\partial R}{\partial \omega} + \frac{1-e^2}{2ea}\dot{a},$$
$$\dot{\iota} = -\frac{1}{\sqrt{\mu a(1-e^2)}\sin\iota}\frac{\partial R}{\partial \Omega} + \frac{\cos\iota}{2a\sin\iota}\dot{a} - \frac{e}{1-e^2}\frac{\cos\iota}{\sin\iota}\dot{e},$$

which are easily solved for $\dot{a}$, $\dot{e}$ and $\dot{\iota}$. We denote the mean motion by

$$n = \sqrt{\mu/a^3}, \tag{9.7}$$

whence

$$\sqrt{\mu a} = a^2 n, \quad \sqrt{\mu/a} = an.$$

The time derivatives of the orbital elements are now:

$$\begin{aligned}
\dot{a} &= \frac{2}{na}\frac{\partial R}{\partial M},\\
\dot{e} &= -\frac{\sqrt{1-e^2}}{na^2 e}\frac{\partial R}{\partial \omega} + \frac{1-e^2}{na^2 e}\frac{\partial R}{\partial M},\\
\dot{\iota} &= -\frac{1}{na^2\sqrt{1-e^2}\sin\iota}\frac{\partial R}{\partial \Omega} + \frac{\cos\iota}{na^2\sqrt{1-e^2}\sin\iota}\frac{\partial R}{\partial \omega},\\
\dot{M} &= n - \frac{2}{na}\frac{\partial R}{\partial a} - \frac{1-e^2}{na^2 e}\frac{\partial R}{\partial e},\\
\dot{\omega} &= \frac{\sqrt{1-e^2}}{na^2 e}\frac{\partial R}{\partial e} - \frac{\cos\iota}{na^2\sqrt{1-e^2}\sin\iota}\frac{\partial R}{\partial \iota},\\
\dot{\Omega} &= \frac{1}{na^2\sqrt{1-e^2}\sin\iota}\frac{\partial R}{\partial \iota}.
\end{aligned} \tag{9.8}$$

These are called the *Lagrangian planetary equations.*

These equations are exact. In a first order theory we assume that the elements vary only very slowly. Thus we can keep them constant when evaluating the expressions of the time derivatives. To get a second order theory, the elements on the right hand sides of the equations can then be expressed as functions of time. As can be guessed, that will lead to rather laborious calculations.

9.3 Three-body perturbing function

Let us now consider the perturbing function in the hierarchical three-body problem. The potential energy of the system is (Fig. 9.2)

$$V = -\frac{Gm_1m_2}{r} - \frac{Gm_1m_3}{r_{13}} - \frac{Gm_2m_3}{r_{23}} \tag{9.9}$$

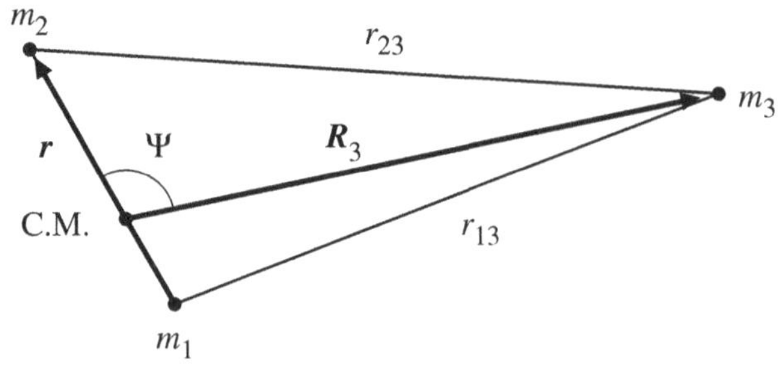

Figure 9.2 A hierarchical three-body system is composed of bodies of masses m_1, m_2 and m_3, with mutual separations r, r_{13} and r_{23}. The centre of mass (C.M.) of the inner binary and the position vector of the third body $\boldsymbol{R}_3$ are also marked.

where r_{23} and r_{13} are the lengths of the vectors

$$\begin{aligned}\boldsymbol{r}_{23} &= \boldsymbol{R}_3 - \frac{m_1}{m_1+m_2}\boldsymbol{r} \\ \boldsymbol{r}_{13} &= \boldsymbol{R}_3 + \frac{m_2}{m_1+m_2}\boldsymbol{r}.\end{aligned} \tag{9.10}$$

In order to evaluate the last two terms in (9.9), we use (3.94), putting

$$\begin{aligned}\boldsymbol{r} &\to \boldsymbol{R}_3 \\ \boldsymbol{r}' &\to \frac{m_1}{m_1+m_2}\boldsymbol{r} \quad \text{or} \quad \boldsymbol{r}' \to -\frac{m_2}{m_1+m_2}\boldsymbol{r} \\ t &= \frac{m_1}{m_1+m_2}\frac{r}{R_3} \quad \text{or} \quad t = -\frac{m_2}{m_1+m_2}\frac{r}{R_3}.\end{aligned} \tag{9.11}$$

Consequently, to the second order,

$$\begin{aligned}V &= -\frac{Gm_1m_2}{r} - \frac{Gm_1m_3}{R_3}\sum_{n=0}^{2}\left(-\frac{m_2}{m_1+m_2}\right)^n\left(\frac{r}{R_3}\right)^n P_n(\cos\psi) \\ &\quad - \frac{Gm_2m_3}{R_3}\sum_{n=0}^{2}\left(\frac{m_1}{m_1+m_2}\right)^n\left(\frac{r}{R_3}\right)^n P_n(\cos\psi) \\ &= -\frac{Gm_1m_2}{r} - \frac{Gm_3}{R_3}\Bigg[(m_1+m_2) \\ &\quad + \left(-\frac{m_1m_2}{m_1+m_2} + \frac{m_1m_2}{m_1+m_2}\right)\frac{r}{R_3}\cos\psi \\ &\quad + \left(\frac{m_1m_2^2}{(m_1+m_2)^2} + \frac{m_2m_1^2}{(m_1+m_2)^2}\right)\left(\frac{r}{R_3}\right)^2\frac{1}{2}(3\cos^2\psi - 1)\Bigg] \\ &= -\frac{Gm_1m_2}{r} - \frac{Gm_3(m_1+m_2)}{R_3} \\ &\quad - \frac{Gm_1m_2m_3}{2(m_1+m_2)R_3}\left(\frac{r}{R_3}\right)^2(3\cos^2\psi - 1).\end{aligned}$$

We recognise the first two terms as the potentials of the two two-body motions, the inner binary and the outer binary. The third term is the lowest order term in the perturbing function:

$$R = \frac{Gm_1m_2m_3}{2(m_1+m_2)R_3}\left(\frac{r}{R_3}\right)^2(3\cos^2\psi - 1). \tag{9.12}$$

9.4 Doubly orbit-averaged perturbing function

In this section we calculate the time average of R, first integrating over one complete cycle of the outer orbit, and then over one cycle of the inner orbit. For the first task, we have to integrate

$$\mathrm{d}V_e = \frac{3\cos^2\psi - 1}{R_3^3}\,\mathrm{d}M_e \tag{9.13}$$

over the mean anomaly M_e from 0 to 2π, and divide the result by 2π. The integration is simplest to carry out using the true anomaly ϕ_e as the variable.

For that purpose we need the transformation between dM and dϕ. As an intermediate variable we use E and the expressions (Eq. (3.37)):

$$\sin E = \sqrt{1-e^2}\,\frac{\sin\phi}{1+e\cos\phi},$$

$$\cos E = \frac{\cos\phi + e}{1+e\cos\phi},$$

and Kepler's equation (3.41)

$$M = E - e\sin E.$$

From Kepler's equation we get

$$\mathrm{d}M = (1 - e\cos E)\,\mathrm{d}E = \frac{r}{a}\,\mathrm{d}E$$

due to Eq. (3.49). Differentiation of the sin E equation above gives

$$\cos E\,\mathrm{d}E = \frac{\sqrt{1-e^2}(\cos\phi + e)}{(1+e\cos\phi)^2}\,\mathrm{d}\phi,$$

and substitution of the cos E equation leads to

$$\mathrm{d}E = \frac{\sqrt{1-e^2}}{1+e\cos\phi}\,\mathrm{d}\phi. \tag{9.14}$$

Since by Eq. (3.35)

$$\frac{r}{a} = \frac{1-e^2}{1+e\cos\phi},$$

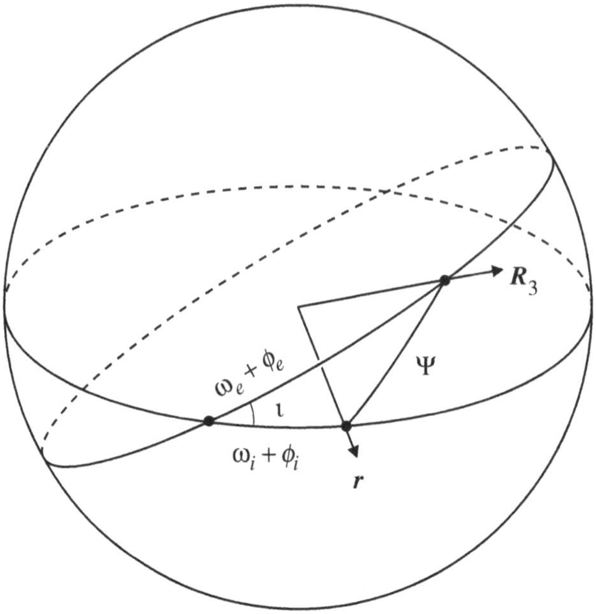

Figure 9.3 Orbital planes of the third body and the binary (horizontal plane) are at an angle ι relative to each other. The current position vectors $\boldsymbol{r}$ of the binary and $\boldsymbol{R}_3$ of the third body are at angles $\omega_i + \phi_i$ and $\omega_e + \phi_e$ from the line of nodes, respectively. The angle between $\boldsymbol{r}$ and $\boldsymbol{R}_3$ is ψ.

(9.14) can be expressed as

$$\mathrm{d}E = \frac{1}{\sqrt{1-e^2}}\left(\frac{r}{a}\right)\,\mathrm{d}\phi.$$

Then we finally get

$$\mathrm{d}M = \frac{1}{\sqrt{1-e^2}}\left(\frac{r}{a}\right)^2\,\mathrm{d}\phi. \tag{9.15}$$

Equipped with this information we write for the outer orbit

$$\begin{aligned} R_3 &= \frac{a_e(1-e_e^2)}{1+e_e\cos\phi_e}, \\ \mathrm{d}M_e &= \left(\frac{R_3}{a_e}\right)^2\frac{\mathrm{d}\phi_e}{\sqrt{1-e_e^2}}, \end{aligned} \tag{9.16}$$

and obtain

$$\mathrm{d}V_e = (3\cos^2\psi - 1)\frac{(1+e_e\cos\phi_e)}{a_e^3(1-e_e^2)^{3/2}}\,\mathrm{d}\phi_e. \tag{9.17}$$

We still have to evaluate $\cos\psi$. With the aid of Fig. 9.3 we write

$$\cos\psi = \cos(\omega_e+\phi_e)\cos(\omega_i+\phi_i) + \sin(\omega_e+\phi_e)\sin(\omega_i+\phi_i)\cos\iota. \tag{9.18}$$

The inclination ι is the relative inclination of the two orbits i.e. $\iota = \iota_i + \iota_e$ in the system based on the invariable plane. At present we are not interested in quantities depending on the variables of the inner binary and can simplify Eq. (9.18) by writing

$$\begin{aligned} C_1 &= \cos(\omega_i + \phi_i), \\ C_2 &= \sin(\omega_i + \phi_i). \end{aligned} \tag{9.19}$$

Then

$$\begin{aligned} \cos^2\psi = &C_1^2(\cos^2\omega_e\cos^2\phi_e + \sin^2\omega_e\sin^2\phi_e \\ &\quad - 2\sin\omega_e\cos\omega_e\sin\phi_e\cos\phi_e) \\ &+ C_2^2(\cos^2\omega_e\sin^2\phi_e + \sin^2\omega_e\cos^2\phi_e \\ &\quad + 2\sin\omega_e\cos\omega_e\sin\phi_e\cos\phi_e)\cos^2\iota \\ &+ 2C_1C_2(\cos\omega_e\sin\omega_e\cos^2\phi_e - \sin^2\omega_e\sin\phi_e\cos\phi_e \\ &\quad + \cos^2\omega_e\sin\phi_e\cos\phi_e - \sin\omega_e\cos\omega_e\sin^2\phi_e)\cos\iota. \end{aligned} \tag{9.20}$$

Now we can start our averaging term by term. Let us denote

$$\langle\cdots\rangle = \frac{1}{2\pi}\int_0^{2\pi}\cdots\,\mathrm{d}\phi_e. \tag{9.21}$$

Then we find easily the following results:

$$\begin{aligned} \langle\cos\phi_e\rangle &= 0, \\ \left\langle\cos^2\phi_e\right\rangle &= \frac{1}{2}, \\ \left\langle\sin^2\phi_e\right\rangle &= \frac{1}{2}, \\ \langle\sin\phi_e\cos\phi_e\rangle &= 0, \\ e_e\left\langle\cos^3\phi_e\right\rangle &= 0, \\ e_e\left\langle\cos\phi_e\sin^2\phi_e\right\rangle &= 0, \\ e_e\left\langle\sin\phi_e\cos^2\phi_e\right\rangle &= 0. \end{aligned} \tag{9.22}$$

After applying these to Eqs. (9.17) and (9.20) we finally get

$$\langle V_e\rangle = \frac{1}{2a_e^3(1-e_e^2)^{3/2}}(3C_1^2 + 3C_2^2\cos^2\iota - 2). \tag{9.23}$$

Since we are now ready to start the second phase of the orbit averaging, we put back the terms containing the variables of the inner binary:

$$\begin{aligned} \langle R\rangle = &\frac{Gm_1m_2m_3}{2(m_1+m_2)}\frac{r^2}{2a_e^3(1-e_e^2)^{3/2}} \\ &\times\left\{3\left[\cos^2(\omega_i+\phi_i) + \sin^2(\omega_i+\phi_i)\cos^2\iota\right] - 2\right\}. \end{aligned} \tag{9.24}$$

The quantity which we need to integrate is

$$dV_i = \left(\frac{r}{a_i}\right)^2 \left(3\left[\cos^2(\omega_i + \phi_i) + \sin^2(\omega_i + \phi_i)\cos^2 \iota\right] - 2\right) dM_i. \tag{9.25}$$

In this case it is simplest to use the eccentric anomaly E for the inner orbit and the following transformations (Eqs. (3.36), (3.49))

$$\begin{aligned}
\sin\phi_i &= \sqrt{1-e_i^2}\frac{\sin E}{1-e_i\cos E},\\
\cos\phi_i &= \frac{\cos E - e_i}{1-e_i\cos E},\\
\frac{r}{a_i} &= 1 - e_i\cos E,
\end{aligned}$$

which lead to

$$\begin{aligned}
\frac{r}{a_i}\sin\phi_i &= \sqrt{1-e_i^2}\sin E,\\
\frac{r}{a_i}\cos\phi_i &= \cos E - e_i.
\end{aligned} \tag{9.26}$$

From (3.41) we get

$$dM_i = (1 - e_i\cos E)\, dE.$$

Some of the terms which we need to average are

$$\begin{aligned}
&\left(\frac{r}{a_i}\right)^2 \cos^2(\omega_i+\phi_i) =\\
&\quad \cos^2\omega_i(\cos E - e_i)^2 + \sin^2\omega_i(1-e_i^2)\sin^2 E\\
&\quad - 2\cos\omega_i\sin\omega_i(\cos E - e_i)\sqrt{1-e_i^2}\sin E,\\
&\left(\frac{r}{a_i}\right)^2 \sin^2(\omega_i+\phi_i) =\\
&\quad \sin^2\omega_i(\cos E - e_i)^2 + \cos^2\omega_i(1-e_i^2)\sin^2 E\\
&\quad + 2\cos\omega_i\sin\omega_i(\cos E - e_i)\sqrt{1-e_i^2}\sin E,\\
&- e_i\cos E\left(\frac{r}{a_i}\right)^2 \cos^2(\omega_i+\phi_i) =\\
&\quad - e_i\cos^2\omega_i\cos E(\cos E - e_i)^2 - e_i\sin^2\omega_i(1-e_i^2)\cos E\sin^2 E\\
&\quad + 2e_i\cos\omega_i\sin\omega_i(\cos E - e_i)\sqrt{1-e_i^2}\sin E\cos E.\\
&- e_i\cos E\left(\frac{r}{a_i}\right)^2 \sin^2(\omega_i+\phi_i) =\\
&\quad - e_i\sin^2\omega_i\cos E(\cos E - e_i)^2 - e_i\cos^2\omega_i(1-e_i^2)\cos E\sin^2 E\\
&\quad - 2e_i\cos\omega_i\sin\omega_i(\cos E - e_i)\sqrt{1-e_i^2}\sin E\cos E.
\end{aligned} \tag{9.27}$$

Therefore we need the following averages

$$\begin{aligned}
\left\langle(\cos E - e_i)^2\right\rangle &= \frac{1}{2} + e_i^2, \\
\left\langle\sin^2 E\right\rangle &= \frac{1}{2}, \\
\langle(\cos E - e_i)\sin E\rangle &= 0, \\
-e_i\left\langle\cos E(\cos E - e_i)^2\right\rangle &= e_i^2, \\
-e_i\left\langle\sin^2 E\cos E\right\rangle &= 0, \\
-e_i\langle(\cos E - e_i)\sin E\cos E\rangle &= 0, \\
\left\langle(1 - e_i\cos E)^3\right\rangle &= 1 + \frac{3}{2}e_i^2.
\end{aligned} \tag{9.28}$$

After using these averages together with Eq. (9.25) we get

$$\begin{aligned}
\langle V_i\rangle = {} & 3\cos^2\omega_i\left(\frac{1}{2} + e_i^2\right) + \frac{3}{2}\sin^2\omega_i(1 - e_i^2) + 3\cos^2\omega_i e_i^2 + 3\sin^2\omega_i e_i^2 \\
& -3\sin^2\omega_i e_i^2\sin^2\iota - 2 - 3e_i^2 + 3\sin^2\omega_i\left(\frac{1}{2} + e_i^2\right) \\
& +\frac{3}{2}\cos^2\omega_i(1 - e_i^2) + \left[-3\sin^2\omega_i\left(\frac{1}{2} + e_i^2\right) - \frac{3}{2}\cos^2\omega_i(1 - e_i^2)\right]\sin^2\iota
\end{aligned} \tag{9.29}$$

(Problem 9.2). Putting all factors together we get the final result for the doubly orbit averaged perturbing function:

$$\langle\langle R\rangle\rangle = \frac{Gm_1m_2m_3a_i^2}{8m_Ba_e^3(1 - e_e^2)^{3/2}}\left[2 + 3e_i^2 - 3\sin^2\iota\left(5e_i^2\sin^2\omega_i + 1 - e_i^2\right)\right]. \tag{9.30}$$

9.5 Motions in the hierarchical three-body problem

We first note that the perturbing function $\langle\langle R\rangle\rangle$ does not depend on the Delaunay elements l_i and l_e since we have averaged our Hamiltonian over these elements. As we have learnt before in Chapter 4, the corresponding canonical momenta L_i and L_e are constant which means that the semi-major axes a_i and a_e remain constant during the orbital motion. In addition, we notice that $\langle\langle R\rangle\rangle$ does not depend on ω_e either which means that the corresponding canonical momentum G_e is a constant of motion. Since $1 - e_e^2 = G_e^2/L_e^2$, we also have the outer orbit eccentricity e_e as a constant of motion. The outer orbit does not vary its size nor its shape in the first approximation. Also, Eq. (4.123) tells us further that the outer inclination i_e remains constant. Without loss of generality, we choose $i_e = 0$, which means that in Eq. (9.30) $\iota = \iota_i + \iota_e = \iota_i$. Then Eq. (4.123) with a constant a_i tells us that

$$\sqrt{1 - e_i^2}\,\cos\iota = \text{constant}. \tag{9.31}$$

Let us then get into the business and start by calculating the partial derivatives of $\langle\langle R\rangle\rangle$ which are needed in the Lagrangian planetary equations.

$$
\begin{aligned}
\frac{\partial\langle\langle R\rangle\rangle}{\partial M_i} &= \frac{\partial\langle\langle R\rangle\rangle}{\partial\Omega_i} = 0,\\
\frac{\partial\langle\langle R\rangle\rangle}{\partial\omega_i} &= -\frac{15}{8}e_i^2\sin 2\omega_i\sin^2\iota\frac{Gm_1m_2m_3a_i^2}{m_Ba_e^3(1-e_e^2)^{3/2}},\\
\frac{\partial\langle\langle R\rangle\rangle}{\partial e_i} &= \frac{3}{4}e_i\left(1+\sin^2\iota-5\sin^2\iota\sin^2\omega_i\right)\frac{Gm_1m_2m_3a_i^2}{m_Ba_e^3(1-e_e^2)^{3/2}},\\
\frac{\partial\langle\langle R\rangle\rangle}{\partial\iota} &= -\frac{3}{4}\sin\iota\cos\iota\left(1-e_i^2+5e_i^2\sin^2\omega_i\right)\frac{Gm_1m_2m_3a_i^2}{m_Ba_e^3(1-e_e^2)^{3/2}}.
\end{aligned}
\tag{9.32}
$$

There is a common factor in all of the non-zero derivatives. Let us call this factor

$$
A = \frac{Gm_1m_2m_3}{m_Ba_e^3(1-e_e^2)^{3/2}}. \tag{9.33}
$$

Outside this factor all variables refer to the inner orbit. Therefore we drop the subscripts i. Then the Lagrangian planetary equations become

$$
\begin{aligned}
\dot{\iota} &= -\frac{15}{8}\frac{e^2}{\sqrt{1-e^2}}\sin 2\omega\sin\iota\cos\iota\,\frac{A}{n},\\
\dot{e} &= \frac{15}{8}e\sqrt{1-e^2}\sin 2\omega\sin^2\iota\,\frac{A}{n},\\
\dot{\omega} &= \frac{3}{4}\frac{1}{\sqrt{1-e^2}}\left[2(1-e^2)+5\sin^2\omega(e^2-\sin^2\iota)\right]\frac{A}{n},\\
\dot{\Omega} &= -\frac{\cos\iota}{4\sqrt{1-e^2}}(3+12e^2-15e^2\cos^2\omega)\frac{A}{n}.
\end{aligned}
\tag{9.34}
$$

At this point it is useful to introduce a new normalised time unit τ:

$$
\tau = \frac{A}{n}t = \frac{Gm_1m_2m_3}{m_Bnb_e^3}t. \tag{9.35}
$$

Here b_e is the semi-minor axis of the outer orbit and n the mean motion of the inner binary. Then the Lagrangian equations are simplified to:

$$\begin{aligned}
\frac{d\iota}{d\tau} &= -\frac{15}{8}\frac{e^2}{\sqrt{1-e^2}}\sin 2\omega \sin\iota\cos\iota, \\
\frac{de}{d\tau} &= \frac{15}{8}e\sqrt{1-e^2}\sin 2\omega \sin^2\iota, \\
\frac{d\omega}{d\tau} &= \frac{3}{4}\frac{1}{\sqrt{1-e^2}}\left[2(1-e^2)+5\sin^2\omega(e^2-\sin^2\iota)\right], \\
\frac{d\Omega}{d\tau} &= -\frac{\cos\iota}{4\sqrt{1-e^2}}(3+12e^2-15e^2\cos^2\omega)).
\end{aligned} \tag{9.36}$$

From here on we neglect the factor $m_1 m_2/m_B$ in the definition of τ which is of the order of unity for nearly equal masses. If one of the bodies, say m_2, is very small, we should start the calculation from a Hamiltonian per unit mass. In this case the mass of the small body does not appear in Eq. (9.35), and the ratio $m_1/m_B = 1$.

It is possible to solve these equations exactly in closed form but the derivation is rather lengthy (Kozai 1962, 2004, Sidlichovsky 1983, Marchal 1990, Kinoshita and Nakai 1999). Heggie and Hut (2003) give a simple description of the solutions. Here we will follow Innanen *et al.* (1997) and consider the special case of the small inner eccentricity e, so small that terms containing e^2 can be neglected. Then the equations are:

$$\begin{aligned}
\frac{d\iota}{d\tau} &= 0, \\
\frac{de}{d\tau} &= \frac{15}{8}e\sin 2\omega \sin^2\iota, \\
\frac{d\omega}{d\tau} &= \frac{3}{4}(2-5\sin^2\omega\sin^2\iota), \\
\frac{d\Omega}{d\tau} &= -\frac{3}{4}\cos\iota,
\end{aligned} \tag{9.37}$$

and in addition,

$$\sqrt{1-e^2}\cos\iota = \text{constant}. \tag{9.38}$$

The first equation (9.37) tells us that ι is constant in this approximation, and therefore the quantity

$$A = 5\sin^2\iota - 2$$

is also constant. The third equation (9.37) is now easily integrated using the known expressions (Petit Bois 1961)

$$\begin{aligned}\int \frac{\mathrm{d}x}{a + b\sin^2 x} &= \frac{1}{\sqrt{a(a+b)}} \arctan \frac{(a+b)\tan x}{\sqrt{a(a+b)}}, \qquad \text{if } a(a+b) > 0, \\ &= \frac{1}{2\sqrt{-a(a+b)}} \log\left[\frac{(a+b)\tan x - \sqrt{-a(a+b)}}{(a+b)\tan x + \sqrt{-a(a+b)}}\right], \qquad \text{if } a(a+b) < 0.\end{aligned} \tag{9.39}$$

For inclinations greater than about 39.23° ($\sin^2 \iota > 0.4$), the solution is

$$\omega = \arctan\left[\sqrt{\frac{2}{A}} \frac{e^{\frac{3}{2}\sqrt{2A}\tau} + 1}{e^{\frac{3}{2}\sqrt{2A}\tau} - 1}\right], \tag{9.40}$$

if $\omega = \pi/2$ when $\tau = 0$ while for inclinations smaller than this limit

$$\omega = \arctan\left[\sqrt{\frac{2}{-A}} \tan\left(\frac{3}{4}\sqrt{-2A\tau}\right)\right] \tag{9.41}$$

if $\omega = 0$ when $\tau = 0$ (Problem 9.3). The former solution approaches a constant value when $\tau \to \infty$, while the latter remains periodic. Therefore, in the former solution we may put $\mathrm{d}\omega/\mathrm{d}\tau = 0$ in the third Eq. (9.37) and find

$$5 \sin^2 \omega \sin^2 \iota = 2. \tag{9.42}$$

Substitution of this into the second Eq. (9.37) gives the first order equation for the evolution of eccentricity:

$$\frac{\mathrm{d}e}{\mathrm{d}\tau} = \frac{15}{4} e \sqrt{\frac{2}{5}\left(\sin^2 \iota - \frac{2}{5}\right)}. \tag{9.43}$$

Depending on the quadrant of ω, $\mathrm{d}e/\mathrm{d}\tau$ could also have a negative sign, but here we consider only the positive case. This is easily integrated:

$$\tau = 0.42 \log(e/e_0)/\sqrt{\sin^2 \iota - 0.4} \tag{9.44}$$

if the eccentricity $e = e_0$ when $\tau = 0$. This is valid when ω is at its asymptotic value. Equation (9.43) implies a rapid growth of the eccentricity. Numerical computations by Innanen *et al.* (1997) show that if we put $e = 1$ we get a fairly good estimate (about 2/3 of the correct value) of the time required to reach the maximum eccentricity when one starts with a very small eccentricity. Our approximation is not valid any more when e has increased considerably, and the condition ι = constant (first Eq. (9.37)) does not hold. However we may estimate what happens then by using Eq. (9.38). It tells us that when e increases, $\cos \iota$ must increase so as to keep $\sqrt{1-e^2}\cos\iota$ constant. However, since $\sin\iota$ must be greater than $2/5$, $\cos\iota$ cannot exceed $\sqrt{3/5}$, and this imposes a maximum value for $e = e_{\max}$. We may write

$$\sqrt{1-e_0^2}\cos\iota_0 = \sqrt{1-e_{\max}^2}\sqrt{\frac{3}{5}} \qquad (9.45)$$

where ι_0 is the initial value of the inclination and e_0 is the initial (small) value of the eccentricity. In accordance with our assumption $1-e_0^2$ must be practically equal to 1. Then

$$e_{\max} = \sqrt{1-\frac{5}{3}\cos^2\iota_0}. \qquad (9.46)$$

Let us take, for example, $e_0 = 0.05$ and $\iota_0 = 60°$. Then $e_{\max} = 0.76$ (Eq. (9.45)). If we replace e by $e_{\max}$ in Eq. (9.44) and ι by ι_0 in the approximation where the inclination is constant, we get a rough estimate for the time it takes the eccentricity to reach $e_{\max}$, starting from a small initial value: $\tau = 1.93$. When e_e is small and $m_1m_2/m_B = 1$, Eq. (9.35) gives

$$t = 1.93\frac{na_e^3}{Gm_3} \approx 0.3\left(\frac{a_e}{a}\right)^3\frac{m_B}{m_3}P \qquad (9.47)$$

where $P = 2\pi\sqrt{a^3/Gm_B}$ is the period of the inner orbit. For a Jupiter-like planet, perturbed by a companion star of $0.4M_\odot$ at $a_e = 1000$ AU from the Sun, Eq. (9.47) gives $t = 6.3\times10^7$ years. Figure 9.4 shows that this underestimates the time of reaching $e_{\max}$ a little which is not surprising since we have solved the equations of motion in a special case of small eccentricities. In general, we may estimate the period of the full cycle to be

$$P_{\text{Kozai}} \approx \left(\frac{a_e}{a}\right)^3\frac{m_B}{m_3}P. \qquad (9.48)$$

The full solution of the equations of motion shows that after obtaining the maximum value $e_{\max}$ the eccentricity decreases to a minimum, and then another eccentricity cycle starts again. This is called a *Kozai cycle* (Kozai 1962). During the cycle the inclination varies so that Eq. (9.31) remains valid. At the same time ω may swing back and forth ('librate') about $\pi/2$ to the extreme positions given by the condition $d\omega/d\tau = 0$, or it may rotate around continuously ('circulate'). Equation (9.46) is valid independent of the initial e_0 for the libration solutions which are found only if $\cos^2\iota < 3/5$, as we deduced above (Kinoshita and Nakai 1999).

If $\cos^2\iota > 3/5$, i.e. $\iota < 39.23°$ or $\iota > 140.73°$, we always have a circulation case. As an example, let us put $2A = -1$ in Eq. (9.41), i.e. $\cos^2\iota = 0.7$, $\iota = 33.2°$. Then for $\tau \lesssim 1$ we may solve Eq. (9.41) to get approximately $\omega \approx 3/2\tau$, and in

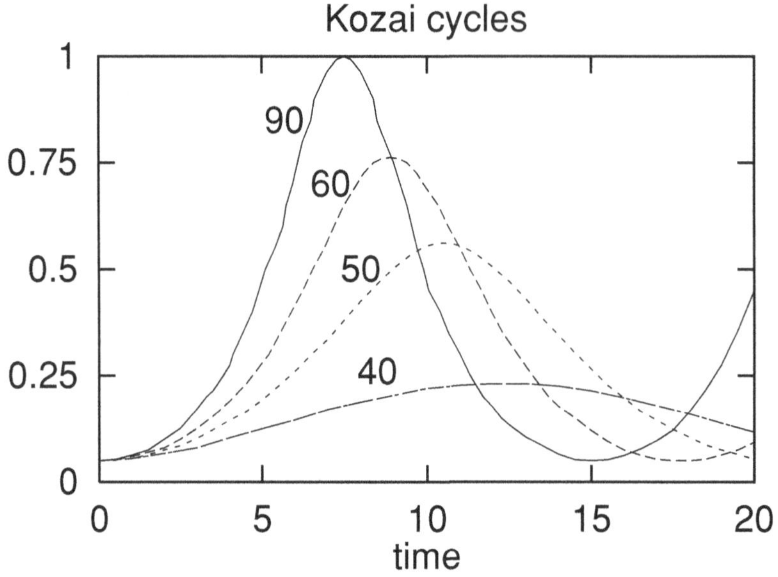

Figure 9.4 The evolution of eccentricity e of the inner orbit as a function of time. The masses of the bodies in the inner orbit are $0.001 M_\odot$ and $1 M_\odot$ and the third-body mass is $0.4 M_\odot$. The initial eccentricity of the inner binary is $e_0 = 0.05$ and its semi-major axis is 5.2 AU. The third body is at the distance of 1000 AU from the binary. Various values of the orbital inclination $\iota = 40°$, $50°$, $60°$ and $90°$ (labelled by the curves) have been studied. The time unit is 10^7 yr. In the first order theory, we expect the binary to reach its maximum eccentricity $e_{\max}$ at $t \approx 10^8$ yr.

this approximation the second equation (9.37) becomes

$$\frac{de}{e} = \frac{3}{16} \sin 2\omega \, d2\omega. \tag{9.49}$$

The solution is

$$e = e_0 \exp\left[-\frac{3}{16} \cos 2\omega\right] \tag{9.50}$$

where e_0 is the value of eccentricity when $\omega = \omega_0 = \pi/4$, say, and e varies sinusoidally on either side of it between $0.83 e_0$ and $1.21 e_0$. In general, the solutions for $\iota < 39.23°$ resemble this in that they possess the periodic term in 2ω, and the amplitude of the variation of e is diminished when $\sin^2 \iota$ becomes small (see Fig. 9.5).

The period of the ω cycle for $\iota > 39.23°$ is obtained approximately from Eq. (9.44) by putting $e = 1$ and by multiplying τ by six. The factor six is required since the typical eccentricity evolution time from a small value to its peak value is about $(3/2)\tau$, and the eccentricity period is twice as long, i.e. $\approx 3\tau$. The ω cycle is twice

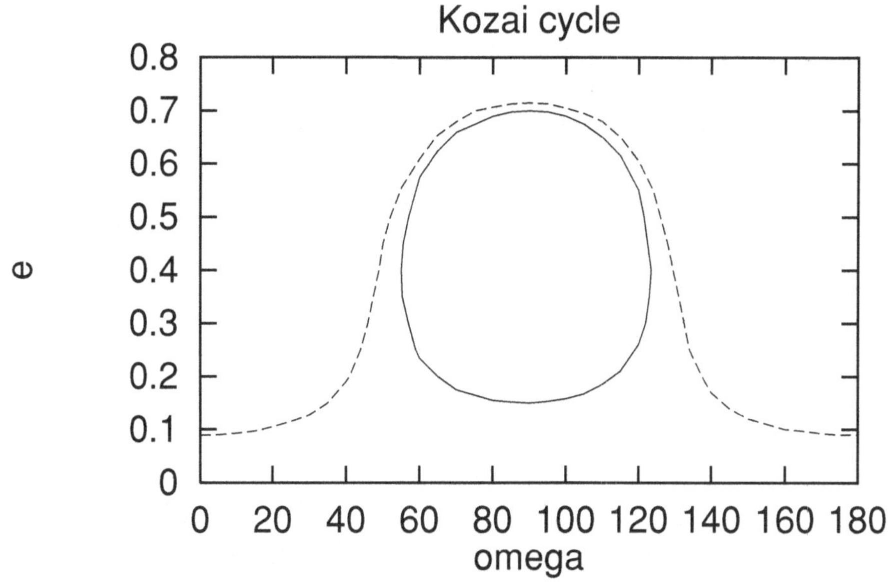

Figure 9.5 The evolution of the eccentricity e as a function of ω of the inner binary. An example of both the libration (solid line) and the circulation (dashed line) orbit is shown. The constant on the right hand side of Eq. (9.38) is 0.55 (Kinoshita and Nakai 1999).

the length of the eccentricity cycle. Thus

$$P_\omega \approx 2.5 \log(1/e_0)/\sqrt{\sin^2 \iota_0 - 0.4}. \tag{9.51}$$

Some numerical examples are displayed in Fig. 9.6 (Kinoshita and Nakai 1999).

Example 9.1 What if the Sun is in reality a binary star? Are the orbits of planets stable in that case? We probably do not have a companion; it would already have been detected even if it were of low luminosity. But this question is relevant to other stellar systems which quite often are binaries.

The Kozai mechanism suggests that as soon as the inclination of the companion is greater than about 40°, oscillations in orbital eccentricities of planets should take place and they should destroy the stability of the planetary system. Qualitatively this should be true for any mass value m_3 of the companion star; only the time of the destruction is pushed further into the future when the companion mass is decreased. Also the influence of the companion should work at different rates on different planets and destroy the planar structure of the system.

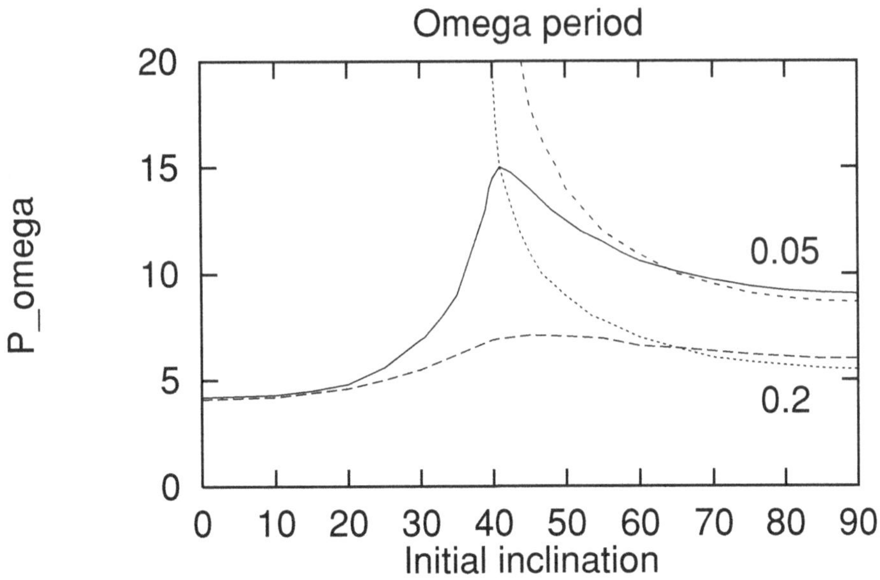

Figure 9.6 The period of the ω cycle expressed in the normalised time units (Eq. (9.35)) as a function of the initial inclination ι_0. The results are shown for $e_0 = 0.05$ and 0.2 both for the circulation (solid and long-dashed lines) and the libration case (dotted and short-dashed lines).

Numerical simulations of the Solar System evolution with companions of different masses and in orbits of different inclinations (Innanen *et al.* 1997) have shown that this is only partly true. It was found that a companion at the distance of 400 AU did not destabilise the Solar System if its inclination was 45° or less. This was independent of the companion mass which was varied between $0.05M_\odot$ and $0.4M_\odot$. To some extent this agrees with what we expect from the Kozai cycles. But what is clearly different from the Kozai theory is the fact that the lowest mass companion ($0.05M_\odot$) cannot destabilise the system at any inclination. Also the Solar System remains 'dynamically rigid', the planetary orbits do not evolve at different rates, but rather the whole system remains planar even though the Solar System plane precesses as a whole. These interesting features can be assigned to the mutual gravitational attraction between the planets since they disappear if the planet masses are taken to be close to zero.

Therefore we cannot rule out on dynamical grounds the possibility that the Sun has a small companion star. More importantly, binary stars can have stable planetary systems like the Solar System as long as the companion is far enough from the primary star, and is small enough in mass. The mutual gravity between the planets gives the system rigidity which protects it from the destructive Kozai cycles.

Problems

Problem 9.1 Consider a hypothetical perturbation, the perturbing function of which is

$$R = \alpha e^2 \cos M + \beta e^2 \sin \iota \cos(M + \Omega).$$

How does this affect the orbital elements? Which changes are secular and which only periodic?

Problem 9.2 Calculate the averages over a complete cycle of ϕ_e given in Eq. (9.22) and the averages over a complete cycle of E given in Eq. (9.28). Then verify Eq. (9.29).

Problem 9.3 Show that the solutions of Eq. (9.37) are given by Eq. (9.40) if $\iota > 39.23^\circ$ and by Eq. (9.41) if $\iota < 39.23^\circ$.

10

Perturbations in strong three-body encounters

The theory of the previous section is applicable to gentle perturbations by a distant companion of a binary. Strong perturbations are often more important in astrophysics where a major change in the binary orbit is achieved in a few orbital periods. Necessarily the techniques have to be different and also the results are more approximate. But we have the advantage that numerical orbit calculations do not require much time and it is possible to refine the analytical theory to suit the numerical results. In this chapter we also study triple stars and the question of stable orbits in such systems.

10.1 Perturbations of the integrals *k* and *e*

The Lagrangian equations used in the previous chapter are the traditional approach to perturbations. We will now study another method for finding the effect of perturbations on the vector elements $\boldsymbol{k}$ and $\hat{\boldsymbol{e}}$. This method can be applied to a great variety of perturbations.

We begin with the Newtonian equation of motion in the form

$$\ddot{\boldsymbol{r}} = -\mu \boldsymbol{r}/r^3 + \boldsymbol{f}, \tag{10.1}$$

where $\boldsymbol{f}$ is a (small) extra term due to perturbations. It can be any vector valued quantity with a dimension of acceleration, and it may be a function of the position and velocity.

First we repeat the definitions of $\boldsymbol{k}$ and $\boldsymbol{e}$:

$$\begin{aligned} \boldsymbol{k} &= \boldsymbol{r} \times \dot{\boldsymbol{r}}, \\ -\mu \boldsymbol{e} &= \boldsymbol{k} \times \dot{\boldsymbol{r}} + \mu \boldsymbol{r}/r. \end{aligned} \tag{10.2}$$

The time derivatives of these are

$$\begin{aligned}\dot{\boldsymbol{k}} &= \dot{\boldsymbol{r}} \times \dot{\boldsymbol{r}} + \boldsymbol{r} \times \ddot{\boldsymbol{r}} = \boldsymbol{r} \times (-\mu \boldsymbol{r}/r^3) + \boldsymbol{r} \times \boldsymbol{f} \\ &= \boldsymbol{r} \times \boldsymbol{f},\end{aligned} \tag{10.3}$$

$$\begin{aligned}-\mu\dot{\boldsymbol{e}} &= \dot{\boldsymbol{k}} \times \dot{\boldsymbol{r}} + \boldsymbol{k} \times \ddot{\boldsymbol{r}} + \mu\frac{\dot{\boldsymbol{r}}}{r} - \mu\frac{\boldsymbol{r}\dot{r}}{r^2} \\ &= \dot{\boldsymbol{k}} \times \dot{\boldsymbol{r}} + (\boldsymbol{r} \times \dot{\boldsymbol{r}}) \times \left(-\mu\frac{\boldsymbol{r}}{r^3}\right) + \boldsymbol{k} \times \boldsymbol{f} + \mu\frac{\dot{\boldsymbol{r}}}{r} - \mu\frac{\boldsymbol{r}\dot{r}}{r^2} \\ &= \dot{\boldsymbol{k}} \times \dot{\boldsymbol{r}} + \boldsymbol{k} \times \boldsymbol{f} + \mu\left(\boldsymbol{r}\frac{\dot{\boldsymbol{r}} \cdot \boldsymbol{r}}{r^3} - \dot{\boldsymbol{r}}\frac{\boldsymbol{r} \cdot \boldsymbol{r}}{r^3} + \frac{\dot{\boldsymbol{r}}}{r} - \frac{\boldsymbol{r}\dot{r}}{r^2}\right) \\ &= \dot{\boldsymbol{k}} \times \dot{\boldsymbol{r}} + \boldsymbol{k} \times \boldsymbol{f}.\end{aligned} \tag{10.4}$$

We know that $\boldsymbol{k}$ is perpendicular to the orbital plane and its length depends on the parameter p of the orbit. Therefore it can be expressed in terms of p and a unit vector $\hat{\boldsymbol{e}}_\zeta$ perpendicular to the orbital plane:

$$\boldsymbol{k} = \sqrt{p\mu}\hat{\boldsymbol{e}}_\zeta. \tag{10.5}$$

Similarly, the length of $\boldsymbol{e}$ is the eccentricity of the orbit, and its direction is the direction of the perihelion. Hence

$$\boldsymbol{e} = e\hat{\boldsymbol{e}}_\xi. \tag{10.6}$$

Using these we find the following expressions for the derivatives of $\boldsymbol{k}$ and $\boldsymbol{e}$:

$$\dot{\boldsymbol{k}} = \tfrac{1}{2}\sqrt{\mu/p}\,\dot{p}\hat{\boldsymbol{e}}_\zeta + \sqrt{p\mu}\,\dot{\hat{\boldsymbol{e}}}_\zeta, \tag{10.7}$$

$$\dot{\boldsymbol{e}} = \dot{e}\hat{\boldsymbol{e}}_\xi + e\dot{\hat{\boldsymbol{e}}}_\xi. \tag{10.8}$$

Next we try to find the change in the eccentricity. Since $\hat{\boldsymbol{e}}_\xi \cdot \dot{\hat{\boldsymbol{e}}}_\xi = 0$, the scalar product of Eq. (10.8) with $\hat{\boldsymbol{e}}_\xi$ gives

$$\hat{\boldsymbol{e}}_\xi \cdot \dot{\boldsymbol{e}} = \dot{e}, \tag{10.9}$$

from which

$$\begin{aligned}\dot{e} = \hat{\boldsymbol{e}}_\xi \cdot \dot{\boldsymbol{e}} &= -\frac{1}{\mu}(\hat{\boldsymbol{e}}_\xi \cdot \dot{\boldsymbol{k}} \times \dot{\boldsymbol{r}} + \hat{\boldsymbol{e}}_\xi \cdot \boldsymbol{k} \times \boldsymbol{f}) \\ &= -\frac{1}{\mu}(\hat{\boldsymbol{e}}_\xi \cdot (\boldsymbol{r} \times \boldsymbol{f}) \times \dot{\boldsymbol{r}} + \hat{\boldsymbol{e}}_\xi \cdot \boldsymbol{k} \times \boldsymbol{f}) \\ &= -\frac{1}{\mu}(\hat{\boldsymbol{e}}_\xi \cdot (\boldsymbol{r} \times \boldsymbol{f}) \times \dot{\boldsymbol{r}} + \hat{\boldsymbol{e}}_\xi \cdot (\sqrt{p\mu}\hat{\boldsymbol{e}}_\zeta \times \boldsymbol{f})) \\ &= \frac{1}{\mu}(\hat{\boldsymbol{e}}_\xi \cdot \dot{\boldsymbol{r}} \times (\boldsymbol{r} \times \boldsymbol{f}) + \sqrt{p\mu}\hat{\boldsymbol{e}}_\eta \cdot \boldsymbol{f}).\end{aligned} \tag{10.10}$$

The change in the semi-latus rectum of the orbit is found by taking the scalar product of Eq. (10.7) and $\hat{\boldsymbol{e}}_\zeta$:

$$\hat{\boldsymbol{e}}_\zeta \cdot \dot{\boldsymbol{k}} = \tfrac{1}{2}\sqrt{\mu/p}\,\dot{p},$$

from which

$$\dot{p} = 2\sqrt{p/\mu}\,\hat{\boldsymbol{e}}_\zeta \cdot (\boldsymbol{r} \times \boldsymbol{f}). \tag{10.11}$$

The effect on the semi-major axis is found from the equation $a = p/(1 - e^2)$, when we know the changes in p and e:

$$\dot{a} = \frac{\dot{p}}{1 - e^2} + \frac{2pe\dot{e}}{(1 - e^2)^2}. \tag{10.12}$$

We may also derive $\dot{a}$ directly as follows. Define the concept of perturbation derivative $\mathrm{d_P}/\mathrm{d}t$. For any quantity Q, the derivative $\mathrm{d}Q/\mathrm{d}t$ can be divided in two parts, the Keplerian part $\mathrm{d_K}Q/\mathrm{d}t$ and the perturbative part $\mathrm{d_P}/\mathrm{d}t$:

$$\frac{\mathrm{d}Q}{\mathrm{d}t} = \frac{\mathrm{d_K}Q}{\mathrm{d}t} + \frac{d_\mathrm{P}Q}{\mathrm{d}t}. \tag{10.13}$$

The Keplerian part gives the change of the quantity along the (unperturbed) Keplerian orbit, while the latter part arises from the perturbing acceleration $\boldsymbol{f}$. By definition, an orbital element (say, a) does not change in a Keplerian orbit, and thus

$$\dot{a} = \frac{\mathrm{d_P}a}{\mathrm{d}t}. \tag{10.14}$$

The position vector $\boldsymbol{r}$ is the same for the Keplerian orbit and for the osculating orbit (Figure 9.1); thus

$$\frac{\mathrm{d_P}\boldsymbol{r}}{\mathrm{d}t} = 0, \quad \frac{\mathrm{d_P}r}{\mathrm{d}t} = 0 \tag{10.15}$$

while

$$\frac{\mathrm{d_P}\boldsymbol{v}}{\mathrm{d}t} = \boldsymbol{f}. \tag{10.16}$$

Let us now apply the perturbative differentiation to Eq. (3.32), one of the basic relations of Keplerian motion:

$$v^2 = \mu\left(\frac{2}{r} - \frac{1}{a}\right). \tag{10.17}$$

This becomes

$$2\boldsymbol{v} \cdot \boldsymbol{f} = \mu\frac{\dot{a}}{a^2}.$$

Using Eq. (3.45)

$$\mu = n^2 a^3, \tag{10.18}$$

and solving for $\dot{a}$, we get

$$\dot{a} = \frac{2}{n^2 a} \boldsymbol{v} \cdot \boldsymbol{f}. \tag{10.19}$$

It is easy to show that this result agrees with Eq. (10.12) (Problem 10.1).

As a summary, we have the following formulae for the time derivatives of the elements:

$$\begin{aligned}
\dot{e} &= \frac{1}{\mu}[\hat{\boldsymbol{e}}_\xi \cdot (\dot{\boldsymbol{r}} \times (\boldsymbol{r} \times \boldsymbol{f})) + \sqrt{p\mu}\,\hat{\boldsymbol{e}}_\eta \cdot \boldsymbol{f}], \\
\dot{p} &= 2\sqrt{\frac{p}{\mu}}\,\hat{\boldsymbol{e}}_\zeta \cdot (\boldsymbol{r} \times \boldsymbol{f}), \\
\dot{a} &= \frac{\dot{p}}{1-e^2} + \frac{2pe\dot{e}}{(1-e^2)^2}.
\end{aligned} \tag{10.20}$$

10.2 Binary evolution with a constant perturbing force

As another application of the previous section we consider first order secular perturbations of the semi-major axis of the inner binary, under a constant perturbing force. This is relevant to highly hierarchical binaries where the ratio of the semi-major axes a_e/a_i is large. Then the outer binary component appears practically stationary in relation to the fast orbital motion of the inner binary. Previously we claimed, without proof, that there is no secular energy exchange between the inner and outer binary components in this situation. Here we establish this claim using the methods of the previous section.

The first order perturbing potential was given in Eq. (9.12); the corresponding perturbing acceleration is (Problem 10.5)

$$\boldsymbol{f} = -\frac{Gm_3}{R_3^3}\left(\boldsymbol{r} - \frac{3\boldsymbol{r} \cdot \boldsymbol{R}_3}{R_3^2}\boldsymbol{R}_3\right). \tag{10.21}$$

This acceleration is substituted in Eqs. (10.20), using a constant perturber at

$$\boldsymbol{R}_3 = R_3 \hat{\boldsymbol{e}}_r \tag{10.22}$$

where $\hat{\boldsymbol{e}}_r$ is the unit vector towards the the third body, together with the standard description of two-body motion:

$$\begin{aligned}
\boldsymbol{r} &= a\cos E\hat{\boldsymbol{e}}_\xi - ae\hat{\boldsymbol{e}}_\xi + b\sin E\hat{\boldsymbol{e}}_\eta,\\
\dot{\boldsymbol{r}} &= -a\dot{E}\sin E\hat{\boldsymbol{e}}_\xi + b\dot{E}\cos E\hat{\boldsymbol{e}}_\eta,\\
\dot{E} &= \frac{\sqrt{Gm_B}a^{-3/2}}{1-e\cos E},\\
\mathrm{d}M &= (1-e\cos E)\,\mathrm{d}E.
\end{aligned} \tag{10.23}$$

Then we have some routine calculations to carry out. We start with the calculation of

$$\boldsymbol{r}\times\boldsymbol{f} = \frac{3Gm_3}{R_3^3}(\boldsymbol{r}\cdot\hat{\boldsymbol{e}}_r)(\boldsymbol{r}\times\hat{\boldsymbol{e}}_r). \tag{10.24}$$

After substituting

$$\begin{aligned}
\boldsymbol{r}\cdot\hat{\boldsymbol{e}}_r &= a(\cos E - e)(\hat{\boldsymbol{e}}_\xi\cdot\hat{\boldsymbol{e}}_r) + b\sin E(\hat{\boldsymbol{e}}_\eta\cdot\hat{\boldsymbol{e}}_r),\\
\boldsymbol{r}\times\hat{\boldsymbol{e}}_r &= a(\cos E - e)(\hat{\boldsymbol{e}}_\xi\times\hat{\boldsymbol{e}}_r) + b\sin E(\hat{\boldsymbol{e}}_\eta\times\hat{\boldsymbol{e}}_r),
\end{aligned}$$

one obtains

$$\begin{aligned}
(\boldsymbol{r}\times\boldsymbol{f})\,\mathrm{d}M = \frac{3Gm_3}{R_3^3}\Big[&\left(a^2\cos^2 E + a^2e^2 - 2a^2e\cos E\right)\\
&\times(\hat{\boldsymbol{e}}_\xi\cdot\hat{\boldsymbol{e}}_r)(\hat{\boldsymbol{e}}_\xi\times\hat{\boldsymbol{e}}_r)\\
&+ b^2\sin^2 E(\hat{\boldsymbol{e}}_\eta\cdot\hat{\boldsymbol{e}}_r)(\hat{\boldsymbol{e}}_\eta\times\hat{\boldsymbol{e}}_r)\\
&+ ab\sin E(\cos E - e)(\hat{\boldsymbol{e}}_\xi\cdot\hat{\boldsymbol{e}}_r)(\hat{\boldsymbol{e}}_\eta\times\hat{\boldsymbol{e}}_r)\\
&+ ab\sin E(\cos E - e)(\hat{\boldsymbol{e}}_\eta\cdot\hat{\boldsymbol{e}}_r)(\hat{\boldsymbol{e}}_\xi\times\hat{\boldsymbol{e}}_r)\Big]\\
&\times(1-e\cos E)\,\mathrm{d}E,
\end{aligned} \tag{10.25}$$

where both sides were multiplied by the differential $\mathrm{d}M$ in preparation for the integration over one orbital cycle. The integration may now be carried out using the eccentric anomaly E as the integration variable. Keeping in mind our earlier results for orbital averaging (Eqs. (9.22) and (9.28)), it becomes obvious that only the first two terms in the square brackets of Eq. (10.25) make a non-zero contribution to the average. Therefore

$$\begin{aligned}
\langle\boldsymbol{r}\times\boldsymbol{f}\rangle = \frac{3Gm_3}{R_3^3}\Big[&\left(\frac{1}{2}a^2 + 2a^2e^2\right)(\hat{\boldsymbol{e}}_\xi\cdot\hat{\boldsymbol{e}}_r)(\hat{\boldsymbol{e}}_\xi\times\hat{\boldsymbol{e}}_r)\\
&+ \tfrac{1}{2}b^2(\hat{\boldsymbol{e}}_\eta\cdot\hat{\boldsymbol{e}}_r)(\hat{\boldsymbol{e}}_\eta\times\hat{\boldsymbol{e}}_r)\Big].
\end{aligned} \tag{10.26}$$

Next, one requires the projection of the above quantity in the $\hat{e}_\zeta$ direction:

$$\hat{e}_\zeta \cdot \langle r \times f \rangle = \frac{3Gm_3}{R_3^3}\left(\frac{1}{2}a^2 + 2a^2e^2 - \frac{1}{2}b^2\right)(\hat{e}_\eta \cdot \hat{e}_r)(\hat{e}_\xi \cdot \hat{e}_r). \tag{10.27}$$

Here we have made use of the well known properties of triple vector products, e.g.

$$\hat{e}_\zeta \cdot (\hat{e}_\xi \times \hat{e}_r) = (\hat{e}_\zeta \times \hat{e}_\xi) \cdot \hat{e}_r = \hat{e}_\eta \cdot \hat{e}_r. \tag{10.28}$$

Since $b^2 = a^2(1 - e^2)$ Eq. (10.27) simplifies to

$$\langle \hat{e}_\zeta \cdot r \times f \rangle = \frac{15}{2}\frac{Gm_3}{R_3^3}a^2e^2(\hat{e}_\xi \cdot \hat{e}_r)(\hat{e}_\eta \cdot \hat{e}_r). \tag{10.29}$$

The next quantity to be calculated is

$$\dot{r} \times (r \times f) = -\frac{3Gm_3}{R_3^3}(r \cdot \hat{e}_r)\left[(\dot{r} \cdot r)\hat{e}_r - (\dot{r} \cdot \hat{e}_r)r\right]. \tag{10.30}$$

We substitute

$$\dot{r} \cdot r = -a^2\dot{E}\sin E\cos E + a^2e\dot{E}\sin E + b^2\dot{E}\sin E\cos E,$$

$$\dot{r} \cdot \hat{e}_r = -a\dot{E}\sin E(\hat{e}_\xi \cdot \hat{e}_r) + b\dot{E}\cos E(\hat{e}_\eta \cdot \hat{e}_r),$$

in Eq. (10.30) and take the projection on the ξ axis. Then (Problem 10.6)

$$\hat{e}_\xi \cdot (\dot{r} \times (r \times f))\,\mathrm{d}M = -\frac{3Gm_3\sqrt{Gm_B}\,a^{-3/2}}{R_3^3}A\,\mathrm{d}E \tag{10.31}$$

where

$$\begin{aligned} A = {} & a^3(1 - e^2)\sin E\cos E(\cos E - e)\left[(\hat{e}_\xi \cdot \hat{e}_r)^2 - (\hat{e}_\eta \cdot \hat{e}_r)^2\right] \\ & + a^2b\cos E\left[1 - 2e^2 - (2 - e^2)\cos^2 E + 2e\cos E\right](\hat{e}_\xi \cdot \hat{e}_r)(\hat{e}_\eta \cdot \hat{e}_r). \end{aligned} \tag{10.32}$$

Averaging over the complete orbital cycle and observing that the only non-zero contribution comes from the last term inside the second square brackets in Eq. (10.32),

$$\langle \hat{e}_\xi \cdot (\dot{r} \times (r \times f)) \rangle = -\frac{3Gm_3\sqrt{Gm_B}}{R_3^3}a^{1/2}be(\hat{e}_\xi \cdot \hat{e}_r)(\hat{e}_\eta \cdot \hat{e}_r). \tag{10.33}$$

Finally,

$$\begin{aligned} (\hat{e}_\eta \cdot f)\mathrm{d}M = -\frac{Gm_3}{R_3^3}\big[& b\sin E - 3a(\cos E - e)(\hat{e}_\xi \cdot \hat{e}_r)(\hat{e}_\eta \cdot \hat{e}_r) \\ & - 3b\sin E(\hat{e}_\eta \cdot \hat{e}_r)^2\big](1 - e\cos E)\,\mathrm{d}E \end{aligned} \tag{10.34}$$

which, after orbital averaging, becomes

$$\langle \hat{\boldsymbol{e}}_\eta \cdot \boldsymbol{f} \rangle = -\frac{9}{2}\frac{Gm_3}{R_3^3} a\, e(\hat{\boldsymbol{e}}_\xi \cdot \hat{\boldsymbol{e}}_r)(\hat{\boldsymbol{e}}_\eta \cdot \hat{\boldsymbol{e}}_r). \tag{10.35}$$

Now we have calculated all the necessary pieces for using Eqs. (10.20). We can easily see that

$$\langle \dot{e} \rangle = -\frac{15}{2}\sqrt{\frac{p}{Gm_B}}\frac{Gm_3}{R_3^3} a\, e(\hat{\boldsymbol{e}}_\xi \cdot \hat{\boldsymbol{e}}_r)(\hat{\boldsymbol{e}}_\eta \cdot \hat{\boldsymbol{e}}_r), \tag{10.36}$$

$$\langle \dot{p} \rangle = 15\sqrt{\frac{p}{Gm_B}}\frac{Gm_3}{R_3^3} a^2 e^2(\hat{\boldsymbol{e}}_\xi \cdot \hat{\boldsymbol{e}}_r)(\hat{\boldsymbol{e}}_\eta \cdot \hat{\boldsymbol{e}}_r), \tag{10.37}$$

which leads to

$$\langle \dot{a} \rangle = \frac{\langle \dot{p} \rangle}{(1-e^2)} + \frac{2pe\,\langle \dot{e} \rangle}{(1-e^2)^2} = 0. \tag{10.38}$$

Thus, in our current approximation, the semi-major axis of the binary does not have secular evolution, unlike for example the eccentricity of the binary. This provides the necessary justification for the use of the orbit averaged perturbing potential in the earlier sections.

Our next problem, in order of increasing difficulty, is the calculation of the energy change of a binary when a third body passes by at a close distance to it. Then it is not possible to claim that the third body is stationary relative to the binary but we must describe its orbital motion. For that purpose the approximate ways of describing elliptic motion as a function of time derived in Section 3.15 are used. It is not as simple as one might expect from the simple geometry of the orbit; one resorts to infinite series, truncating the series at suitable points for practical calculations.

10.3 Slow encounters

The problem of the energy change of a binary (mass m_B) caused by a passing third body (mass m_3) is rather complicated (Walters 1932a, b, Lyttleton and Yabushita 1965, Yabushita 1966, Heggie 1975, Heggie and Hut 1993, Roy and Haddow 2003). It is obvious that it should be so since the two orbits may be oriented in many different ways relative to each other, with different eccentricities, semi-major axes, closest approach distances etc. For this reason we limit ourselves to the rather simple case of a circular binary and a third-body orbit of fixed eccentricity. The outer orbit is taken to be elliptic and thus we are not really dealing with three-body scattering. The approximation considered here is called *adiabatic* since the perturbing potential varies slowly in comparison with the orbital frequency of the binary.

The inclination between the two orbital planes is called ι. The other parameters describing the relative orientations of the inner and the outer orbits Ω and ω, are considered less essential, and their influence is averaged over in the end.

The eccentricity of the outer orbit is called e. It has been shown by numerical orbit integration that the energy change is not very sensitive to the eccentricity of the outer orbit near the parabolic case (Saslaw *et al.* 1974). In the calculation below we give the eccentricity a fixed value $e = 0.265$. It is low enough that the series presented in Section 3.15 converge well, and the particular value of 0.265 is chosen for convenience since the value of mean anomaly $M = \pi/3$ corresponds to the true anomaly $\phi_e = \pi/2$ at nearly this eccentricity. The calculation could be carried out for any other low value of e without significant change in the final result.

The encounter is assumed to be effective only between $\phi_e = -\pi/2$ and $\phi_e = \pi/2$ which is where most of the action takes place especially in highly eccentric or parabolic orbits. Therefore we expect that the derived model is applicable to parabolic or even mildly hyperbolic encounters.

The circular binary is rather special but much simpler than the general case because the orientation of the major axis in its orbital plane does not need to be specified. The special symmetry due to the zero inner eccentricity simplifies the derivation considerably.

We divide the integration in two parts: the approaching branch, from $M = -M_0$ to $M = 0$, and the receding branch from $M = 0$ to $M = +M_0$. M_0 is defined so that the inner binary executes exactly one revolution in its initial orbit while the third body progresses from $M = 0$ (pericentre) to $M = M_0$. In our example below M_0 is close to $\pi/3$. We ignore the effects of the subsequent revolutions since they typically happen while the third body is outside the range

$$-\frac{\pi}{2} \le \phi_e \le \frac{\pi}{2}. \tag{10.39}$$

We start by integrating the relevant functions between $M = 0$ and $M = M_0$. Numerical experiments have shown that the effects of the whole encounter may be well estimated by using the receding branch.

The energy change may be calculated using Eq. (9.8) with the perturbing function of Eq. (9.12). We get

$$n^2 \dot{a} = \frac{2}{a}\frac{\partial R}{\partial t} = \frac{2}{a}\frac{\partial R}{\partial \boldsymbol{r}} \cdot \frac{\partial \boldsymbol{r}}{\partial t} = \frac{2}{a}\nabla R \cdot \dot{\boldsymbol{r}},$$

where the gradient of the perturbing function R is related to the perturbing acceleration of Eq. (10.21) by

$$\nabla R = \boldsymbol{f}.$$

Thus the rate of change of the binary semi-major axis becomes

$$\dot{a} = 2\sqrt{\frac{a}{\mu}}\left(\frac{1}{n}\boldsymbol{f}\cdot\dot{\boldsymbol{r}}\right).$$

The product $\boldsymbol{f}\cdot\dot{\boldsymbol{r}}$ includes the term $\boldsymbol{r}\cdot\dot{\boldsymbol{r}}$ which equals zero as long as the binary remains circular. Therefore, using Eqs. (10.21)–(10.23) with $e = 0$, $\dot{E} = n$ and $a = a_i$,

$$\frac{1}{n}\boldsymbol{f}\cdot\dot{\boldsymbol{r}} = \frac{3Gm_3}{R_3^3}a_i^2\Big[\cos^2\phi(\hat{\boldsymbol{e}}_\xi\cdot\hat{\boldsymbol{e}}_r)(\hat{\boldsymbol{e}}_\eta\cdot\hat{\boldsymbol{e}}_r) - \sin^2\phi(\hat{\boldsymbol{e}}_\xi\cdot\hat{\boldsymbol{e}}_r)(\hat{\boldsymbol{e}}_\eta\cdot\hat{\boldsymbol{e}}_r)$$
$$- \sin\phi\,\cos\phi(\hat{\boldsymbol{e}}_\xi\cdot\hat{\boldsymbol{e}}_r)^2 + \sin\phi\,\cos\phi(\hat{\boldsymbol{e}}_\eta\cdot\hat{\boldsymbol{e}}_r)^2\Big].$$

Alternatively, we may derive the energy change by following the steps taken in Section 10.2. Putting the inner eccentricity $e_i = 0$, replacing the inner eccentric anomaly E by ϕ and writing a_i for the inner semi-major axis, Eqs. (10.20) and (10.25) lead to $\dot{a} = 2\sqrt{a/\mu}$ times the vector product

$$\begin{aligned}\hat{\boldsymbol{e}}_\zeta\cdot(\boldsymbol{r}\times\boldsymbol{f}) = \frac{3Gm_3}{R_3^3}a_i^2[&\cos^2\phi(\hat{\boldsymbol{e}}_\xi\cdot\hat{\boldsymbol{e}}_r)\hat{\boldsymbol{e}}_\zeta\cdot(\hat{\boldsymbol{e}}_\xi\times\hat{\boldsymbol{e}}_r)\\ &+\sin^2\phi(\hat{\boldsymbol{e}}_\eta\cdot\hat{\boldsymbol{e}}_r)\hat{\boldsymbol{e}}_\zeta\cdot(\hat{\boldsymbol{e}}_\eta\times\hat{\boldsymbol{e}}_r)\\ &+\sin\phi\cos\phi(\hat{\boldsymbol{e}}_\xi\cdot\hat{\boldsymbol{e}}_r)\hat{\boldsymbol{e}}_\zeta\cdot(\hat{\boldsymbol{e}}_\eta\times\hat{\boldsymbol{e}}_r)\\ &+\sin\phi\cos\phi(\hat{\boldsymbol{e}}_\eta\cdot\hat{\boldsymbol{e}}_r)\hat{\boldsymbol{e}}_\zeta\cdot(\hat{\boldsymbol{e}}_\xi\times\hat{\boldsymbol{e}}_r)].\end{aligned} \tag{10.40}$$

After calculating the vector products of the unit vectors, we realise that they correspond to the products in the previous equation, and that the right hand sides of the two equations are identical. Thus the two routes of calculating $\dot{a}$ give the same answer, as they should.

We choose the orientation of the outer orbit such that the projection of its major axis lies along the ξ axis, with the pericentre direction on the negative side of the axis, and the line of nodes is taken to be on the η axis ($\omega = \Omega = \pi/2$ for the outer orbit; see Fig. 10.1). This makes the calculation less cumbersome; the general case of arbitrary relative orientations will be discussed later (Section 10.4). Let the unit vector $\hat{\boldsymbol{e}}_A$ point towards the pericentre of the outer orbit and let the unit vector $\hat{\boldsymbol{e}}_B$ be the corresponding perpendicular unit vector. Then

$$\hat{\boldsymbol{e}}_r = \frac{a}{R_3}(\cos E - e)\hat{\boldsymbol{e}}_A + \frac{b}{R_3}\sin E\hat{\boldsymbol{e}}_B. \tag{10.41}$$

Here the symbols a, b, e and E (the semi-axes, eccentricity and the eccentric anomaly, respectively) refer to the outer orbit. The vector products needed in

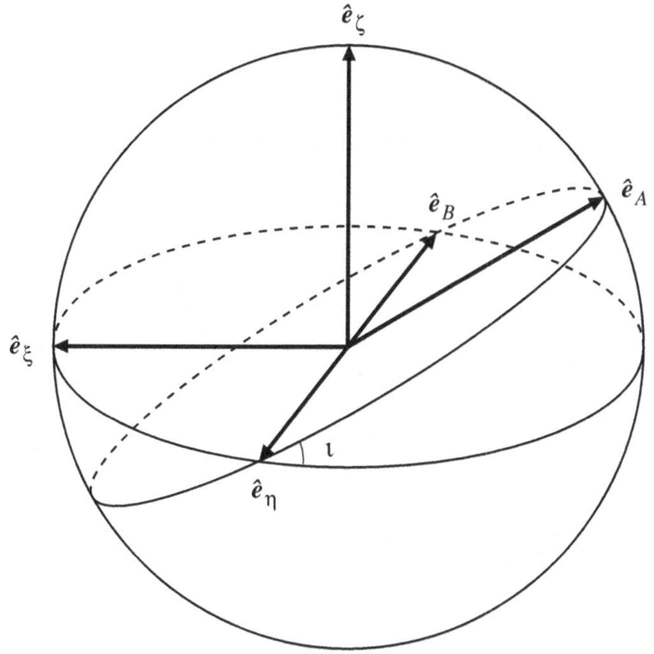

Figure 10.1 Coordinate base vectors $\hat{\boldsymbol{e}}_\xi$, $\hat{\boldsymbol{e}}_\eta$ of the binary orbital plane and the base vectors $\hat{\boldsymbol{e}}_A$, $\hat{\boldsymbol{e}}_B$ of the third-body orbit, see text.

Eq. (10.40) are

$$\begin{aligned}
\hat{\boldsymbol{e}}_\xi \cdot \hat{\boldsymbol{e}}_r &= -\frac{a}{R_3}(\cos E - e)\cos\iota, \\
\hat{\boldsymbol{e}}_\zeta \cdot (\hat{\boldsymbol{e}}_\xi \times \hat{\boldsymbol{e}}_r) = \hat{\boldsymbol{e}}_\eta \cdot \hat{\boldsymbol{e}}_r &= -\frac{b}{R_3}\sin E, \\
\hat{\boldsymbol{e}}_\eta \cdot \hat{\boldsymbol{e}}_r &= -\frac{b}{R_3}\sin E, \\
\hat{\boldsymbol{e}}_\zeta \cdot (\hat{\boldsymbol{e}}_\eta \times \hat{\boldsymbol{e}}_r) = -\hat{\boldsymbol{e}}_\xi \cdot \hat{\boldsymbol{e}}_r &= \frac{a}{R_3}(\cos E - e)\cos\iota,
\end{aligned} \tag{10.42}$$

where ι is the angle of inclination between the two orbital planes. Use was made here of the identity $\hat{\boldsymbol{e}}_B = -\hat{\boldsymbol{e}}_\eta$ which is applicable for our chosen relative orientations.

Substituting these products in Eq. (10.40) and making use of the identities

$$\begin{aligned}
\cos^2\phi - \sin^2\phi &= \cos 2\phi, \\
2\sin\phi\cos\phi &= \sin 2\phi,
\end{aligned}$$

leads to

$$\begin{aligned}\hat{\boldsymbol{e}}_\zeta \cdot (\boldsymbol{r} \times \boldsymbol{f}) = &\frac{3Gm_3a_i^2}{a^3}\left(\frac{a}{R_3}\right)^5 \\ &\times \left[\sqrt{1-e^2}\sin E(\cos E - e)\cos\iota\cos 2\phi \right.\\ &- \frac{1}{2}(\cos E - e)^2\cos^2\iota\sin 2\phi \\ &\left. + \frac{1}{2}(1-e^2)\sin^2 E\sin 2\phi\right].\end{aligned} \tag{10.43}$$

The factor $(a/R_3)^5$ decreases monotonically from its greatest value $(1-e)^{-5}$ at the pericentre ($\phi_e = 0$) to $(1-e^2)^{-5}$ at the end of the latus rectum ($\phi_e = \pi/2$). For an orbit of high eccentricity the decrease is by a factor of almost 32, while at the low eccentricity of $e = 0.265$ it is still by a factor of ≈ 3.2. Therefore we expect that the integral of $\hat{\boldsymbol{e}}_\zeta \cdot (\boldsymbol{r} \times \boldsymbol{f})$ over M receives its greatest contribution from the second term in the square brackets which is also at its maximum at the pericentre. In the following we ignore the other two terms in the square brackets of Eq. (10.43) which go to zero at $\phi_e = 0$.

Next, E and R_3 are expressed as functions of the mean anomaly M of the outer orbit. Using Eq. (3.90) with $e = 0.265$ we get

$$\frac{a}{R_3} = 1 + 0.263\cos M + 0.0686\cos 2M + 0.0201\cos 3M \tag{10.44}$$

which is raised to the fifth power. After some labour we get approximately, ignoring terms of higher order than $\cos 5M$:

$$\begin{aligned}\left(\frac{a}{R_3}\right)^5 = &\,1.416(1 + 1.193\cos M + 0.586\cos 2M \\ &+ 0.272\cos 3M + 0.108\cos 4M + 0.0473\cos 5M).\end{aligned} \tag{10.45}$$

Similarly, expressions (3.85) for $\cos E$ and $\cos 2E$ become

$$\begin{aligned}\cos E &= -0.1325 + 0.974\cos M + 0.126\cos 2M \\ &\quad + 0.0246\cos 3M + 0.00568\cos 4M, \\ \cos 2E &= -0.263\cos M + 0.931\cos 2M \\ &\quad + 0.245\cos 3M + 0.0636\cos 4M + 0.0173\cos 5M,\end{aligned} \tag{10.46}$$

and

$$\frac{1}{2}(\cos E - e)^2 = 0.25\cos 2E - 0.265\cos E + 0.285$$
$$= 0.320 - 0.324\cos M + 0.199\cos 2M \qquad (10.47)$$
$$+ 0.0546\cos 3M + 0.0143\cos 4M$$

(Problem 10.7).

The final step, before integration of Eq. (10.40), is to write ϕ in terms of M. Since we are dealing with a circular binary, the true anomaly is simply a linear function of time t:

$$\phi = n_i t + \phi_0 \qquad (10.48)$$

where n_i is the mean motion of the binary and ϕ_0 is the true anomaly at the time of the pericentre passage ($t = 0$). The corresponding relation for the mean anomaly of the outer orbit is

$$M = n_e t. \qquad (10.49)$$

Equating the times in these two expressions gives

$$t = \frac{M}{n_e} = \frac{\phi - \phi_0}{n_i} \qquad (10.50)$$

or

$$\phi = \frac{n_i}{n_e} M + \phi_0. \qquad (10.51)$$

The ratio n_i/n_e depends on the normalised pericentre distance Q, defined as

$$Q = a(1-e)/a_i = 0.735\frac{a}{a_i}. \qquad (10.52)$$

Instead of Q it is often more practical to use the parameter

$$k \equiv 2\frac{n_i}{n_e} = \sqrt{2}\left(\frac{2m_B}{m_B + m_3}\right)^{1/2}\left(\frac{a}{a_i}\right)^{3/2}$$
$$= 8.87\left(\frac{2m_B}{m_B + m_3}\right)^{1/2}\left(\frac{Q}{2.5}\right)^{3/2}. \qquad (10.53)$$

With these definitions

$$2\phi = kM + 2\phi_0. \qquad (10.54)$$

Using the identity

$$\sin 2\phi = \sin kM \cos 2\phi_0 + \cos kM \sin 2\phi_0$$

the integral over Eq. (10.40) may now be formally written as

$$\int_0^{M_0} \hat{\boldsymbol{e}}_\zeta \cdot (\boldsymbol{r} \times \boldsymbol{f})\,\mathrm{d}M = \frac{3Gm_3}{a^3} a_i^2 \left(I^{(1)} \sin 2\phi_0 + I^{(2)} \cos 2\phi_0\right) \cos^2 \iota \qquad (10.55)$$

where

$$I^{(1)} = -\frac{1}{2} \int_0^{M_0} \cos kM \left(\frac{a}{R_3}\right)^5 (\cos E - e)^2\,\mathrm{d}M \qquad (10.56)$$

and

$$I^{(2)} = -\frac{1}{2} \int_0^{M_0} \sin kM \left(\frac{a}{R_3}\right)^5 (\cos E - e)^2\,\mathrm{d}M. \qquad (10.57)$$

Let us start with the integral $I^{(2)}$. Substituting Eqs. (10.45) and (10.47) into Eq. (10.57) and transforming the products $\sin aM \cos bM$ (where a and b are arbitrary real numbers) by

$$\sin aM \cos bM = \frac{1}{2}\left[\sin(a+b)M + \sin(a-b)M\right],$$

one obtains

$$\begin{aligned} I^{(2)} = &-\int_0^{M} [0.279 + 0.183 \cos M + 0.291 \cos 2M + 0.229 \cos 3M \\ &+ 0.139 \cos 4M + 0.0733 \cos 5M + 0.0358 \cos 6M \\ &+ 0.0163 \cos 7M + 0.0072 \cos 8M] \sin kM\,\mathrm{d}M \\ = &\ \frac{1}{200}\Bigg|_0^{M_0} \sum_{n=-8}^{+8} \frac{B_n}{k-n} \cos(k-n)M. \end{aligned} \qquad (10.58)$$

The last step can be verified by taking the derivative $\mathrm{d}/\mathrm{d}M$ of the right hand side; by doing so, one also finds the values of the coefficients B_n:

$$\begin{aligned} B_0 &= 55.8, \\ B_1 &= B_{-1} = 18.3, \\ B_2 &= B_{-2} = 29.1, \\ B_3 &= B_{-3} = 22.9, \\ B_4 &= B_{-4} = 13.9, \\ B_5 &= B_{-5} = 7.33, \\ B_6 &= B_{-6} = 3.58, \\ B_7 &= B_{-7} = 1.63, \\ B_8 &= B_{-8} = 0.72. \end{aligned} \qquad (10.59)$$

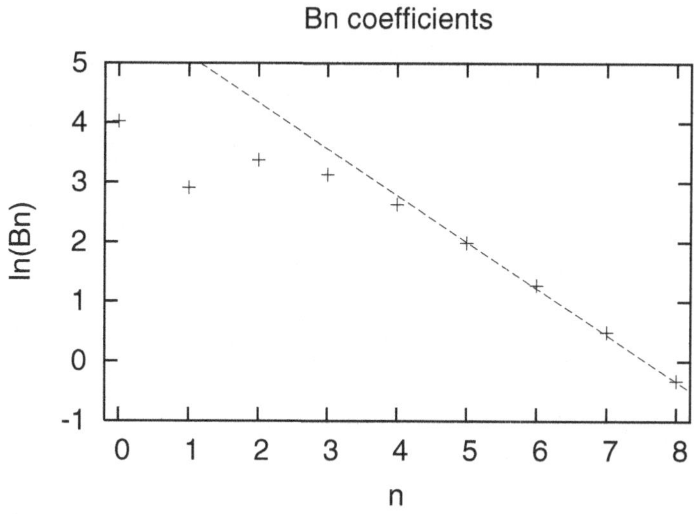

Figure 10.2 The values of the coefficients B_n in Eq. (10.60) as a function of n. The line shows an exponential fit.

We see that the values of the coefficients decrease with increasing n approximately as

$$B_n \simeq 365\mathrm{e}^{-0.78n} \tag{10.60}$$

for $n \geq 3$ (see Fig. 10.2). Qualitatively this result holds also for eccentricities other than $e = 0.265$.

Using the identity

$$\cos aM \cos bM = \frac{1}{2}\left[\cos(a+b)M + \cos(a-b)M\right]$$

the other integral becomes

$$I^{(1)} = -\frac{1}{200}\Bigg|_0^{M_0} \sum_{n=-8}^{+8} \frac{B_n}{k-n} \sin(k-n)M. \tag{10.61}$$

The value of M_0 is obtained from Eq. (10.54) by putting $\phi - \phi_0 = 2\pi$:

$$M_0 = \frac{4\pi}{k}. \tag{10.62}$$

This is used to evaluate $I^{(1)}$ and $I^{(2)}$.

In the above sums the term $n = 0$ gives no contribution since at the upper limit $\sin kM_0 = \sin 4\pi = 0$ and $\cos kM_0 = \cos 4\pi = 1$, and the values are the same at the lower limit. The remaining terms may be combined pairwise; for $I^{(1)}$ and $I^{(2)}$

the pairs are

$$\begin{aligned} &-\tfrac{1}{200}B_n\left[\tfrac{1}{k+n}\sin(k+n)\tfrac{4\pi}{k}+\tfrac{1}{k-n}\sin(k-n)\tfrac{4\pi}{k}\right], \\ &\tfrac{1}{200}B_n\left[\tfrac{1}{k+n}\left(\cos(k+n)\tfrac{4\pi}{k}-1\right)+\tfrac{1}{k-n}\left(\cos(k-n)\tfrac{4\pi}{k}-1\right)\right], \end{aligned} \tag{10.63}$$

respectively. Since

$$\begin{aligned} \sin(k+n)\frac{4\pi}{k} &= -\sin(k-n)\frac{4\pi}{k} = \sin 4\pi\frac{n}{k}, \\ \cos(k+n)\frac{4\pi}{k} &= \cos(k-n)\frac{4\pi}{k} = \cos 4\pi\frac{n}{k}, \end{aligned}$$

these pairs become

$$\begin{aligned} &-\frac{1}{200}B_n\sin 4\pi\frac{n}{k}\left[\frac{1}{k+n}-\frac{1}{k-n}\right] \\ &\quad = \frac{1}{100}\frac{nB_n}{k^2}\left[1-\left(\frac{n}{k}\right)^2\right]^{-1}\sin 4\pi\frac{n}{k}, \\ &\frac{1}{200}B_n\left(\cos\frac{4\pi n}{k}-1\right)\left[\frac{1}{k+n}+\frac{1}{k-n}\right] \\ &\quad = \frac{1}{100}\frac{B_n}{k}\left[1-\left(\frac{n}{k}\right)^2\right]^{-1}\left(\cos\frac{4\pi n}{k}-1\right). \end{aligned} \tag{10.64}$$

Therefore the sums for $I^{(1)}$ and $I^{(2)}$ become

$$\begin{aligned} I^{(1)} &= \frac{1}{100}\frac{1}{k^2}\sum_{n=1}^{8} nB_n\left[1-\left(\frac{n}{k}\right)^2\right]^{-1}\sin\left(\frac{4\pi n}{k}\right) \\ I^{(2)} &= \frac{1}{100}\frac{1}{k}\sum_{n=1}^{8} B_n\left[1-\left(\frac{n}{k}\right)^2\right]^{-1}\left[\cos\left(\frac{4\pi n}{k}\right)-1\right]. \end{aligned} \tag{10.65}$$

For example, let us take $k = 12$. Equations (10.51) and (10.53) show that with $k = 12$, $\phi - \phi_0 = 2\pi$ when $M = \pi/3$, i.e. the binary completes one revolution while the third body moves from the pericentre to the end of the latus rectum. This value of k corresponds to $Q \simeq 3.06$ if $m_B = m_3$ (Eq. (10.53)). We get

$$\begin{aligned} I^{(1)} &= \frac{1}{14400}(15.98+51.82-54.05-38.41+14.93+8.98) \\ &\simeq -0.00005 \\ I^{(2)} &= -\frac{1}{1200}(9.22+44.88+48.90+23.41+4.43+1.23+1.94) \\ &\simeq -0.11. \end{aligned} \tag{10.66}$$

Since $I^{(1)}$ is much smaller than $I^{(2)}$ we will not consider it further. By numerical evaluation of the sums in Eq. (10.65) one may show that this is true also for other values of k, $k > 8$.

The integral $I^{(2)}$ depends on k in two ways: first, there is the obvious k^{-1} factor in front of the summation, but also the sum itself depends on k. It is easy to verify that at small k ($k \lesssim 15$, corresponding to $Q \lesssim 3.5$ when $m_B = m_3$) this dependence is weak and in practice may be neglected. For higher k the sum decreases exponentially. The exponential factor has its origin in the exponential decay of the B_n coefficients as a function of n (Eq. (10.60)) together with the connection $n = k/4$ at the first maximum of the absolute value of the periodic factor $[\cos(4\pi n/k) - 1]$. The exponential factor is of great importance at large Q (Heggie 1975, Roy and Haddow 2003). At first we will ignore it, and come back to it later.

At $k = 12$ Eq. (10.55) thus becomes

$$\int_0^{M_0} \hat{\boldsymbol{e}}_\zeta \cdot (\boldsymbol{r} \times \boldsymbol{f})\,\mathrm{d}M \approx -0.3\frac{Gm_3}{a^3}a_i^2 \cos^2 \iota \cos 2\phi_0. \tag{10.67}$$

For any other value of k ($9 \lesssim k \lesssim 15$), the right hand side should be multiplied by $12/k$ (Eq. (10.68)). From Eq. (10.53) we get

$$\frac{12}{k} = 1.35\left(\frac{2m_B}{m_B + m_3}\right)^{-1/2}\left(\frac{Q}{2.5}\right)^{-3/2} \tag{10.68}$$

and consequently

$$\begin{aligned}\int_0^{M_0} \hat{\boldsymbol{e}}_\zeta \cdot (\boldsymbol{r} \times \boldsymbol{f})\,\mathrm{d}M \simeq &-0.4\frac{Gm_3}{a^3}a_i^2\left(\frac{2m_B}{m_B + m_3}\right)^{-1/2}\\ &\times\left(\frac{Q}{2.5}\right)^{-3/2} \cos^2 \iota \cos 2\phi_0.\end{aligned} \tag{10.69}$$

Note that we recover Eq. (10.67) if $Q = 3.06$ (corresponding to $k = 12$) and $m_B = m_3$.

Now the change in the semi-major axis Δa_i may be calculated from Eq. (10.20) when the inner eccentricity $e_i = 0$:

$$\Delta a_i = \frac{2a_i^{1/2}}{\sqrt{Gm_B}}\int_0^{t_0} \hat{\boldsymbol{e}}_\zeta \cdot (\boldsymbol{r} \times \boldsymbol{f})\,\mathrm{d}t = \frac{2a_i^{1/2}}{n_e\sqrt{Gm_B}}\int_0^{M_0} \hat{\boldsymbol{e}}_\zeta \cdot (\boldsymbol{r} \times \boldsymbol{f})\,\mathrm{d}M. \tag{10.70}$$

Here the integration over time t, through one period t_0 of the inner orbit, was replaced by integration over the mean anomaly of the outer orbit M using $M = n_e t$, and the mean motion of the outer orbit

$$n_e = \left(\frac{G(m_B + m_3)}{a^3}\right)^{1/2}. \tag{10.71}$$

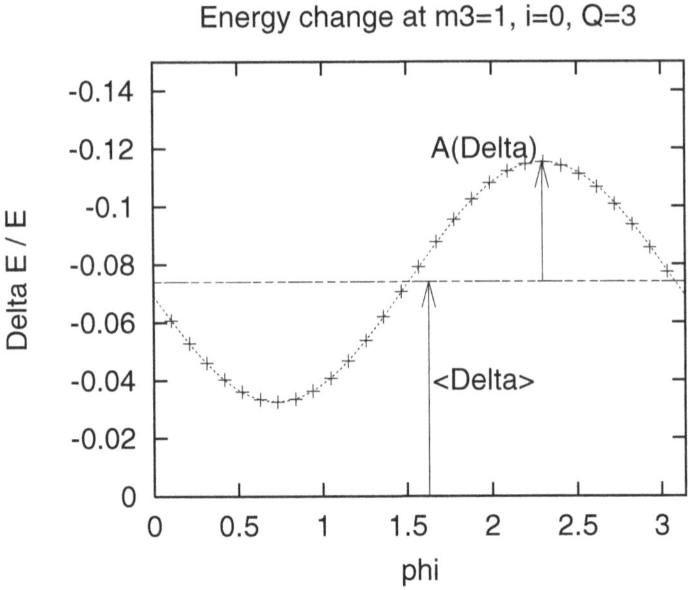

Figure 10.3 The relative energy change $-\Delta E/E = \Delta$ of the binary in a parabolic encounter with a third body. Computer experiments (+) are compared with a sinusoidal function of ϕ_0. The values of the mean energy change $\langle\Delta\rangle$ and the amplitude $A(\Delta)$ are indicated. The case of $\iota = 0°$, $m_3 = m_B$, $e_i = 0$ and $Q = 3$.

Substituting Eq. (10.69) into Eq. (10.70) and making use of Eqs. (10.52) and (10.71), one obtains

$$\Delta = \frac{\Delta a_i}{a_i} = -0.09\frac{m_3}{m_B}\left(\frac{Q}{2.5}\right)^{-3}\cos^2\iota\cos 2\phi_0. \tag{10.72}$$

Note that in this chapter we reverse the sign of Δ relative to its definition in Eq. (8.1).

This expression may be compared with results from numerical orbit calculations with parabolic third-body orbits. Figure 10.3 shows an example of how Δ varies with the phase angle ϕ_0. We see that in fact Δ is of the form

$$\Delta = \langle\Delta\rangle + A(\Delta)\cos 2\phi_0, \tag{10.73}$$

where $A(\Delta)$ is the amplitude of the variation and $\langle\Delta\rangle$ is the mean level. Equation (10.72) suggests $\langle\Delta\rangle = 0$; why this is not true becomes clearer in Section 10.5.

Let us now consider the exponential factor. The sum of $I^{(2)}$ (Eq. (10.65)) can be evaluated numerically and approximated by $-230\mathrm{e}^{-0.04k}$ in the interval $15 \le k \le 75$ (Problem 10.8). Using Eq. (10.68) it becomes

$$\exp[-0.5(Q/Q_1)^{3/2}],$$

where the scale factor Q_1 is

$$Q_1 = 2.5(1 + m_3/m_B)^{1/3}. \tag{10.74}$$

Therefore the Q dependence in $A(\Delta)$ should be of the form

$$Q^{-3} \exp[-0.5(Q/Q_1)^{3/2}]. \tag{10.75}$$

In many applications it is easier to use a pure power-law rather than this combination of power-law and exponential. The exponential can always be modelled locally by a power-law, with a power which increases with increasing Q. Thus we have to specify the range of Q we are interested in, and then make the proper choice of the approximate power. We do this first at the smallest values of Q where our theory may still be applicable. This range is approximately

$$Q_{\rm st} < Q < 1.5 Q_{\rm st}$$

where $Q_{\rm st}$ signifies the *stability boundary*. It is defined as the minimum value Q where the original binary survives the encounter at all phase angles ϕ_0 and at all values of ω and Ω.

It is possible to derive an approximate expression for $Q_{\rm st}$ from Eq. (10.77) by requiring that the sinusoidally varying component of Δ has a specific value at the stability boundary, e.g. $A(\Delta) = 0.09$. Our argument would be that greater amplitudes of Δ would lead to an exchange of a binary member with the third body. Then

$$Q_{\rm st}^{(a)} = 2.5(m_3/m_B)^{1/3}(\cos^2 \iota)^{1/3}. \tag{10.76}$$

Taken literally, this would imply that the stability boundary goes to zero at $\iota = 90^\circ$ which is not reasonable. In the next section we will see that generally one should replace $\cos \iota$ by $1 + \cos \iota$ in expressions like this. In Section 10.5 we will learn that we should expect a functional form of a constant + $(1 + \cos \iota)^2$ instead of $\cos^2 \iota$. In anticipation of these results we write numerical fitting functions

$$\begin{aligned} Q_{\rm st}^{(d)} &= 2.52[(1 + m_3/m_B)/2]^{0.45}[(0.1 + (1 + \cos \iota)^2)/4]^m, \\ Q_{\rm st}^{(r)} &= 2.75[(1 + m_3/m_B)/2]^{0.225}[(0.4 + (1 + \cos \iota)^2)/4]^{0.4}. \end{aligned} \tag{10.77}$$

The first one is for direct orbits ($\cos \iota_0 \le \cos \iota \le 1$), and the second one for retrograde orbits ($-1 \le \cos \iota \le \cos \iota_0$). The power law index m is given by

$$m = 0.06 + 0.08(1 + m_3/m_B),$$

and the direct/retrograde border $\cos \iota_0$ is defined as

$$\cos \iota_0 = 1.52[(1 + m_3/m_B)/2] - 1.28.$$

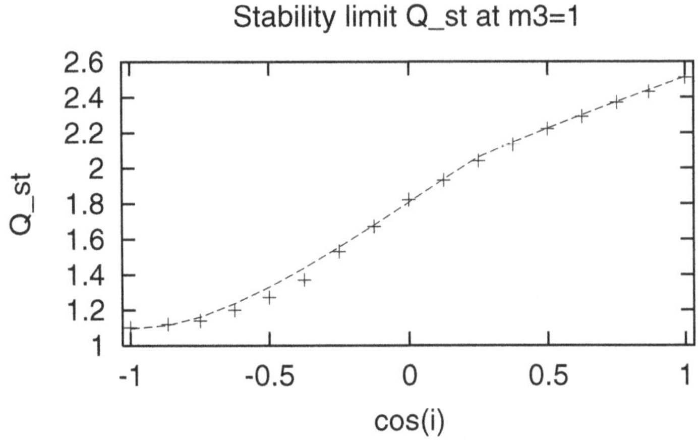

Figure 10.4 The stability limit in numerical experiments (+) as a function of $\cos\iota$. The line follows Eq. (10.77). The case of $m_3 = m_B$.

This gives a good representation of the stability boundary when the masses are not too unequal, i.e. in the mass range $0.2 \leq m_3/m_B \leq 2.0$. Note that (m_3/m_B) of $Q_{\rm st}^{(a)}$ is replaced by $(1 + m_3/m_B)$ in the more accurate $Q_{\rm st}^{(d)}$ and $Q_{\rm st}^{(r)}$. Because of this the stability boundary scales as a power of the exponential scale factor Q_1 (Eq. (10.74)). A fit of these functions to experimental data is shown in Fig. 10.4.

Hills (1992) determined the stability boundary numerically over the mass range $0.15 \leq m_3/m_B \leq 5000$. He used orbits of random inclinations and obtained the result

$$Q_{\rm st}(\text{Hills}) = 2.1(1 + m_3/m_B)^{1/3}. \tag{10.78}$$

This is in some ways a compromise between $Q_{\rm st}^{(d)}$ and $Q_{\rm st}^{(r)}$, since the power $1/3$ of the mass factor is intermediate between the corresponding powers of 0.45 and 0.225 in Eq. (10.77). It is simpler to use Eq. (10.78) than (10.77) if the inclinations are random (or unknown).

We may now return to Eq. (10.72) which may be written in two parts as

$$\begin{aligned} \Delta &= A(\Delta)\cos 2\phi_0, \\ A(\Delta) &= -0.09\frac{m_3}{m_B}\left(\frac{Q}{2.5}\right)^{-3}\cos^2\iota. \end{aligned} \tag{10.79}$$

As in the case of the stability boundary, numerical experiments show that in reality the functional form of $A(\Delta)$ is more complicated than this.

Experimentally we find that the power n of $(Q/2.5)^{-n}$ is steeper than $n = 3$. It is a function of both m_3/m_B and $\cos\iota$. It may be represented by a piecewise function

$$\begin{aligned}
&n_1 = 7, \quad 0.5 \le \cos\iota \le 1,\\
&n_1 = 6 + 2\cos\iota, \quad 0 \le \cos\iota \le 0.5,\\
&n_1 = 6 + 5\cos\iota, \quad -0.3 \le \cos\iota \le 0,\\
&n_1 = 4.5, \quad -0.933 \le \cos\iota \le -0.3,\\
&n_1 = 5.5 - 15(1 + \cos\iota), \quad -1 \le \cos\iota \le -0.933,
\end{aligned} \tag{10.80}$$

and

$$n = n_1 + 0.25\left(1 - \frac{m_3}{m_B}\right) \Big/ (m_3/m_B). \tag{10.81}$$

Since the range of applicability of the numerical result is $0.2 \le m_3/m_B \le 2.0$, the index n has the range $4.25 \le n \le 8$. Thus it is always greater than the power $n = 3$ of Eq. (10.72). This is at least partly because of the exponential factor (Eq. (10.75)) which has now been transformed into power-law form and its influence has been effectively absorbed by raising the power-law index n.

The mass factor in Eq. (10.72) is similarly influenced. The reference value $Q = 2.5$ is replaced by $Q_1 = 2.5(1 + m_3/m_B)^{1/3}$ which is the scale factor of the exponential (Eq. (10.74)). Therefore the mass factor is of the general form

$$f(m_3) = \left(\frac{m_3}{m_B}\right)^{m^{(1)}} \left(1 + \frac{m_3}{m_B}\right)^{m^{(2)}}. \tag{10.82}$$

The values of $m^{(1)}$ and $m^{(2)}$ depend on the inclination and they may be expressed in piecewise manner:

$$\begin{aligned}
&m^{(1)} = 1, \quad m^{(2)} = n_1/7, \quad -0.3 \le \cos\iota \le 1,\\
&m^{(1)} = 1.15, \quad m^{(2)} = 0, \quad -0.96 \le \cos\iota \le -0.3,\\
&m^{(1)} = 1.75 - 15(1 + \cos\iota), \quad m^{(2)} = 0, \quad -1 \le \cos\iota \le -0.96.
\end{aligned} \tag{10.83}$$

Since $A(\Delta)$ does not go to zero at $\iota = 90^\circ$, $\cos^2\iota$ is not a suitable inclination factor. Instead for direct orbits we may define

$$\mathrm{cs}(\iota) = 0.75\,[(1 + \cos\iota)/2]^2, \quad 0 \le \cos\iota \le 1. \tag{10.84}$$

This factor goes to zero at $\iota = 180^\circ$ which cannot be correct either. Thus for retrograde orbits the inclination factor has to be more complicated. We may use

$$\begin{aligned}
\mathrm{cs}(\iota) &= 0.75\,[(3+2\cos\iota)/5]^2\,, \quad -0.865 \le \cos\iota \le 0,\\
\mathrm{cs}(\iota) &= 0.75\,[0.015+0.37(1+\cos\iota)]\,, \quad -0.94 \le \cos\iota \le -0.865,\\
\mathrm{cs}(\iota) &= 0.75\left[0.0068+8(1+\cos\iota)^2\right], \quad -1 \le \cos\iota \le -0.94.
\end{aligned} \tag{10.85}$$

Note that these forms are fitting functions to numerically calculated orbits. To a certain extent they are justified by the previous theory. But the last three equations are clear examples of functional forms where a simple justification is not available.

After defining these rather complicated functions the final result may be written in a simple form

$$A(\Delta) = 0.09 f(m_3)\left(\frac{Q}{2.5}\right)^{-n} \mathrm{cs}(\iota). \tag{10.86}$$

Notice that Eq. (10.86) does not contain the mass ratio m_a/m_b of the binary members. Numerical experiments by Hills (1984) for close encounters between a star–planet system and a stellar intruder show that indeed we may neglect this parameter in the first approximation.

At low inclinations, $0.5 \le \cos\iota \le 1$, the Q-dependence may also be represented by a function of the form $Q^{-3}\exp[-0.5(Q/Q_1)^{3/2}]$, as an alternative to the Q^{-7} power-law, in accordance with Eq. (10.75). In the Q-range $2.5 \le Q/Q_1 \le 4$, $(Q/Q_1)^{-7} \simeq 0.18(Q/Q_1)^{-3}\exp[-0.5(Q/Q_1)^{3/2}]$.

10.4 Inclination dependence

Let us now briefly discuss the general case where the relative alignment of the two orbits is arbitrary. Then we may describe the direction of the unit vector $\hat{\boldsymbol{e}}_A$ relative to the unit vector $\hat{\boldsymbol{e}}_\xi$ by three angles Ω, ω and ι, with their usual meanings (see Fig. 10.5). The case calculated above corresponds to $\omega = \Omega = \pi/2$; here we initially allow any value of ω and Ω but in the end take an average result for all possible values of these 'less important' parameters.

It is easy to show (Problem 10.9) using spherical trigonometry that the unit vectors are

$$\begin{aligned}
\hat{\boldsymbol{e}}_\xi &= \cos\Omega\hat{\boldsymbol{i}} - \sin\Omega\hat{\boldsymbol{j}},\\
\hat{\boldsymbol{e}}_\eta &= \sin\Omega\hat{\boldsymbol{i}} + \cos\Omega\hat{\boldsymbol{j}},\\
\hat{\boldsymbol{e}}_A &= \cos\omega\hat{\boldsymbol{i}} + \sin\omega\cos\iota\hat{\boldsymbol{j}} + \sin\omega\sin\iota\hat{\boldsymbol{k}},\\
\hat{\boldsymbol{e}}_B &= -\sin\omega\hat{\boldsymbol{i}} + \cos\omega\cos\iota\hat{\boldsymbol{j}} + \cos\omega\sin\iota\hat{\boldsymbol{k}},
\end{aligned} \tag{10.87}$$

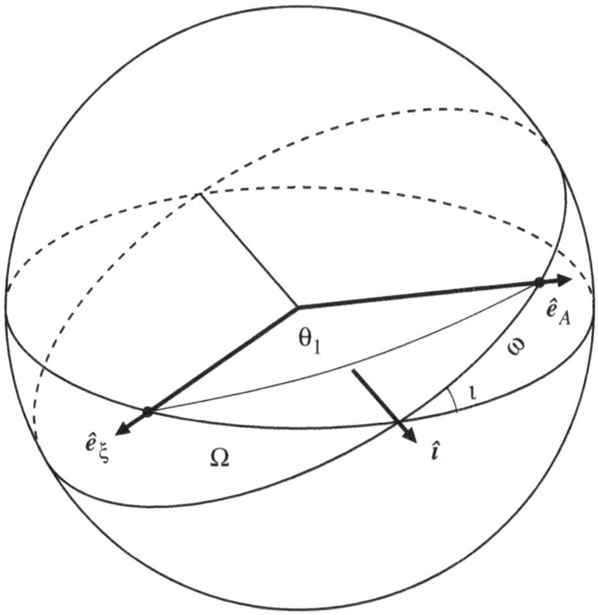

Figure 10.5 The coordinate base vectors $\hat{\boldsymbol{e}}_\xi$ and $\hat{\boldsymbol{e}}_A$ of the orbital planes of the binary and the third body. The planes are at an inclination ι relative to each other. The line of nodes is at an angle Ω relative to $\hat{\boldsymbol{e}}_\xi$, and $\hat{\boldsymbol{e}}_A$ is at an angle ω relative to the line of nodes. The angle between the two base vectors is θ_1.

where $\hat{\boldsymbol{i}}$, $\hat{\boldsymbol{j}}$ and $\hat{\boldsymbol{k}}$ are the unit vectors of a rectangular coordinate system. Unit vector $\hat{\boldsymbol{i}}$ points along the line of nodes (Fig. 10.5).

Now we go back to Eqs. (10.40) and (10.41). Vector products of Eq. (10.42) have to be calculated again using Eq. (10.87). We start by calculating

$$
\begin{aligned}
\hat{\boldsymbol{e}}_A \cdot \hat{\boldsymbol{e}}_\xi = \cos\theta_1 &= \cos\omega\cos\Omega - \sin\omega\sin\Omega\cos\iota \\
&= \cos(\omega-\Omega) - \sin\omega\sin\Omega(1+\cos\iota), \\
\hat{\boldsymbol{e}}_B \cdot \hat{\boldsymbol{e}}_\xi = \cos\theta_2 &= -\sin\omega\cos\Omega - \cos\omega\sin\Omega\cos\iota \\
&= -\sin(\omega-\Omega) - \cos\omega\sin\Omega(1+\cos\iota), \\
\hat{\boldsymbol{e}}_A \cdot \hat{\boldsymbol{e}}_\eta = \cos\theta_3 &= \cos\omega\sin\Omega + \sin\omega\cos\Omega\cos\iota \\
&= -\sin(\omega-\Omega) + \sin\omega\cos\Omega(1+\cos\iota), \\
\hat{\boldsymbol{e}}_B \cdot \hat{\boldsymbol{e}}_\eta = \cos\theta_4 &= -\sin\omega\sin\Omega + \cos\omega\cos\Omega\cos\iota \\
&= -\cos(\omega-\Omega) + \cos\omega\cos\Omega(1+\cos\iota).
\end{aligned}
$$

The θ are the angles between the unit vectors. For example, θ_1 is the angle between vectors $\hat{\boldsymbol{e}}_A$ and $\hat{\boldsymbol{e}}_\xi$, $\hat{\boldsymbol{e}}_A \cdot \hat{\boldsymbol{e}}_\xi = \cos\theta_1$ (Fig. 10.5).

With the help of these products and Eq. (10.41) we write the expressions equivalent to Eq. (10.42):

$$\begin{aligned}
\hat{\boldsymbol{e}}_\xi \cdot \hat{\boldsymbol{e}}_r &= \frac{a}{R_3}(\cos E - e)\cos\theta_1 + \frac{b}{R_3}\sin E\cos\theta_2, \\
\hat{\boldsymbol{e}}_\zeta \cdot (\hat{\boldsymbol{e}}_\xi \times \hat{\boldsymbol{e}}_r)\hat{\boldsymbol{e}}_\eta \cdot \hat{\boldsymbol{e}}_r &= \frac{a}{R_3}(\cos E - e)\cos\theta_3 + \frac{b}{R_3}\sin E\cos\theta_4, \\
\hat{\boldsymbol{e}}_\eta \cdot \hat{\boldsymbol{e}}_r &= \frac{a}{R_3}(\cos E - e)\cos\theta_3 + \frac{b}{R_3}\sin E\cos\theta_4, \\
\hat{\boldsymbol{e}}_\zeta \cdot (\hat{\boldsymbol{e}}_\eta \times \hat{\boldsymbol{e}}_r) = -\hat{\boldsymbol{e}}_\xi \cdot \hat{\boldsymbol{e}}_r &- \frac{a}{R_3}(\cos E - e)\cos\theta_1 - \frac{b}{R_3}\sin E\cos\theta_2.
\end{aligned} \tag{10.88}$$

As before we assume that the terms containing $\sin E$ may be neglected in comparison with the terms containing $\cos E - e$ since the $(a/R_3)^5$ factor in our integral peaks strongly at $E = 0$ where the $\sin E$ terms go to zero. This effectively drops the second terms out on the right hand sides of Eq. (10.88). Thus Eq. (10.40) is modified to

$$\begin{aligned}
\hat{\boldsymbol{e}}_\zeta \cdot (\boldsymbol{r} \times \boldsymbol{f}) = -\frac{3Gm_3a_i^2}{a^3}\left(\frac{a}{R_3}\right)^5 \frac{1}{2}(\cos E - e)^2 \\
\times [A_1\cos 2\phi + A_2\sin 2\phi],
\end{aligned} \tag{10.89}$$

where

$$\begin{aligned}
A_1 &= -2\cos\theta_1\cos\theta_3 \\
&= \sin[2(\omega - \Omega)] + \sin^2\omega\sin 2\Omega(1 + \cos\iota)^2 \\
&\quad - 2\sin\omega\cos(\omega - 2\Omega)(1 + \cos\iota), \\
A_2 &= \cos^2\theta_1 - \cos^2\theta_3 \\
&= \cos[2(\omega - \Omega)] - \sin^2\omega\cos 2\Omega(1 + \cos\iota)^2 \\
&\quad + 2\sin\omega\sin(\omega - 2\Omega)(1 + \cos\iota).
\end{aligned} \tag{10.90}$$

Putting $2\phi = kM + 2\phi_0$ Eq. (10.55) becomes

$$\begin{aligned}
\int_0^{M_0} \hat{\boldsymbol{e}}_\zeta \cdot (\boldsymbol{r} \times \boldsymbol{f})\,\mathrm{d}M = &\frac{3Gm_3}{a^3}a_i^2 \\
&\times \big(I^{(1)}[A_1\cos 2\phi_0 + A_2\sin 2\phi_0] \\
&\quad + I^{(2)}[-A_1\sin 2\phi_0 + A_2\cos 2\phi_0]\big).
\end{aligned} \tag{10.91}$$

Since $\left|I^{(1)}\right| \ll \left|I^{(2)}\right|$, the terms proportional to $I^{(1)}$ may be ignored. Therefore the product $\cos^2\iota\cos 2\phi_0$ in Eq. (10.72) is replaced by

$$-A_1\sin 2\phi_0 + A_2\cos 2\phi_0.$$

Rather than using A_1 and A_2 with their full dependence on ω, Ω and ι, let us do some averaging. We may take the attitude that we do not really care which

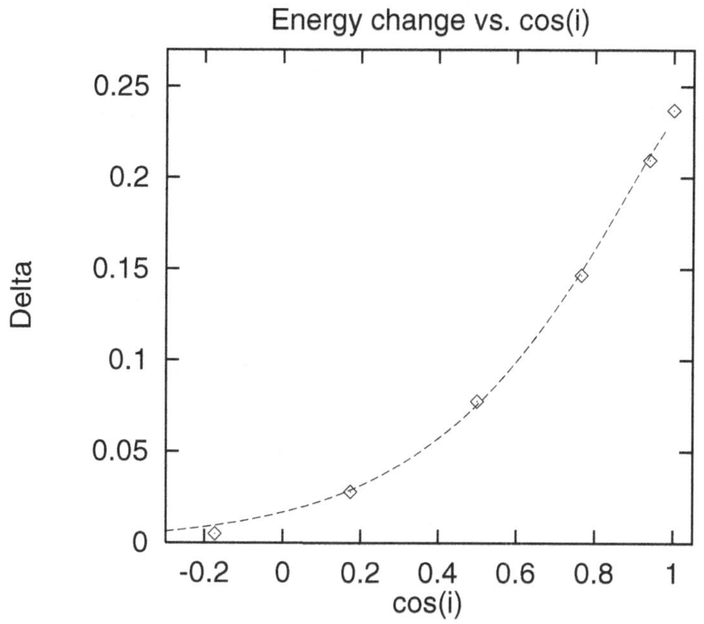

Figure 10.6 A comparison of the expected $|\Delta| \approx (1 + \cos \iota)^4$ dependence with experimental data from Huang and Valtonen (1987) at $Q = 2.5$.

way the major axis of the outer orbit points, and similarly we consider the line of nodes to be randomly placed in the binary orbital plane. Then the average values of $\sin 2(\omega - \Omega)$, $\cos 2(\omega - \Omega)$, $\sin \omega \sin(\omega - 2\Omega)$ and $\sin \omega \cos(\omega - 2\Omega)$ are zero while $\langle \sin^2 \omega \rangle = \frac{1}{2}$. What remains is

$$- \langle A_1 \rangle \sin 2\phi_0 + \langle A_2 \rangle \cos 2\phi_0 = -\frac{1}{2} \cos 2(\phi_0 - \Omega)(1 + \cos \iota)^2. \qquad (10.92)$$

(A different calculation is given as an exercise, Problem 10.10.) To be exact, also the $\cos 2(\phi_0 - \Omega)$ factor averages to zero when Ω is arbitrary. We have kept it, however, to show that the inclination dependence is more likely to be a function of $(1 + \cos \iota)^2$ rather than of $\cos^2 \iota$. It is reasonable that our result should depend on $\phi_0 - \Omega$ since both angles ϕ_0 and Ω are measured from the ξ axis which itself has no physical significance for a circular binary. The only physically significant quantity is their difference $\phi_0 - \Omega$. In the second order theory we have to square this factor, as we will see in the next section, and thus we expect the energy change in a slow encounter to be proportional to $(1 + \cos \iota)^4$. Figure 10.6 shows that this is indeed a good approximation (Huang and Valtonen 1987). However, the amplitude of the sinusoidal variation of Δ comes from the first order theory and therefore it is of the order of $(1 + \cos \iota)^2$ only (Eq. (10.84)).

10.5 Change in eccentricity

The change in the binary eccentricity e_i may be calculated by using Eq. (9.8). Since the binary orbit precesses slowly in comparison with the orbital motion, we may write

$$\frac{\partial R}{\partial \omega} \ll \frac{\partial R}{\partial M}$$

in Eq. (9.8). Therefore

$$\dot{e}_i = \frac{1}{2}\frac{(1-e_i^2)}{a_i e_i}\dot{a}_i \tag{10.93}$$

or

$$\Delta e_i = \frac{1}{2}\frac{1-e_i^2}{e_i}\Delta.$$

Initially $e_i = 0$; therefore its typical value is $e_i \approx \frac{1}{2}\Delta e_i$. As long as it is small, $1 - e_i^2 \approx 1$. Thus

$$(\Delta e_i)^2 \approx |\Delta|. \tag{10.94}$$

This expectation agrees well with numerical experiments (Huang and Valtonen 1987). They show that $(\Delta e_i)^2$ varies sinusoidally (with period π) about a non-zero mean value (Valtonen 1975b), as expected.

To be more quantitative, let us move onto the calculation of the $\dot{e}$ term in Eq. (10.20). It leads to a theory for the mean energy change $\langle\Delta\rangle$. First we need an expression for Δe_i. The change in eccentricity Δe_i is calculated in a manner similar to the calculation of $\Delta a_i/a_i$. We use Eqs. (10.20), (10.31)–(10.32) and (10.34), substitute ϕ for E, a_i for a and b and put $e = 0$. In addition, we use Eq. (10.88) where we neglect the $\sin E$ terms as small in comparison with the $\cos E - e$ terms, as we did previously. Then

$$\begin{aligned}
\dot{e}_i = &-3\frac{m_3}{m_B}\frac{\sqrt{Gm_B}}{R_3^3}a_i^{3/2} \\
&\times\left[\left(\frac{a}{R_3}\right)^2(\cos E - e)^2(\cos^2\theta_1 - \cos^2\theta_3)\sin\phi\cos^2\phi\right. \\
&- 2\left(\frac{a}{R_3}\right)^2(\cos E - e)^2\cos\theta_1\cos\theta_3\cos^3\phi \\
&\left.- \left(\frac{a}{R_3}\right)^2(\cos E - e)^2\cos^2\theta_3\sin\phi + \frac{1}{3}\sin\phi\right].
\end{aligned}$$

The integral of $\dot{e}_i$ over the whole pericentre passage is

$$\Delta e_i = \int_{-t_0}^{t_0} \dot{e}_i \,\mathrm{d}t = \frac{a^{3/2}}{\sqrt{G(m_B + m_3)}} \int_{-M_0}^{M_0} \dot{e}_i \,\mathrm{d}M.$$

The integral is split into several terms by using the identities

$$\begin{aligned}
\sin\phi\cos^2\phi &= \frac{1}{4}(\sin\phi + \sin 3\phi),\\
\cos^3\phi &= \frac{1}{4}(3\cos\phi + \cos 3\phi),\\
\sin\phi &= \sin\frac{k}{2}M\cos\phi_0 + \cos\frac{k}{2}M\sin\phi_0,\\
\cos\phi &= \cos\frac{k}{2}M\cos\phi_0 - \sin\frac{k}{2}M\sin\phi_0,\\
\sin 3\phi &= \sin\frac{3}{2}kM\cos 3\phi_0 + \cos\frac{3}{2}kM\sin 3\phi_0,\\
\cos 3\phi &= \cos\frac{3}{2}kM\cos 3\phi_0 - \sin\frac{3}{2}kM\sin 3\phi_0.
\end{aligned}$$

However, we know from the experience gained in Section 10.3 that the terms proportional to $\cos\frac{1}{2}kM$ and $\cos\frac{3}{2}kM$ are going to be small compared with the $\sin\frac{1}{2}kM$ and $\sin\frac{3}{2}kM$ terms, and the former may be neglected. Thus

$$\begin{aligned}
\Delta e_i =& \frac{3}{2\sqrt{2}}\sqrt{\frac{2m_B}{m_B+m_3}\frac{m_3}{m_B}}\left(\frac{a_i}{a}\right)^{3/2}\int_{-M_0}^{M_0}\left[-\frac{2}{3}\left(\frac{a}{R_3}\right)^3\sin\frac{k}{2}M\cos\phi_0\right.\\
&-\frac{1}{2}\left(\frac{a}{R_3}\right)^5(\cos E - e)^2(\cos^2\theta_1 - 5\cos^2\theta_3)\sin\frac{k}{2}M\cos\phi_0\\
&-\frac{1}{2}\left(\frac{a}{R_3}\right)^5(\cos E - e)^2(\cos^2\theta_1 - \cos^2\theta_3)\sin\frac{3}{2}kM\cos 3\phi_0\\
&-3\left(\frac{a}{R_3}\right)^5(\cos E - e)^2\cos\theta_1\cos\theta_3\sin\frac{k}{2}M\sin\phi_0\\
&\left.-\left(\frac{a}{R_3}\right)^5(\cos E - e)^2\cos\theta_1\cos\theta_3\sin\frac{3}{2}kM\sin 3\phi_0\right]\mathrm{d}M.
\end{aligned}$$

The integrals consist of two segments: from $-M_0$ to 0 and from 0 to M_0. Between the two segments ϕ_0 changes its sign in order that the two orbit solutions match each other at $M = 0$. The integrals are antisymmetric relative to $M = 0$; that means that the terms containing $\cos\phi_0$ or $\cos 3\phi_0$ are antisymmetric, and add up to zero when integrated from $-M_0$ to $+M_0$. The terms containing $\sin\phi_0$ and $\sin 3\phi_0$ are symmetric and make equal contributions to both segments (Fig. 10.7).

The integrals which we need to evaluate are the familiar $I^{(2)}$ (Eq. (10.57)), now calculated at $k/2$ and $3k/2$ instead of k. Let us call these values $I^{(2)}(k/2)$ and

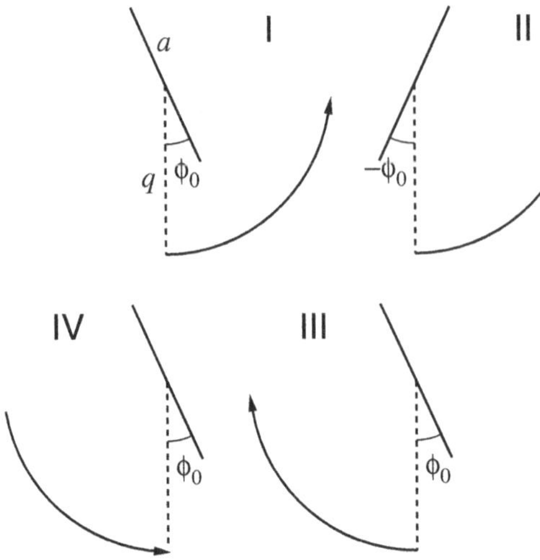

Figure 10.7 The outward path (I) and the inward path (IV) and how they are obtained from each other by three transformations. Transformation I→II changes $\phi_0 \to -\phi_0$ which leaves $\cos\phi_0$ unchanged. Transformation II→III is simply a left-to-right flip, while in the transformation III→IV the time is reversed, i.e. $M \to -M$. This leads to $I^{(2)} \to -I^{(2)}$, Eq. (10.58). Thus for terms proportional to $\cos\phi_0$ or $\cos 3\phi_0$ the eccentricity change Δe_i in the inward path is equal in magnitude to the eccentricity change in the outward path but opposite in sign. In contrast, the terms proportional to $\sin\phi_0$ or $\sin 3\phi_0$ add up from the two paths.

$I^{(2)}(3k/2)$, respectively. Then

$$I^{(2)}(k/2) = \frac{1}{50}\frac{1}{k}\sum_{n=1}^{5} B_n\left[1-\left(\frac{2n}{k}\right)^2\right]^{-1}\left[\cos\left(\frac{8\pi n}{k}\right)-1\right],$$
$$I^{(2)}(3k/2) = \frac{1}{150}\frac{1}{k}\sum_{n=1}^{8} B_n\left[1-\left(\frac{2n}{3k}\right)^2\right]^{-1}\left[\cos\left(\frac{8\pi n}{3k}\right)-1\right].$$

At $k = 12$ a straightforward calculation gives $I^{(2)}(k/2) \approx -0.25$ and $I^{(2)}(3k/2) \approx -0.06$. The latter value is small enough that it may be ignored in comparison with the former. Numerical evaluation of the sums shows that this is reasonable for $k > 8$, and especially for higher values of k, e.g. for $k > 27$, $I^{(2)}(k/2) > 10I^{(2)}(3k/2)$. The value of $I^{(2)}(k/2)$ may be taken as a constant in the range $8 < k \leq 30$.

Then

$$\Delta e_i = \frac{3 \cdot 0.25}{\sqrt{2}}\sqrt{\frac{2m_B}{m_B+m_3}\frac{m_3}{m_B}}\left(\frac{a_i}{a}\right)^{3/2} 6\cos\theta_1\cos\theta_3\sin\phi_0$$
$$\approx -3\sqrt{\frac{2m_B}{m_B+m_3}\frac{m_3}{m_B}}\left(\frac{a_i}{a}\right)^{3/2} A_1\sin\phi_0.$$

The coefficient A_1 is given by Eq. (10.90); when averaged over ω

$$A_1 = \frac{1}{2}\sin 2\Omega(1+\cos\iota)^2.$$

Finally we need $\langle(\Delta e_i)^2\rangle$, averaged over all Ω and ϕ_0:

$$\langle(\Delta e_i)^2\rangle = 0.025\left(\frac{m_3}{m_B}\right)^2\left(\frac{Q}{2.5}\right)^{-6}(1+\cos\iota)^4, \tag{10.95}$$

where we have made use of Eq. (10.52) for a_i/a and have multiplied the right hand side by $12/k$ (Eq. (10.68)) in order to make the expression valid for all Q.

Combining numerical data with theory we get finally

$$\langle(\Delta e_i)^2\rangle = 0.016\left(\frac{m_3}{m_B}\right)^2\left(1+\frac{m_3}{m_B}\right)^{2/3}\left(\frac{Q}{2.5}\right)^{-8}(1+\cos\iota)^4. \tag{10.96}$$

Note that the power n in $(Q/2.5)^{-n}$ is higher than what is expected from Eq. (10.95). The semi-experimental expression of Eq. (10.96) is valid only in the range $0^\circ \le \iota \le 60^\circ$; at other inclinations it becomes more complicated.

We may now return to the second term in the change of the binary energy (Eq. (10.20))

$$\langle\Delta\rangle = \left\langle\frac{\Delta a_i}{a_i}\right\rangle = \left\langle\frac{2e_i\Delta e_i}{1-e_i^2}\right\rangle \approx \langle(\Delta e_i)^2\rangle$$

if $e_i \approx \frac{1}{2}\Delta e_i$. This is a second order term since in the first order $e_i = 0$.

There is a complication in the the second order theory. The mean energy change varies its sign between direct and retrograde orbits. While $\langle\Delta\rangle$ is negative for direct orbits, it becomes positive for retrograde orbits. At intermediate inclinations $\langle\Delta\rangle$ is positive at small Q and at some value $Q = Q_0$ it crosses the $\langle\Delta\rangle = 0$ line and becomes negative. Typically $Q_0 \approx 2/3Q_1$. This behaviour may be modelled by using a $(Q - Q_0)$ factor as a multiplier in the expression for $\langle\Delta\rangle$.

The results based on numerical experiments are best described in piecewise manner in different ranges of inclination. In the range $90^\circ \le \iota \le 150^\circ$ the amplitude $A(\Delta)$ always dominates over $\langle\Delta\rangle$ and the values of $\langle\Delta\rangle$ are relatively small. In this range we may thus put effectively $\langle\Delta\rangle = 0$. In the remainder of the retrograde range we may write

$$\langle\Delta\rangle = 0.11 f_2(m_3)\left(\frac{Q}{2.62}\right)^{-6}\mathrm{cs}^2(\iota), \tag{10.97}$$

where

$$f_2(m_3) = \left(\frac{m_3}{m_B}\right)^{2+7.5(1+\cos\iota)} \Bigg/ \left[\left(1+\frac{m_3}{m_B}\right)\Big/2\right]^{10(1+\cos\iota)}, \quad (10.98)$$

$$\mathrm{cs}^2(\iota) = 0.136\left[0.38\left(1+\frac{m_3}{m_B}\right)^{-0.45} - (1+\cos\iota)\right]^2. \quad (10.99)$$

For direct orbits the equation is of the form

$$\langle\Delta\rangle = -0.11 f_2(m_3)(Q - Q_0)\left(\frac{Q}{2.62}\right)^{-n_2} \mathrm{cs}^2(\iota), \quad (10.100)$$

where

$$\begin{aligned} Q_0 &= 1.98\left(\frac{m_3}{m_B}\right)^{0.11}\left(1+\frac{m_3}{m_B}\right)^{0.155}, \quad 0.5 \le \cos\iota \le 1, \\ Q_0 &= (1.54 + 0.21\cos\iota)\left(1+\frac{m_3}{m_B}\right)^{0.4}, \quad 0 \le \cos\iota \le 0.5, \end{aligned} \quad (10.101)$$

and

$$\begin{aligned} n_2 &= 10, \quad 0.5 \le \cos\iota \le 1, \\ n_2 &= 9 + 2\cos\iota, \quad 0 \le \cos\iota \le 0.5, \end{aligned} \quad (10.102)$$

$$\begin{aligned} f_2(m_3) &= \left(\frac{m_3}{m_B}\right)^{2}\left(1+\frac{m_3}{m_B}\right)^{-0.5}, \quad 0.5 \le \cos\iota \le 1, \\ f_2(m_3) &= \left(\frac{m_3}{m_B}\right)^{4/3}\left(1+\frac{m_3}{m_B}\right)^{7/8}, \quad 0 \le \cos\iota \le 0.5, \end{aligned} \quad (10.103)$$

$$\begin{aligned} \mathrm{cs}^2(\iota) &= 1.4\left[0.5(m_3/m_B)^{1/2} + ((1+\cos\iota)/2)^2\right]^{20/7}, \quad 0.5 \le \cos\iota \le 1, \\ \mathrm{cs}^2(\iota) &= 0.55[0.27 + 1.35\cos\iota] \\ &\quad \times \left[0.5(m_3/m_B)^{1/2} + ((1+\cos\iota)/2)^2\right]^{2n_2/7}, \quad 0 \le \cos\iota \le 0.5. \end{aligned} \quad (10.104)$$

A comparison of numerical data with the above equations is shown in Figures 10.8 and 10.9.

10.6 Stability of triple systems

Up to now the notion of stability has been used in relation to only one pericentre passage. Often it is more interesting to know what happens after many pericentre passages when the third body approaches the binary repeatedly. Obviously, a more

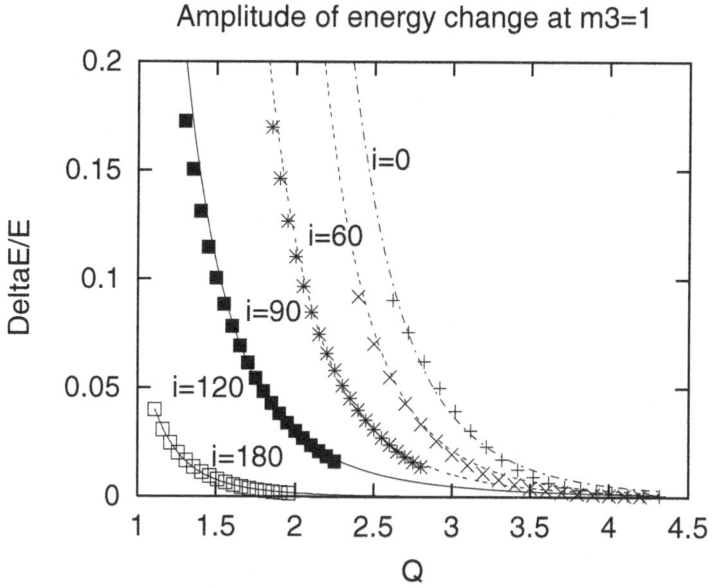

Figure 10.8 The amplitude $A(\Delta)$ of the energy change as a function of the pericentre distance Q at several inclinations: $\iota = 0°$ (+), 60° (×), 90° (∗), 120° (■), and at 180° (□). The curves follow Eqs. (10.80)–(10.86). The case of $m_3 = m_B$.

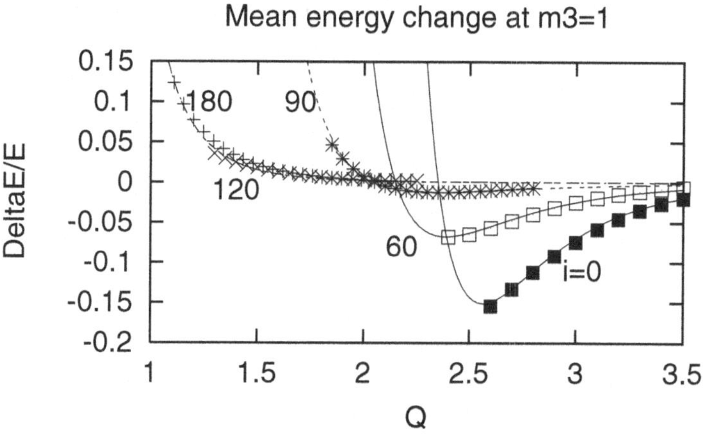

Figure 10.9 The mean energy change $\langle\Delta\rangle$ as a function of the pericentre distance Q at several inclinations: $\iota = 0°$ (■), 60° (□), 90° (∗), 120° (×) and at 180° (+). The curves follow Eqs. (10.97)–(10.104). The case of $m_3 = m_B$.

stringent stability limit, i.e. a greater value of Q, is needed to guarantee stability. The stability may also be defined in different ways which give slightly different results.

Let us start by defining stability so that for a stable orbit the relative energy change should be no greater than 10^{-3} in either direction during a single pericentre passage.

This corresponds to the survival of the triple system for 10^5 revolutions of the outer binary, according to the following estimate. If the destabilising level of accumulated relative energy change is $10^{-1/2}$, and the energy change accumulates in the manner of a random walk (Huang and Innanen 1983), then the destabilising level is achieved after $\left(10^{-1/2}/10^{-3}\right)^2 = 10^5$ steps. The random walk type behaviour of the energy changes is due to the phase factor $\cos 2\phi_0$ in Eq. (10.73). Generally successive encounters take place with different values of ϕ_0, the latter being distributed more or less randomly. Equation (10.73) has also a constant (independent of ϕ_0) drift factor which may be dominant depending on the inclination. However, numerical experiments (Saslaw *et al.* 1974) have shown that once the eccentricity e_i has increased to about $e_i = 0.2$, the drift becomes insignificant in comparison with the random walk. Equation (10.96) shows that the eccentricity goes over this limit quite easily at moderate values of Q.

For example, $\left\langle(\Delta e_i)^2\right\rangle \approx 0.0016$ at $Q = Q_{\rm st}(A(\Delta) = 10^{-3})$, when $m_3 = m_B$. Then it takes about 25 steps for e_i^2 to go over the limit of $e_i^2 = 0.04$.

When we extend the theory to small values of Δ, i.e. to large Q, we have to take account of the exponential factor. It makes the power law Q^{-n} steeper the greater is the value of Q. For the relative energy change $\Delta = 10^{-3}$ the suitable effective power is 11 (Valtonen 1975b).

Therefore the power of the mass and inclination factors in Eq. (10.76) should be lowered from $1/3$ to ≈ 0.09. Using numerical experiments we further refine the expression and get a new stability limit, suitable for all inclinations:

$$Q_{\rm st}(A(\Delta) = 10^{-3}) = 3.62\,[(1 + m_3/m_B)/2]^{0.23}\left(\frac{m_3}{m_B}\right)^{0.09} \times [1.035 + \cos\iota]^{0.18}\,. \qquad (10.105)$$

Figure 10.10 shows that this gives a good description of the stability boundary. Figure 10.10 also demonstrates that the stability limit based on the drift $|\langle\Delta\rangle|$ at the level of 10^{-3} is about equal to or less than the limit derived above from the amplitude $A(\Delta)$. Thus the contribution of the drift to the stability boundary can generally be ignored. Only at small inclinations and close to $\iota = 180°$, and as long as e_i stays small, does the drift become important.

Less stringent stability criteria have also been used. One may require that in 100 revolutions of the outer orbit there is no major orbital change (Mardling & Aarseth 1999), or that within some specific number of revolutions of the original outer orbit there should be neither exchanges of the binary members nor escapes of any of the bodies (Huang and Innanen 1983 use the revolution number $N = 62$, Eggleton and Kiseleva 1995 use $N = 100$); sometimes the survival through an even smaller number (10–20) of revolutions has been considered to be enough for

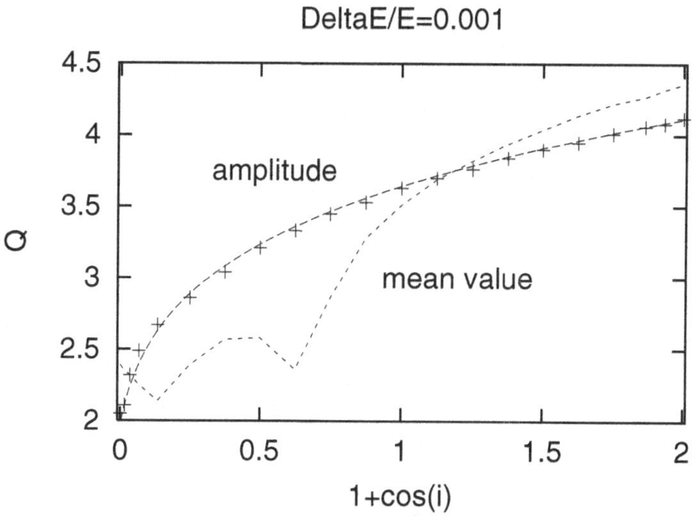

Figure 10.10 The Q-boundary of $A(\Delta) = 10^{-3}$ (+) and of $|\langle\Delta\rangle| = 10^{-3}$ (dotted line). The analytical function for the former is drawn as a dashed line. The mean value changes its sign from positive to negative at $\cos\iota \approx -0.375$ when going from -1 to $+1$ along the $\cos\iota$ axis. The case of $m_3 = m_B$.

stability (Harrington 1972, 1975). However, the results are not very different even though the chosen revolution number N varies a lot. This is because in a random walk with a constant energy step, in order to cover a standard magnitude change in energy, $\sqrt{N} = \text{const}\, Q^{11}$, i.e. the stability limit Q_{st} varies only as the $1/22$nd power of N.

The stability limit also depends on the strength of binding of the outer binary to the inner binary. If the outer binary is initially only very loosely bound, then even a small positive energy increase at the pericentre may set it loose. The degree of relative binding is best described by the axial ratio a_i/a, or since $Q = (a/a_i)(1-e)$, by $(1-e)/Q$. Putting this relative binding equal to the relative energy change $|\Delta| \propto Q^{-11}$, i.e. setting $|\Delta| = \Delta E_B/E_B \approx E_{\text{outer}}/E_B \approx (1-e)/Q$, we find that the stability limit varies as

$$Q_{\text{st}} \propto (1-e)^{-\alpha} \tag{10.106}$$

where $\alpha = 0.1$. Actually, putting α as 0.3–0.4 gives a better agreement with some experiments (see Fig. 10.11; Huang and Innanen 1983, Mardling and Aarseth 1999) while in others $\alpha \simeq 0.0$ (Eggleton and Kiseleva 1995); the results depend on the definition of stability and on the masses of the bodies.

The experimental value for the stability limit for equal masses $m_1 = m_2 = m_3$, $e = 0$ and $\cos\iota = 1$ is $Q = 2.7$ after $N = 62$ revolutions (Huang and Innanen 1983). With these parameters $A(\Delta)$ gives the value 0.03 which becomes

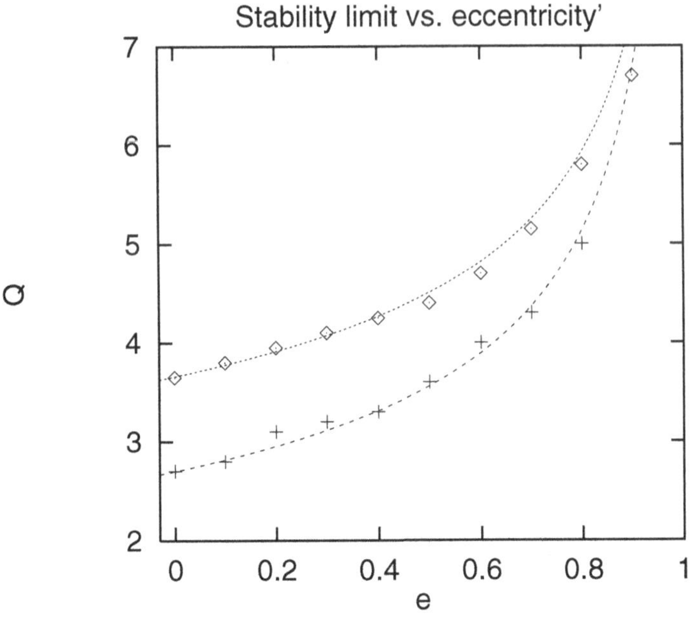

Figure 10.11 A comparison of Eq. (10.107) with experimental data from Huang and Innanen 1983 (+) and Mardling and Aarseth 1999 (◇).

$\Delta E_B/E_B = 0.23$ after multiplication by $\sqrt{62}$. This is just at the relative energy change usually associated with instability (i.e. ≈ 0.2) and thus the stability limit of Huang and Innanen (1983) is as expected. The stability limit of Mardling and Aarseth (1999) for the same case is $Q = 3.65$ which seems contradictory. At this pericentre distance we expect the average energy change per revolution to be 0.003 which is multiplied by $\sqrt{100}$ and adds up to 0.03 after 100 revolutions. This is only about 10% of the value at the stability boundary. However, in Mardling and Aarseth (1999) the stability criterion was such that two orbits initially differing by one part in 10^5 in the eccentricity should remain close after 100 orbits. $|\Delta E_B/E_B|$ being at about 10% of the stability limit could well be used as a definition for two nearly identical orbits not to have evolved too far apart from each other and for the system to be stable. The corresponding stability limit of Bailyn (1984), $Q = 3.1$, lies between the previous two, and it appears that the definition of instability is also intermediate between those of Huang and Innanen (1983) and Mardling and Aarseth (1999).

So far we have not considered the possibility that the inner binary orbit may be eccentric. In the first approximation we may take the time averaged mean separation $r = a_i[1 + 0.5e_i^2]$ in place of a_i in our perturbation equations. Then the stability limits obtained earlier are simply multiplied by r/a_i since the inner binary is effectively this much greater in extent, and the encounter has to be more distant by the same factor for stability. In practice it appears that this method gives reasonable agreement with numerical experiments (Bailyn 1984).

Even though there are obviously many different ways to define stability, and correspondingly many possible stability limits, it appears safest to use $Q_{\rm st}$ as defined by orbit calculations:

$$Q_{\rm st}^{(2)} = 2.8\left(1+\frac{m_3}{m_B}\right)^{1/6}(1-e)^{-0.1} \times\left(1+\frac{1}{2}e_i^2\right)\left[\frac{1}{3}+((1+\cos\iota)(1.97-\cos\iota))^{0.8}\right]^{1/3} \qquad (10.107)$$

This expression has been found to be quite satisfactory in numerical experiments in the 'random walk range' $-0.9 \le \cos\iota \le 0.6$ (Valtonen *et al.* 2006) for the stability of $N = 1000$ revolutions when the stability is defined in the manner of Huang and Innanen (1983).

Here we must remember the Kozai resonance which operates effectively at inclination angles close to $\iota = \pi/2$. The inner eccentricity grows up to values close to $e_i = 1$ which means that the factor $1 + 0.5e_i^2$ is best replaced by 1.5 in Eq. (10.107) for these inclinations, independent of the original eccentricity e_i (Miller and Hamilton 2002, Wen 2003).

Example 10.1 A large number of stars are found in triple systems many of which appear to be stable in the sense that the pericentre of the outer orbit is far beyond $Q_{\rm st}^{(2)}$. How do they originate?

Besides the obvious possibility that triple stars were born like this, we may consider evolutionary processes which add a third star to a binary. A star passing by a binary in a retrograde orbit is likely to lose energy and may become bound to it. However, its relative pericentre distance Q remains small, and subsequent encounters are likely to reduce it further until a resonance forms and leads to an escape. Is there anything that might alter this evolution and make a stable triple star? In the capture process the eccentricity of the binary is increased and it may happen that the increase is enough to make two stars of the inner binary touch or almost touch each other at the pericentre of the orbit. Then tidal dissipation of the orbital energy makes the inner binary shrink, and effectively Q increases for the outer binary to the extent that the triple is stable. Some triples must have followed this evolutionary path (Bailyn and Grindlay 1987, Bailyn 1987, 1989). A more efficient process is a binary–binary interaction. Then we may regard the more compact binary as a single point which interacts with a wider binary. One of the members of the wide binary is likely to escape and what remains is a hierarchical three-body system. It may be stable or unstable. Numerical simulations by Mikkola (1983, 1984a, b) indicate a large probability for forming stable triples in this way.

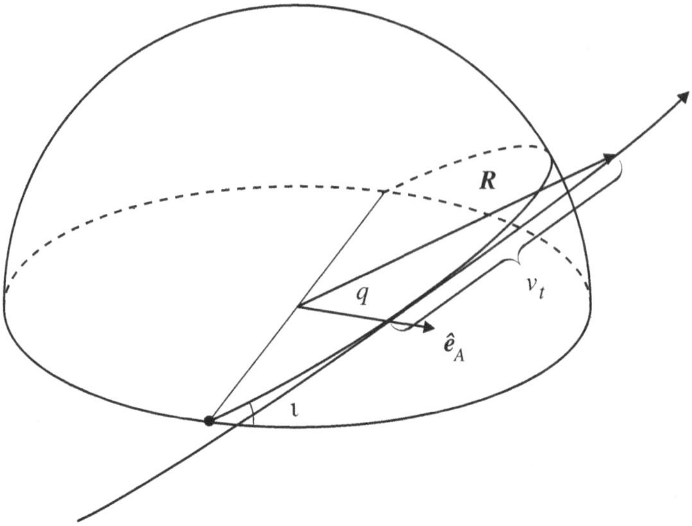

Figure 10.12 The third body passes the binary centre of mass at the distance q with a constant speed V and is located at the position $\boldsymbol{R}$ at time t after the pericentre passage. The orbital plane of the third body is at an angle ι relative to the orbital plane of the binary (horizontal plane). The unit vector $\hat{\boldsymbol{e}}_A$ points from the binary centre of mass towards the pericentre.

10.7 Fast encounters

Fast encounters imply that the outer orbit is so strongly hyperbolic that we may use a straight line approximation for the relevant segment of the third-body orbit. To simplify the calculation further, assume that the binary is at a constant phase angle ϕ during the third-body passage. Since most of the action takes place close to the pericentre this *impulsive approximation* may be quite appropriate. The geometry of the encounter is illustrated in Fig. 10.12. Let q be the closest approach distance of the third body to the binary centre of mass, and let V be the uniform speed of the third body in its straight line orbit. Then

$$\boldsymbol{R} = q\hat{\boldsymbol{e}}_A + Vt\hat{\boldsymbol{e}}_B. \tag{10.108}$$

Here the unit vectors $\hat{\boldsymbol{e}}_A$ and $\hat{\boldsymbol{e}}_B$ point from the binary centre of mass towards the pericentre point of closest approach and along a line parallel to the orbital motion, respectively. As before, t is the time, with $t = 0$ at the pericentre. The magnitude of $\boldsymbol{R}$ is

$$R = \sqrt{q^2 + V^2t^2} \tag{10.109}$$

and the corresponding unit vector is

$$\hat{\boldsymbol{e}}_r = \frac{q}{R}\hat{\boldsymbol{e}}_A + \frac{Vt}{R}\hat{\boldsymbol{e}}_B. \tag{10.110}$$

In order to calculate the relative energy change during the pericentre passage we proceed as before. The required quantities are:

$$
\begin{aligned}
\hat{\boldsymbol{e}}_\xi \cdot \hat{\boldsymbol{e}}_r &= \frac{q}{R}\cos\theta_1 + \frac{Vt}{R}\cos\theta_2,\\
\hat{\boldsymbol{e}}_\eta \cdot \hat{\boldsymbol{e}}_r &= \frac{q}{R}\cos\theta_3 + \frac{Vt}{R}\cos\theta_4,\\
\hat{\boldsymbol{e}}_\zeta \cdot (\hat{\boldsymbol{e}}_\xi \times \hat{\boldsymbol{e}}_r) = \hat{\boldsymbol{e}}_\eta \cdot \hat{\boldsymbol{e}}_r &= \frac{q}{R}\cos\theta_3 + \frac{Vt}{R}\cos\theta_4,\\
\hat{\boldsymbol{e}}_\zeta \cdot (\hat{\boldsymbol{e}}_\eta \times \hat{\boldsymbol{e}}_r) = -\hat{\boldsymbol{e}}_\xi \cdot \hat{\boldsymbol{e}}_r &= -\frac{q}{R}\cos\theta_1 - \frac{Vt}{R}\cos\theta_2,
\end{aligned}
\tag{10.111}
$$

which are substituted into Eq. (10.40). In writing this equation it was assumed that the inner eccentricity $e_i = 0$, and that the binary separation is a_i. In the case of an eccentric binary we could use the time averaged mean separation $a_i(1 + 0.5e_i^2)$ instead of a_i; this would lead to a correction factor $(1 + 0.5e_i^2)$ on the right hand side of the next equation and in subsequent equations. We get

$$
\begin{aligned}
\hat{\boldsymbol{e}}_\zeta \cdot (\boldsymbol{r} \times \boldsymbol{f}) = \frac{3Gm_3a_i^2}{R^5}\Big[&\big(q^2\cos\theta_1\cos\theta_3 + V^2t^2\cos\theta_2\cos\theta_4\\
&+ qVt(\cos\theta_2\cos\theta_3 + \cos\theta_1\cos\theta_4)\big)\cos 2\phi\\
&+ \Big(\frac{1}{2}q^2(\cos^2\theta_3 - \cos^2\theta_1) + \frac{1}{2}V^2t^2(\cos^2\theta_4 - \cos^2\theta_2)\\
&+ qVt(\cos\theta_3\cos\theta_4 - \cos\theta_1\cos\theta_2)\Big)\sin 2\phi\Big].
\end{aligned}
\tag{10.112}
$$

To integrate from $t = -\infty$ to $t = \infty$, the integrals needed are

$$
\begin{aligned}
I_1 &= \int_{-\infty}^{\infty} \frac{qVt\,\mathrm{d}t}{(q^2 + V^2t^2)^{5/2}} = -\Big|_{-\infty}^{\infty} \frac{q}{3V(q^2 + V^2t^2)^{3/2}} = 0,\\
I_2 &= \int_{-\infty}^{\infty} \frac{q^2\,\mathrm{d}t}{(q^2 + V^2t^2)^{5/2}} = \Big|_{-\infty}^{\infty} \frac{1}{2}\left(1 + \frac{2}{3}\frac{V^2t^2}{q^2}\right)\frac{t}{(q^2 + V^2t^2)^{3/2}}\\
&= \frac{2}{3q^2V},\\
I_3 &= \int_{-\infty}^{\infty} \frac{V^2t^2\,\mathrm{d}t}{(q^2 + V^2t^2)^{5/2}} = \Big|_{-\infty}^{\infty} \frac{V^2t^3}{3q^2(q^2 + V^2t^2)^{3/2}} = \frac{2}{3q^2V} = I_2.
\end{aligned}
\tag{10.113}
$$

Therefore

$$\int_{-\infty}^{\infty} \hat{\boldsymbol{e}}_\zeta \cdot (\boldsymbol{r} \times \boldsymbol{f})\,\mathrm{d}t = -\frac{3Gm_3a_i^2}{2}\big[(A_1I_2 + A_3I_3)\cos 2\phi + (A_2I_2 + A_4I_3)\sin 2\phi\big] = -\frac{Gm_3a_i^2}{q^2V}\big[(A_1 + A_3)\cos 2\phi + (A_2 + A_4)\sin 2\phi\big]. \tag{10.114}$$

The coefficients A_1 and A_2 were previously defined (Eq. (10.90)):

$$\begin{aligned} A_1 &= -2\cos\theta_1\cos\theta_3 \\ &= \sin 2(\omega - \Omega) + \sin^2\omega \sin 2\Omega(1 + \cos\iota)^2 \\ &\quad - 2\sin\omega\cos(\omega - 2\Omega)(1 + \cos\iota), \\ A_2 &= \cos^2\theta_1 - \cos^2\theta_3 \\ &= \cos 2(\omega - \Omega) - \sin^2\omega\cos 2\Omega(1 + \cos\iota)^2 \\ &\quad + 2\sin\omega\sin(\omega - 2\Omega)(1 + \cos\iota). \end{aligned} \tag{10.115}$$

The two other coefficients are

$$\begin{aligned} A_3 &= -2\cos\theta_2\cos\theta_4 \\ &= -\sin 2(\omega - \Omega) + \cos^2\omega\sin 2\Omega(1 + \cos\iota)^2 \\ &\quad + 2\cos\omega\sin(\omega - 2\Omega)(1 + \cos\iota), \\ A_4 &= \cos^2\theta_2 - \cos^2\theta_4 \\ &= -\cos 2(\omega - \Omega) - \cos^2\omega\cos 2\Omega(1 + \cos\iota)^2 \\ &\quad + 2\cos\omega\cos(\omega - 2\Omega)(1 + \cos\iota). \end{aligned} \tag{10.116}$$

As before, we may simplify the result considerably by averaging over ω and by keeping only the terms proportional to $(1 + \cos\iota)^2$ which contain the $\sin^2\omega$ or $\cos^2\omega$ coefficients explicitly. Then

$$\begin{aligned} \langle A_1 + A_3\rangle &= \sin 2\Omega(1 + \cos\iota)^2, \\ \langle A_2 + A_4\rangle &= -\cos 2\Omega(1 + \cos\iota)^2. \end{aligned} \tag{10.117}$$

We use Eq. (10.70) and obtain finally

$$\frac{\Delta a_i}{a_i} = \frac{2\sqrt{Gm_B}a_i^{3/2}}{q^2V}\left(\frac{m_3}{m_B}\right)(1 + \cos\iota)^2\sin 2(\phi - \Omega). \tag{10.118}$$

Averaged over all phases $\phi - \Omega$, to the first order there is no net change in a_i during the pericentre passage.

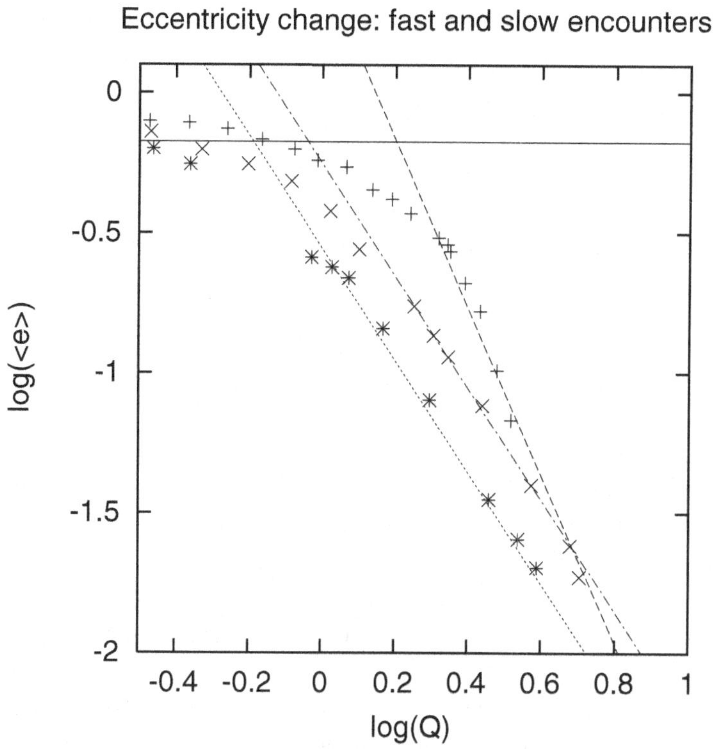

Figure 10.13 The average change of eccentricity as a function of Q. Experimental points are from Hills (1984): $V/v_0 = 0.034$ (+), 3.19 (×) and 10.37 (∗). The lines refer to Eq. (10.95) (slope −3; $\cos\iota = 0.44$) and to Eq. (10.121) (slope −2). The horizontal line corresponds to the average value from Eq. (7.14), $\langle e \rangle = 0.67$.

What about the contribution of the second term in Eq. (10.12)? It gives

$$\frac{\Delta E_B}{E_B} = -\frac{2e_i \Delta e_i}{1 - e_i^2} \approx -(\Delta e_i)^2 \tag{10.119}$$

if e_i is small and typically $e_i = \frac{1}{2}\Delta e_i$.

After a somewhat lengthy calculation we find, averaging over inclinations and binary phases,

$$\langle(\Delta e_i)^2\rangle^{1/2} \approx \frac{\sqrt{2Gm_B}a_i^{3/2}}{q^2 V}\left(\frac{m_3}{m_B}\right). \tag{10.120}$$

Note that Δe_i now depends on $Q^{-2} \equiv (q/a_i)^{-2}$, which is less steep than the Q^{-3} slope in slow encounters (Eq. (10.95), see Fig. 10.13). Using $v_0^2 = Gm_B/a_i$

(Eq. (8.8)), $V = v_\infty$, the definition of v from Eq. (8.11) and $Q = q/a_i$ we get

$$(\Delta e_i)^2 = 0.10 \frac{m_3^3}{2Mm_a m_b} \frac{1}{v^2} \left(\frac{Q}{2.5}\right)^{-4}, \tag{10.121}$$

which gives the relative energy change by Eq. (10.119). Notice that the sign of $\Delta E_B/E_B$ is now negative, since $\Delta E_B/E_B \approx -(\Delta e_i)^2$. This of course is true only on average, but for the remainder of this section we are satisfied to work at the level of averages.

Then

$$\Delta = -\frac{\Delta E_B}{E_B} \approx \frac{2m_3^3}{Mm_a m_b} \frac{1}{v^2 Q^4}. \tag{10.122}$$

Again Eq. (10.122) has a different Q-dependence from the slow encounters: the -6 power of Eq. (10.95) has changed to a -4 power of Q. Also the sign of Δ is different from $\langle\Delta\rangle$ of the slow encounter case at direct orbits. This is shown in Fig. 10.14 where the theory is compared with experiments.

In order to arrive at cross-sections, the average result of many encounters with different impact parameters Q must be calculated. Assume that the passing orbits impact an area $\pi Q_{\max}^2$ randomly where $Q_{\max}$ is the most distant encounter that interests us. Its value is a matter of definition; from numerical work we know that taking $Q_{\max} \approx \sqrt{2}$ will give a rather complete coverage of the energy change $|\Delta|$ of the order of 0.1 or greater at high incoming velocity ($v \approx 4 - 8$, Hut and Bahcall 1983). In this case the geometrical cross-section $\Sigma = 2\pi a_i^2$. The distribution of Q-values within this area is

$$f(Q) = 2Q \tag{10.123}$$

since the annuli of width Q have areas proportional to Q.

At this point a simplifying assumption, not strictly true, is made that at each encounter distance Q energy changes of a definite value Δ are produced according to Eq. (10.122). Remember that this result only applies to average energy changes, and therefore we are replacing the distribution of Δ at each Q by its average value. However, this opens up a simple way forward, and the final result can be checked by comparison with numerical experiments. Adopting this approach,

$$f(\Delta) = f(Q)\frac{\mathrm{d}Q}{\mathrm{d}\Delta} \approx \frac{1}{2}\left(\frac{2m_3^3}{Mm_a m_b}\right)^{1/2} \frac{1}{v\Delta^{3/2}}. \tag{10.124}$$

When multiplied by the geometrical cross-section Σ, the differential cross-section follows:

$$\frac{\mathrm{d}\sigma}{\mathrm{d}\Delta} \approx \sqrt{2}\left(\frac{m_3^3}{Mm_a m_b}\right)^{1/2} \frac{\pi a_i^2}{v\Delta^{3/2}}. \tag{10.125}$$

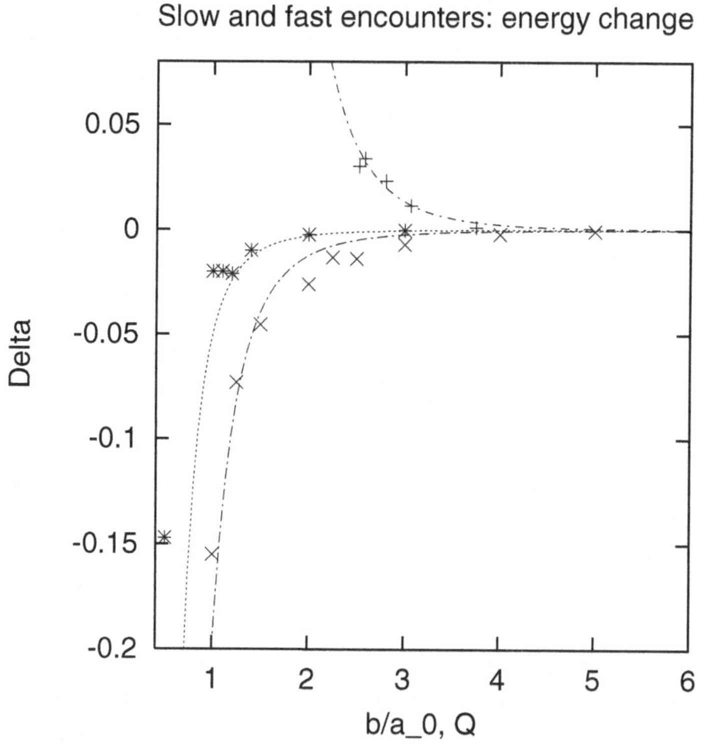

Figure 10.14 The relative energy change $-\Delta$ in slow encounters (positive side of the y axis) and in fast encounters (negative side of the y axis). The values of V/v_0 are 0.034 (+), 3.18 (×) and 10.37 (∗). The data are from Hills (1984) and the theoretical curves from Eq. (10.96) ($\cos \iota = 0.5$) and from Eq. (10.122). The x axis is b/a_0 for fast encounters which is slightly different from Q due to focussing. For slow encounters Q is used as the x axis.

In Fig. 10.15 a comparison of this result with experiments (Heggie and Hut 1993) shows a very good agreement.

The present discussion is based on an impulsive encounter between the third body and the binary as a whole, assuming that the third body does not come very close to either of the binary members. If it does come close we should use the theory of Section 6.8 (Eq. (6.79)) which gives

$$\frac{d\sigma}{d\Delta} = \frac{32}{9} \left(\frac{3m_3^3}{Mm_a m_b} \right) \frac{\pi a_i^2}{v^2 |\Delta|^3}. \tag{10.126}$$

At small $|\Delta|$ this is valid for both positive and negative Δ, and therefore the cross-sections to either direction are equal and the cross-section for the net change Δ is zero. At $|\Delta| > 1$ positive changes dominate and eventually at $|\Delta| \gg 1$ Eq. (10.126) represents the cross-section for the net change (Heggie 1975, Heggie and Hut 1993).

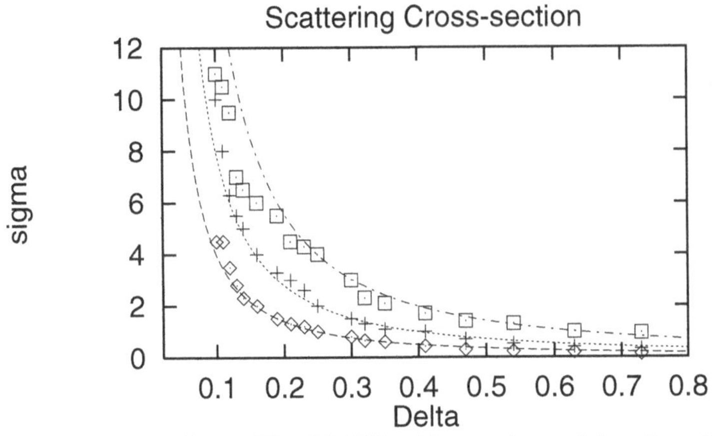

Figure 10.15 A comparison of Eq. (10.125) with experimental data from Heggie and Hut (1993). ($e = 0$, positive energy changes). The values of normalised speed are $v = 2$ ($\square$), $v = 4$ ($+$) and $v = 8$ ($\diamond$). The ordinate is $(\mathrm{d}\sigma/\mathrm{d}\Delta)/\pi a_i^2$.

Therefore we expect that at some point along the Δ axis there should be a transition from the $|\Delta|^{-3/2}$ power-law derived from the perturbation theory to the $|\Delta|^{-3}$ power-law which comes from close encounters. Figure 10.16 shows that the $|\Delta|^{-3/2}$ power-law holds well up to $|\Delta| \approx 1$ if we consider the cross-section for the net change (negative–positive). Above $|\Delta| = 1$ the negative changes agree with Eq. (10.126). For positive changes $|\Delta| > 1$ means dissociations of the binaries (ionisations), and if we discount them, the cross-section for negative Δ represents the net change.

Another way of looking at the scattering is to study the change in v as a result of the encounter. This is especially useful in the case of ionisation. Then it is more significant to speak of the change in the speed of the third body, relative to the centre of mass of the two others, than of the change in the binary energy. Let the final value of the normalised speed be v_f. Then it is easy to see that

$$\Delta = v^2 - v_f^2 \approx 2v(v - v_f) \tag{10.127}$$

since typically v_f is not very different from v. The cross-section for v_f ending up in the interval $[v_f, v_f + \mathrm{d}v_f]$ is (in the regime of Eq. (10.126))

$$\frac{\mathrm{d}\sigma}{\mathrm{d}v_f} = \frac{\mathrm{d}\sigma}{\mathrm{d}\Delta}\frac{\mathrm{d}\Delta}{\mathrm{d}v_f} \approx \frac{8}{9}\frac{\pi a_i^2}{v^4(v - v_f)^3}. \tag{10.128}$$

This gives a good description of numerical experiments where $|\Delta| \gtrsim 1$ (Valtonen and Heggie 1979; see Fig. 10.17).

At $v \approx 1$ the experimental cross-section is lower than predicted when Δ is low (see Fig. 10.15). The fit is improved if the data from scattering with eccentric

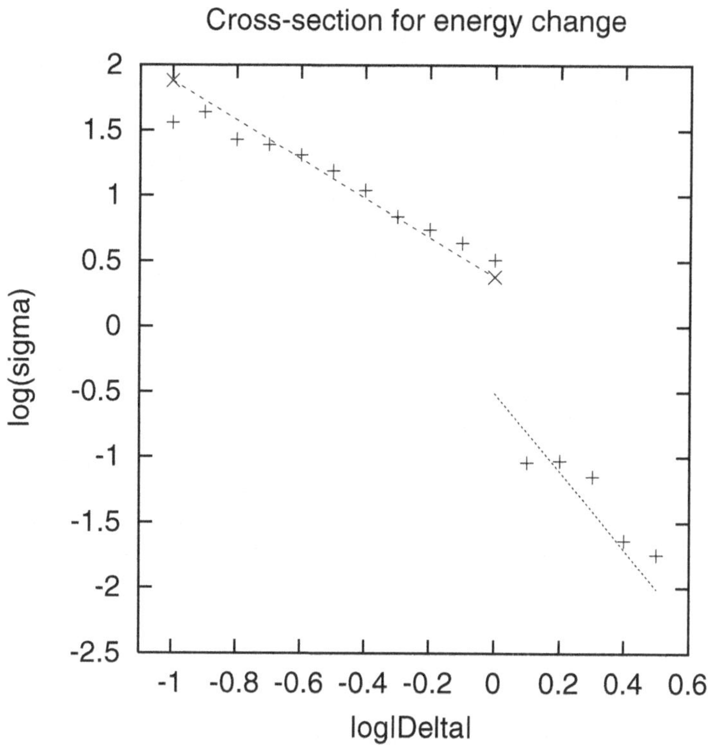

Figure 10.16 The normalised cross-section $(d\sigma/d|\Delta|)/\pi a_i^2$ in scattering experiments by Heggie and Hut (1993) at $v \approx 3.4$. At $|\Delta| < 1$, the cross-section is the difference between the cross-sections for positive Δ and for negative Δ while for $|\Delta| > 1$ only negative Δ are included since $\Delta > 1$ corresponds to ionisation. The lines are based on Eqs. (10.125) ($|\Delta| < 1$) and (10.126) ($|\Delta| > 1$).

binaries are used rather than the zero eccentricity data (Hut and Bahcall 1983). This is not surprising since our approximation of the constant phase angle ϕ is best justified for eccentric binaries. A very eccentric binary spends most of the time close to the apocentre of the orbit when the line joining the two components is very close to the major axis of the orbit.

10.8 Average energy exchange

In Section 8.5, the range of energy transfer between binaries and single stars was discussed, the important quantity being

$$\langle \sigma \Delta \rangle = \int_{-\infty}^{\infty} \Delta \frac{d\sigma}{d\Delta} \, d\Delta . \tag{10.129}$$

We call this quantity the *average energy exchange*.

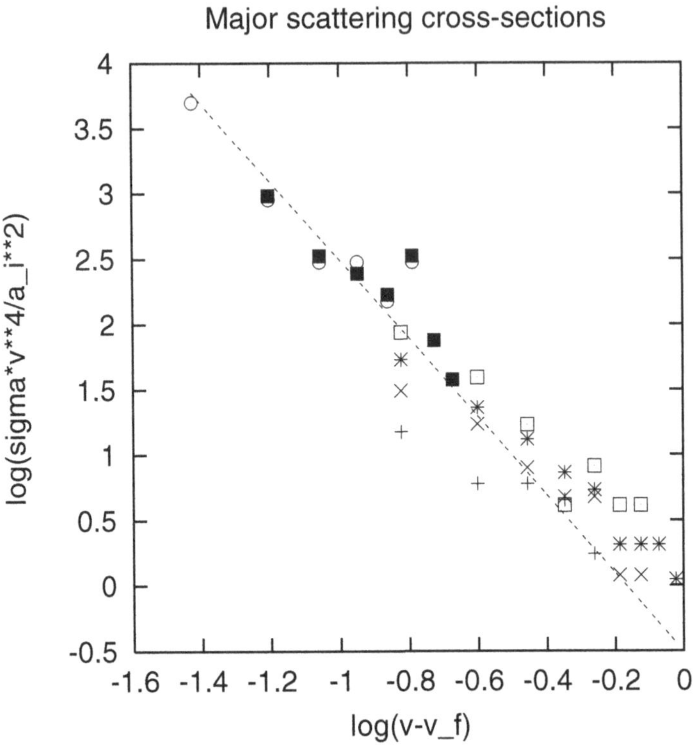

Figure 10.17 A comparison of Eq. (10.128) with experimental data from Valtonen and Heggie (1979). These data correspond to larger values of $|\Delta|$ than in Fig. 10.16 and thus they may be called major scattering. The comparison is presented for equal mass systems and $v_\infty/v_0 = 1.73$ (+), 2.45 ($\times$), 3.54 ($*$), 5 ($\square$), 7.07 ($\blacksquare$) and 10 ($\circ$). The observational points have been raised by 0.6 units in the log scale for $v_\infty/v_0 = 2.45$, by 2×0.6 for the next speed, etc. in order that a single theoretical line of slope -3 applies to all cases. A shallower slope of -1.5 begins at the low end of the experimental range of $v - v_f$, roughly where $v - v_f$ is equivalent to $\Delta = -1$ according to Eq. (10.127).

In Section 8.1, $\mathrm{d}\sigma/\mathrm{d}\Delta$ was derived for slow approach speeds v (Eq. (8.12) with a correction factor of Eq. (8.15)). At these speeds, the limits may be taken from $\Delta = 0$ to $\Delta = \infty$. The integration over Δ is carried out with the help of Eq. (7.54). For equal masses, the result is

$$\langle \sigma \Delta \rangle_{\text{slow}} \approx 2.5(1 - v^2)^{7/2} \frac{\pi a_0^2}{v^2}. \tag{10.130}$$

For high speeds v, Eq. (10.125) is used and integration is carried out from $\Delta = 0$ to $\Delta = 1/3$ since the region of validity of our theory does not generally extend to higher than $\Delta = 1/3$. The integration then yields for equal masses

$$\langle \sigma \Delta \rangle_{\text{fast}} \approx -\frac{\pi a_0^2}{v}. \tag{10.131}$$

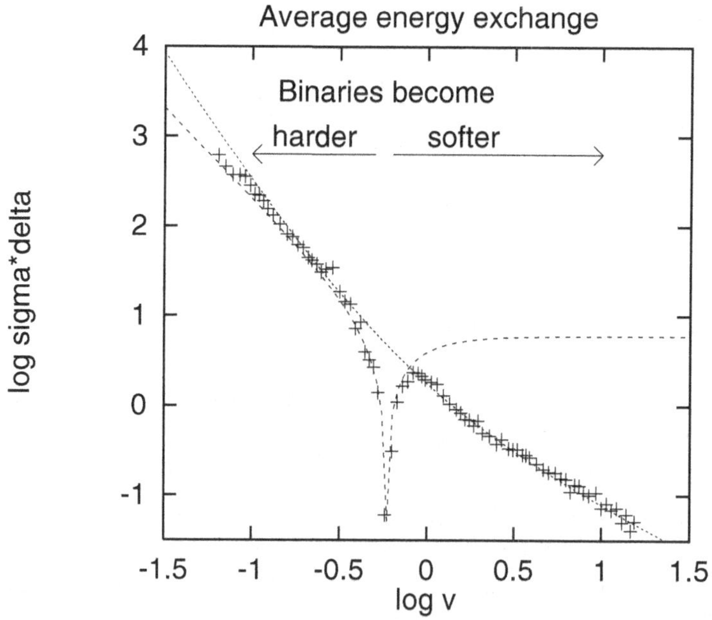

Figure 10.18 Experimental values of $\langle \sigma \Delta \rangle / (\pi a_0^2)$ from Hut and Bahcall (1983) are compared with the predictions of Eq. (10.131). The dashed line should be followed up to $v \approx 0.8$ and the dotted line for higher v values. The theoretical lines include a factor for gravitational focussing. The average energy exchange reverses its sign at $v \approx 0.6$ and therefore goes through zero at this point. At low values of v, scattering tends to make binaries harder (more strongly bound) while at high values of v they tend to become softer (more weakly bound).

In Fig. 10.18 we show numerical data (Hut and Bahcall 1983) which are in fairly good agreement with the expectations of the theory if the two regions are joined at $v \sim 1$. In the range $0.6 \lesssim v \lesssim 1.6$, the theoretical energy exchange rate is not very accurate because neither of the two theoretical approaches is strictly valid. The theoretical expressions may be improved by adding a correction factor $0.7(1 + 0.59/v)^2$ for gravitational focussing (Hut and Bahcall 1983) in Eq. (10.131)

This is not the same focussing factor which was derived previously in Section 6.2. It is less sensitive to the incoming velocity v_∞ than the standard expression which gives the focussing from a very large distance to the binary area. Our expression is used for comparison with numerical orbit calculations which typically start from relatively close to the binary; thus here we are concerned with near range focussing only.

Equation (10.130) is modified also so that the experimentally observed reversal of the sign of $\langle \sigma \Delta \rangle$ at $v \approx 0.6$ is included. This can be achieved by replacing $2.5(1 - v^2)^{7/2}$ by $2.09\left(1 - (v/0.59)^2\right)$ in Eq. (10.130). It includes also a small correction in the asymptotic value as $v \to 0$. Then

$$\langle\sigma\Delta\rangle = -6\left(1-\left(\frac{0.59}{v}\right)^2\right)\pi a_0^2, \quad v \leq 0.8,$$
$$\langle\sigma\Delta\rangle = -0.7\left(1+\frac{0.59}{v}\right)^2\frac{\pi a_0^2}{v}, \quad v \geq 0.8. \tag{10.132}$$

These functions have been drawn in Fig. 10.18, and they improve the agreement with the experimental data points considerably.

The significance of the 'watershed' value $v \approx 0.6$ is as follows. The angular speed of the binary motion is $\Omega_i = v_0/a_i$ and the angular speed of the third body at the pericentre (distance q from the binary centre of mass) is $\Omega_0 = v_\infty/q$ assuming a uniform passage speed v_∞. For equal masses, the latter is $\Omega_0 = \sqrt{3/4}vv_0/q$. The two angular speeds are equal when $v = \sqrt{4/3}Q$ where $Q = q/a_i$. At a 'grazing' encounter $Q = 1/2$, and therefore $v = 1/\sqrt{3} \approx 0.6$. At speeds less than 0.6 the binary turns around faster than the passing third body while the opposite is true for $v > 0.6$. The two cases correspond to different orbital senses in a coordinate system where the binary is stationary. Therefore the average energy exchange reverses its sign at the 'watershed' value. For $v \lesssim 0.6$ the binary contracts (becomes 'harder') while for $v \gtrsim 0.6$ the binary expands and becomes 'softer'. Correspondingly, the third body gains speed when $v \lesssim 0.6$ and loses speed during the encounter when $v \gtrsim 0.6$. Note, however, that these conclusions apply only to the average energy exchange. In individual cases the passing body may lose speed even below the 'watershed' $v = 0.6$ and gain speed above $v = 0.6$, even though the likelihood is greater for the opposite outcome.

The above discussion can be obviously extended to systems with different masses. We have previously learnt that different systems are conveniently scaled according to their stability limit; outside the stability limit the energy change tends to be in one direction only (positive or negative, depending on the inclination). Thus we generalise the 'watershed' speed by using, instead of $v_\infty/v_0 = Q = \frac{1}{2}$, a value which scales with $Q_{\mathrm{st}}^{(a)}$ for the mean energy change $\langle\Delta\rangle$ (Eq. (10.100), putting $\langle\Delta\rangle = 0.11$ and solving for Q as a function of m_3/m_B when $m_a/m_B \gg 1$ and $Q \gg Q_0$):

$$\frac{v_\infty}{v_0} \approx \left(\frac{2m_3}{m_B}\right)^{1/6}. \tag{10.133}$$

Numerical experiments by Hills (1990) show that this is a good approximation for $2m_3 \geq m_B$; for smaller values of m_3, v_∞/v_0 may be taken as a constant.

Problems

Problem 10.1 Show that Eq. (10.12) follows from Eq. (10.19).

Problem 10.2 A satellite circles a planet in an elliptic orbit. Its engine produces a radial acceleration $\mathbf{f} = \alpha\mathbf{r}/r$, where α is a constant. Find the effect of this perturbation during one revolution.

Problem 10.3 A binary radiates energy in the form of gravitational radiation, and it is possible to describe the back-reaction of this radiation as a perturbing force on the binary. In the case of zero eccentricity, the coefficient of 'drag' is

$$\alpha = \frac{32}{5}\frac{G^3 m_1 m_2(m_1 + m_2)}{c^5 a^4},$$

where m_1 and m_2 are the binary component masses and a is the orbital radius. The speed of light is c. Show that the lifetime of the binary is

$$T = \frac{5}{256}\frac{c^5}{G^3}\frac{a^4}{m_1 m_2(m_1 + m_2)},$$

if its initial orbital radius is a. Express the lifetime in years if a is in parsecs and masses in units of $10^9 M_\odot$.

Problem 10.4 The drag coefficient in dynamical friction is (Section 3.14)

$$\alpha = 4\pi G^2 m_s m_a n \ln\Lambda/u^3,$$

where m_s is the mass of a star (say $1 M_\odot$), m_a is the mass of the secondary body (say $10^{10} M_\odot$) and n is the number density of stars of the primary body at the position of the secondary, at a distance a from the centre of the primary. The parameter $\Lambda \approx au^2/Gm_a$. Assume that the orbit is circular, and that the orbital speed u is constant, independent of a (say, $u = 300$ km/s). Show first that in a spherically symmetric galaxy with a constant circular rotation speed u the local number density of stars n at distance a is

$$n = \frac{u^2}{4\pi G m_s a^2}$$

and that for $m_a = 10^{10} M_\odot$ and $r = 10$ kpc

$$\ln\Lambda \approx 3.$$

Then show that the lifetime of the binary (= binary galaxy) is

$$t_{\text{fric}} \approx \frac{a^2 u}{6Gm_a}$$

if a is the original orbital radius. Substitute a in units of 30 kpc, m_a in units of $10^{10} M_\odot$ to show that

$$t_{\mathrm{fric}} \approx 10^9 \text{ yr} \left(\frac{a}{30 \text{ kpc}}\right)^2 \left(\frac{m_a}{10^{10} M_\odot}\right)^{-1} \left(\frac{u}{300 \text{ km/s}}\right)$$

and compare it with the orbital period at the original distance a.

Problem 10.5 Show that the force in Eq. (10.21) is derived from the potential of Eq. (9.12).

Problem 10.6 Show that Eq. (10.31) results from Eq. (10.30).

Problem 10.7 Show that the approximations (10.45)–(10.47) are valid when $e = 0.265$. Note that the series expansions of Section 3.15 give somewhat different coefficients due to truncation errors.

Problem 10.8 Plot the sums $I^{(1)}$ and $I^{(2)}$ in Eq. (10.65), and show that $I^{(2)} \gg I^{(1)}$ when $k > 8$. Show that $I^{(2)} \approx -(2.3/k)\mathrm{e}^{-0.04k}$ when $15 \le k \le 75$.

Problem 10.9 Derive the expressions of Eq. (10.87) for the unit vectors.

Problem 10.10 Apply Eq. (10.105) to the Sun–Earth–Moon system. Is the orbit of the Moon stable? Data: masses, the Sun 2×10^{30} kg, the Earth 6×10^{24} kg, the Moon 7×10^{22} kg; mean distances from the Earth, the Sun 1.5×10^{11} m, the Moon 4×10^8 m.

Problem 10.11 Show that the inclination factor in Eq. (10.92) would be

$$1 - \cos^2 \iota$$

if the averaging is carried out over ω only. Why is this factor less useful than the $(1 + \cos \iota)^2$ factor?

Problem 10.12 Starting from the expression for the eccentricity change in a fast encounter,

$$\begin{aligned}
\Delta e_i = & -3\sqrt{G m_B a_i^{3/2}} \frac{m_3}{m_B} \left\{ \frac{1}{4} (\sin \phi + \sin 3\phi) \big[I_2 (\cos^2 \theta_1 - \cos^2 \theta_3) \right. \\
& + I_3 (\cos^2 \theta_2 - \cos^2 \theta_4) + 2 I_1 (\cos \theta_1 \cos \theta_2 - \cos \theta_3 \cos \theta_4) \big] \\
& - \frac{1}{2} (3 \cos \phi + \cos 3\phi) \big[I_2 \cos \theta_1 \cos \theta_3 + I_3 \cos \theta_2 \cos \theta_4 \\
& + I_1 (\cos \theta_1 \cos \theta_4 + \cos \theta_2 \cos \theta_3) \big] \\
& \left. - \sin \phi \big[I_2 \cos^2 \theta_3 + I_3 \cos^2 \theta_4 + 2 I_1 \cos \theta_3 \cos \theta_4 - I_4 \big] \right\},
\end{aligned}$$

where I_4 is

$$I_4 = \frac{1}{3} \int_{-\infty}^{\infty} \frac{\mathrm{d}t}{(q^2 + V^2 t^2)^{3/2}} = \frac{2}{3q^2 V} = I_2,$$

show that after averaging over ω and Ω and keeping only the term proportional to $(1 + \cos \iota)^2$, Δe_i becomes

$$\Delta e_i = \frac{\sqrt{G m_B a_i^{3/2}}}{q^2 V} \frac{m_3}{m_B} (1 + \cos \iota)^2 \sin \phi,$$

and that after further averaging over ϕ and ι Eq. (10.120) follows.

Problem 10.13 The cross-section for ionisation, i.e. for a complete disruption of the three-body system has been found to be

$$\sigma_{\mathrm{ion}} = \frac{40}{3} \frac{m_3^3}{M m_a m_b} \frac{\pi a_i^2}{v^2}$$

(Hut 1983, 1993). Use Eq. (10.126) to derive this result (except for a constant factor).

11

Some astrophysical problems

11.1 Binary black holes in centres of galaxies

Galactic nuclei are regions of high star densities as well as sites of very massive black holes, at least in many galaxies. For example, the central black hole in our own Galaxy is thought to be about $2 \times 10^6 M_\odot$ while the giant elliptical galaxy M87 possesses a dark central body of $3 \times 10^9 M_\odot$. The observed masses of these central objects are typically 1.2×10^{-3} times the mass of the spheroidal stellar component of their host galaxies (Merritt and Ferrarese 2001).

It is quite likely that there are also supermassive binary black holes in the centres of some galaxies, based both on theoretical (Saslaw *et al.* 1974, Begelman *et al.* 1980) and observational grounds (Komberg 1967, Sillanpää *et al.* 1988, Lehto and Valtonen 1996). They result most likely from mergers of galaxies. While the stars and gas of the two merged galaxies intermingle and form a new single galaxy, the central black holes remain separate for a long time, perhaps as long as the Hubble time (Milosavljevic and Merritt 2001). In the currently popular cold dark matter (CDM) model of cosmology it is believed that merging of galaxies is a common process (e.g. Frenk *et al.* 1988). Therefore binary black holes must also be common. How common they are depends on the interaction of the binary with the surrounding stars and gas clouds. This interaction may be modelled by a three-body process, the two binary members plus one star or one gas cloud.

Before the binary black hole stage has been reached, the separate nuclei of the two galaxies approach each other. In the process their stars are gradually stripped away until only the black holes with a small retinue of stars bound to them survive. At this time the black holes (of $10^9 M_\odot$ category) are about $a_0 = 5$ pc apart, and their relative orbit is practically circular, with eccentricity $e \approx 0.1$ (Milosavljevic and Merritt 2001). The orbital speed of the black hole binary is then about 2000 km/s, an order of magnitude greater than the observed typical speed of a star in the centre of a galaxy.

Since the three-body process depends sensitively on the speeds of the stars as well as on their numbers, let us focus for a detailed model on a definite galaxy M87 which has been well studied (Young *et al.* 1978). The spheroidal luminous mass of this galaxy is about $4.5 \times 10^{11} M_\odot$ and it is surrounded by a much greater dark matter halo. The latter component is not important here since we are considering only the centre of the galaxy. The luminous spheroidal mass M_L is a useful scaling parameter; we will use the luminous mass of M87 as a standard of reference: $M_L^* = 4.5 \times 10^{11} M_\odot$. The radial velocity dispersion in M87 has been observed to be $\sigma_v = 278$ km/s in its core of 690 pc in radius while the average mass density in the 690 pc core has been estimated to be $\rho_0 = 26 M_\odot/\mathrm{pc}^3$ (Young *et al.* 1978).

The galaxies of different sizes may be parametrised in different ways. In the system of Young (1976), which is based purely on observations, the length scale R_L of luminous matter and the luminous mass M_L scale as $M_L \propto R_L^2$. Thus the density scales as $\rho \propto M_L/R_L^3 \propto M_L^{-1/2}$, i.e.

$$\rho_0 = 26 M_\odot/\mathrm{pc}^3 (M_L/M_L^*)^{-1/2}. \tag{11.1}$$

The velocity scales as $\sigma_v \approx (M_L/R_L)^{1/2} \propto M_L^{1/4}$, i.e.

$$\sigma_v = 278 \text{ km/s } (M_L/M_L^*)^{1/4}. \tag{11.2}$$

Since the black hole mass M_{BH} also scales with M_L, we have

$$M_{BH} = 6 \times 10^8 M_\odot (M_L/M_L^*). \tag{11.3}$$

Note that this is a 'typical' black hole mass; there is considerable scatter in the M_{BH} values. For example, the actually observed black hole mass in M87 is five times greater than the 'typical' value.

The three-body interaction generally leads to the escape of the less massive bodies (the stars) from the neighbourhood of the binary. The typical escape speed is the binary orbital speed v_0 (Chapter 7). As stars are ejected, the central stellar density in the galaxy steadily decreases. The evolution is best estimated from N-body simulations with large N (Milosavljevic and Merritt 2001) but the result may be justified as follows. An escaped star carries away the energy $\approx \frac{1}{2} m_s v_0^2$, where m_s is the mass of the star. This energy is extracted from the binary which has binding energy $|E_B| = \frac{1}{2}\mathcal{M} v_0^2$, $\mathcal{M}$ being the reduced mass of the binary. Therefore the relative change in the binding energy of the binary per escaped star is

$$-\frac{\mathrm{d}|E_B|}{|E_B|} = \frac{\mathrm{d}a_B}{a_B} \approx \frac{\frac{1}{2} m_s v_0^2}{\frac{1}{2}\mathcal{M} v_0^2} = \frac{m_s}{\mathcal{M}}. \tag{11.4}$$

Here a_B is the semi-major axis of the black hole binary.

To see how the escape of stars affects the mass density, let us focus on a volume centred on the binary black hole which contains stellar mass $2\mathcal{M}$ (roughly the

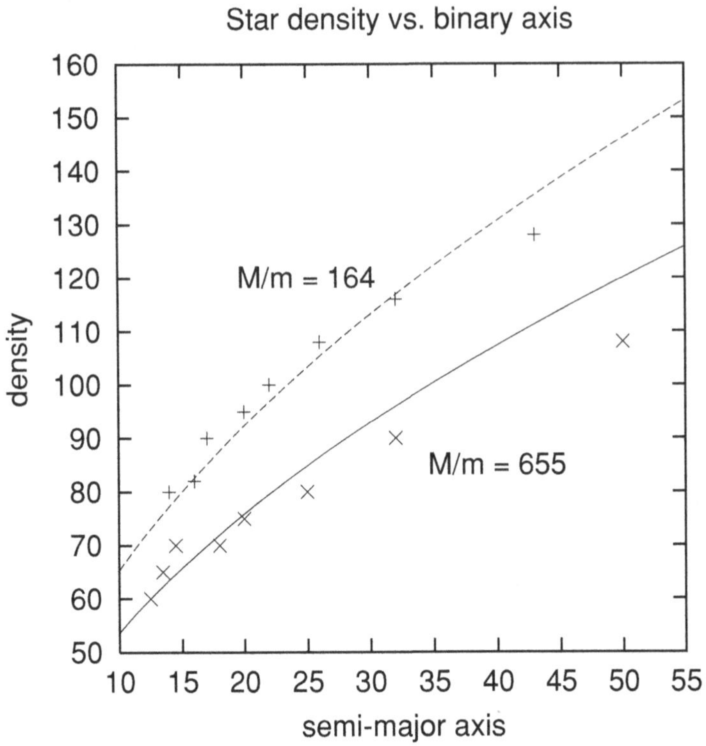

Figure 11.1 The evolution of the density of stars in the neighbourhood of a binary, plotted as a function of the semi-major axis of the binary. Data points come from numerical N-body simulations by Milosavljevic and Merritt (2001; Fig. 8) while the curves trace the ρ proportional to $\sqrt{a}$ evolution. Two mass ratios of the binary mass to the stellar mass have been used. The evolution starts from approximately $a_0 = 100$ units and $\rho = 150$ units.

black hole mass). The radius of this volume is about 70 pc. We thus consider the interaction between the black holes and a similar amount of stellar matter. The escape of a single star changes the average stellar mass density of the volume by

$$\frac{\mathrm{d}\rho}{\rho} = \frac{m_s}{2\mathcal{M}} \approx \frac{1}{2}\frac{\mathrm{d}a_B}{a_B}. \tag{11.5}$$

The solution of this equation is

$$\rho = \frac{\rho_0}{a_0^{1/2}} a_B^{1/2}, \tag{11.6}$$

where a_0 and ρ_0 are the initial values of a_B and ρ, respectively. This density evolution has been found in numerical calculations (Fig. 11.1; Milosavljevic and Merritt 2001) which gives us confidence in using it in the following.

In Chapter 6 (Eq. (6.27)) an expression for the relative change of the semi-major axis a_B was derived:

$$\frac{da_B}{a_B} = R_a \pi a_B \frac{G\rho}{u} dt. \tag{11.7}$$

We may put $u = \sqrt{3}\sigma_v$ for the speeds of the stars (Quinlan 1996), and since $v_0 \gg \sigma_v$, $u/v_0 \approx 0$ and $R_a \approx -6.5$. Consequently the coefficient $R_a\pi \approx -20$. This dimensionless number is usually called the hardening rate. The remaining factor on the right hand side

$$a_B \frac{G\rho}{u} = a_B^{3/2} \frac{G\rho_0}{a_0^{1/2} u} \tag{11.8}$$

has a constant part

$$\begin{aligned} \frac{G\rho_0}{a_0^{1/2} u} &= \frac{4.3 \times 10^{-3} (\mathrm{km/s})^2\ \mathrm{pc}/M_\odot \times 26 M_\odot/\mathrm{pc}^3 \times 10}{(a_0/\mathrm{pc})^{1/2}\ \mathrm{pc}^{1/2} \times 278\ \mathrm{km/s} \times \sqrt{3}} \left(\frac{M_L}{M_L^*}\right)^{-3/4} \\ &= \frac{23 (a_0/\mathrm{pc})^{-1/2}}{\mathrm{pc}^{3/2} \times 10^{10}\ \mathrm{yr}} \left(\frac{M_L}{M_L^*}\right)^{-3/4}. \end{aligned} \tag{11.9}$$

We have multiplied ρ_0 by 10 since the average density within a radial distance of 70 pc is this much greater than within the radial distance of 690 pc, in the expected density law of $\rho \propto r^{-1}$ (Milosavljevic and Merrit 2001). Also the $\sqrt{3}$ factor multiplies the observed σ_v to obtain u. Thus the equation becomes

$$\frac{da_B}{a_B^{5/2}} = -\frac{464 (a_0/\mathrm{pc})^{-1/2}}{\mathrm{pc}^{3/2} \times 10^{10}\ \mathrm{yr}} \left(\frac{M_L}{M_L^*}\right)^{-3/4} dt \tag{11.10}$$

which is easily integrated:

$$-\frac{2}{3}\Big|_{a_0}^{a} a_B^{-3/2} = \frac{-464 (a_0/\mathrm{pc})^{-1/2}}{\mathrm{pc}^{3/2} \times 10^{10}\ \mathrm{yr}} \left(\frac{M_L}{M_L^*}\right)^{-3/4} (t - t_0). \tag{11.11}$$

Here a_0 and a are the initial and final values of a_B at times t_0 and t, respectively. Assuming that $a_0 \gg a$,

$$a = \frac{1}{45} \left(\frac{a_0}{\mathrm{pc}}\right)^{1/3} \left(\frac{M_L}{M_L^*}\right)^{1/2} \left(\frac{10^{10}\ \mathrm{yr}}{t - t_0}\right)^{2/3} \mathrm{pc}. \tag{11.12}$$

The eccentricity of the binary remains zero if it is initially zero, and increases only moderately if it starts from a small value (Chapter 6, Eqs. (6.38), (6.39)). For example, the typical value $e_0 \approx 0.1$ in the N-body experiments (Milosavljevic and Merritt 2001) becomes $e \approx 0.27$ by the time the semi-major axis has fallen by a factor of 10. Therefore the eccentricity evolution is generally not of prime importance.

However, gravitational radiation from the black hole binary may become quite important. It carries away binary energy and eventually leads to a collapse of the binary. The interesting question is the time scale: is it as fast or faster than the energy loss by three-body interactions, and is it significant within the lifetime of a galaxy?

The total collapse of a circular binary of initial orbital radius a takes place in time (Peters 1964, see Problem 10.4)

$$T_c \approx 60 \times 10^{10}\ \mathrm{yr} \left(\frac{a}{\mathrm{pc}}\right)^4 \left[\frac{(10^9 M_\odot)^3}{\mathcal{M}(m_1 + m_2)^2}\right] \tag{11.13}$$

due to gravitational radiation. Putting $m_1 + m_2 = 6 \times 10^8 M_\odot (M_L/M_L^*)$

$$T_c \approx 1.8 \times 10^{10}\ \mathrm{yr} \left(\frac{a}{0.2\ \mathrm{pc}}\right)^4 \left(\frac{m_1 + m_2}{4\mathcal{M}}\right) \left(\frac{M_L}{M_L^*}\right)^{-3}. \tag{11.14}$$

This may be compared with the evolution time scale $t - t_0$ due to three-body interactions:

$$t - t_0 \approx 0.036 \times 10^{10}\ \mathrm{yr} \left(\frac{a}{0.2\ \mathrm{pc}}\right)^{-3/2} \left(\frac{a_0}{5\ \mathrm{pc}}\right)^{1/2} \left(\frac{M_L}{M_L^*}\right)^{3/4}. \tag{11.15}$$

The two time scales are equal if

$$a = 0.1\ \mathrm{pc} \left(\frac{a_0}{5\ \mathrm{pc}}\right)^{0.09} \left(\frac{4\mathcal{M}}{m_1 + m_2}\right)^{0.18} \left(\frac{M_L}{M_L^*}\right)^{0.68}. \tag{11.16}$$

This equality happens at

$$t - t_0 \approx 10^9\ \mathrm{yr} \left(\frac{a_0}{5\ \mathrm{pc}}\right)^{0.365} \left(\frac{4\mathcal{M}}{m_1 + m_2}\right)^{-0.27} \left(\frac{M_L}{M_L^*}\right)^{-0.27}. \tag{11.17}$$

Putting $a_0 = 5$ pc, $m_1 = m_2$ and $M_L = M_L^*$, we find $t - t_0 \approx 10^9$ yr, i.e. after this period gravitational radiation takes over and causes a total collapse of the binary at $t - t_0 \approx 2 \times 10^9$ yr. This is close to the age of a galaxy ($\approx 10^{10}$ yr). One only needs to lower the black hole masses a little below the mean relation in order to make them survive through the Hubble time. On the other hand, binaries well over the mean value for the black hole masses may live less than 10^9 yr. In the absence of processes other than those discussed above, it appears that there are both robust binaries which live through the age of the galaxy, and binaries which coalesce. But as always in astrophysics, there are additional factors which may be important. These include the gravitational field of the stars, loss cone depletion, gas inflow and multiple mergers of galaxies.

The binary evolution is not a pure three-body process since stars interact among each other as well as with the binary. A simple way to include the star–star interactions is to modify the hardening rate πR_a if necessary. N-body simulations show that the hardening rate should be effectively lowered by about a factor of three due to the star–star effects (Quinlan 1996, Milosavljevic and Merritt 2001). Carrying this factor through the calculations (Problem 11.1) shows that the time scale of coalescence of the black holes is increased by more than a factor of two.

Not all the stars in the core interact with the binary. Those stars which do interact will be eventually removed either further out or to escape orbits (Zier and Biermann 2001). These are called loss cone stars since the directions of their approach velocities lie inside a cone whose base is the three-body interaction area (see Fig. 7.3). In this way a low density region develops in the distribution of stars inside the core. Since the black hole binary is able to wander around a bit due to Brownian motion (from encounters with stars) it effectively covers a region of about one parsec in radius, and the underdense region in the stellar distribution is about twice this size. In the end it is the density of stars inside the hole, not the mean density of the core, which determines the response of the binary to the surrounding stars. N-body experiments have not yet been carried out far enough in time to determine reliably how much the density drops in the central hole relative to the mean density in the core. The experiments by Quinlan and Hernquist (1997) suggest that the effective mean density may be lowered by a factor of 10–50 due to the central hole; if this is true then the lifetimes of binaries before coalescence are increased by an order of magnitude, i.e. the coalescence of a binary is not expected within the lifetime of a galaxy unless the binary mass is well above the 'typical' black hole mass. Yu (2002) finds that the central hole in stellar distribution could be filled in highly flattened or triaxial galaxies but not in nearly spherical systems. Therefore it is expected that black hole binaries survive primarily in the latter type of galaxies.

Another obstacle to the coalescence of binary black holes may come from gas flow to the binary components. Each component presumably possesses a disk of gas, its accretion disk. Infall of gas from outside the nucleus will replenish the gas while some of the disk matter falls into the black holes, thus increasing their mass. Let us consider how a binary responds to this process.

The angular momentum of the infalling gas has the same origin as the angular momentum of the black hole binary: the orbital angular momentum of the two galaxies prior to their merger. Therefore we may assume that the angular momenta are aligned with each other, and we may restrict ourselves to a planar problem.

Let the two binary components have masses m_1 and m_2, $m_1 > m_2$. Numerical simulations of gas accretion (Bate and Bonnell 1997, Bate 2000) have shown that the less massive (and therefore more distant from the centre of mass) black hole

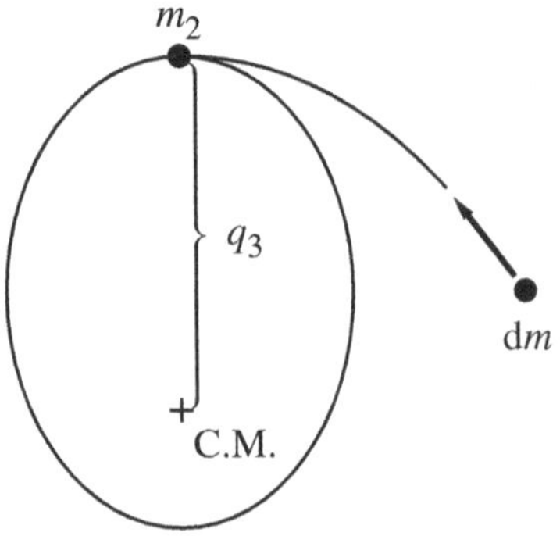

Figure 11.2 A gas cloud of mass dm arrives in a parabolic orbit of pericentre distance $q_3 = (\mathcal{M}/m_2)a(1+e)$ relative to the binary centre of mass and collides with the component of mass m_2 when the latter is at the apocentre of its orbit.

receives most of the infalling gas. This tends to equalise the black hole masses if the amount of accreted gas is comparable to the binary mass.

We assume that a collision between a gas cloud (mass dm) and the less massive binary component (mass m_2) takes place at the apocentre of the binary orbit which is at the same time the pericentre of the gas cloud orbit. As a result of this, the gas cloud starts to follow the binary orbit. We further assume that the total mass, excess energy and excess angular momentum of the gas cloud are passed on to the binary. We then ask what is the resulting change in the binary orbit (see Fig. 11.2).

The initial orbital energy of the gas cloud is taken to be zero by the definition of the parabolic orbit. After the collision, the gas cloud takes the orbit of the component with mass m_2. Therefore the change in its orbital energy is

$$\mathrm{d}E \approx -\frac{1}{2}\,\mathrm{d}m\, v_0^2,$$

where v_0 is the mean orbital speed of the binary (Problem 11.2). The binary energy is changed by the amount $\mathrm{d}V - \mathrm{d}E$, where $\mathrm{d}V$ is the change in the potential energy of the binary. After a little calculation (Problem 11.3) we find

$$\frac{\mathrm{d}|E_B|}{|E_B|} \approx \frac{\frac{1}{2}\,\mathrm{d}m\, v_0^2}{\frac{1}{2}\mathcal{M}v_0^2} = \frac{\mathrm{d}m}{\mathcal{M}}.$$

The semi-major axis

$$a = \frac{Gm_1m_2}{2|E_B|}$$

changes accordingly:

$$\frac{\mathrm{d}a}{a} = -\frac{\mathrm{d}|E_B|}{|E_B|} + \frac{\mathrm{d}m_2}{m_2} \approx -\frac{\mathrm{d}m}{\mathcal{M}} + \frac{\mathrm{d}m}{m_2} \approx 0,$$

when $m_2 \ll m_1$ and $e = 0$. The semi-major axis is expected to stay constant or to increase during the accretion.

The angular momentum L of the binary is (Eq. (7.11))

$$L^2 = Gm_1m_2\mathcal{M}a(1-e^2)$$

while the corresponding angular momentum L_2 for mass dm in the binary orbit is

$$L_2^2 = Gm_1(\mathrm{d}m)^2a(1-e^2).$$

In the parabolic orbit, the angular momentum L_1 of mass dm is

$$L_1^2 = 2G(m_1+m_2)(\mathrm{d}m)^2\frac{\mathcal{M}}{m_2}a(1+e).$$

In the last equation we have put the pericentre distance of the outer orbit equal to the apocentre distance of the binary:

$$q_3 = \frac{\mathcal{M}}{m_2}a(1+e), \tag{11.18}$$

and have replaced $a_3(1-e_3^2)$ for the outer orbit by $2q_3$.

Since $e^2 = 1-(m_1+m_2)L^2/(Gm_1^2m_2^2a)$,

$$\mathrm{d}(e^2) = (1-e^2)\left(-2\frac{\mathrm{d}L}{L} + \left(\frac{2m_1+m_2}{m_1+m_2}\right)\frac{\mathrm{d}m}{m_2} + \frac{\mathrm{d}a}{a}\right).$$

The change in the angular momentum of the binary $\mathrm{d}L = L_1 - L_2$ is (Problem 11.4)

$$\frac{\mathrm{d}L}{L} = \sqrt{\frac{\mathcal{M}}{m_2}}\left(\sqrt{\frac{2}{(1-e)}}-1\right)\frac{\mathrm{d}m}{\mathcal{M}}. \tag{11.19}$$

When substituted above we obtain d(e^2). The expression is rather complicated (Problem 11.5). We get a simpler result by putting $\mathrm{d}a/a = 0$ which is justified by numerical simulations (Fig. 11.3). Then

$$\mathrm{d}(e^2) = -2(1-e^2)\sqrt{\frac{\mathcal{M}}{m_2}}$$
$$\times\left[\sqrt{\frac{2}{(1-e)}}-1-\frac{1}{2}\left(1+\frac{m}{m_2}\right)\sqrt{\frac{\mathcal{M}}{m_2}}\right]\frac{\mathrm{d}m}{\mathcal{M}}$$

which is < 0 for large values of e and > 0 for small e). Therefore $e \to e$ final ≈ 0.15 independently of its initial value. If $\mathrm{d}a/a > 0$, the eccentricity can grow bigger, but in no case to a higher value than $e \approx 0.8$ (Problem 11.5).

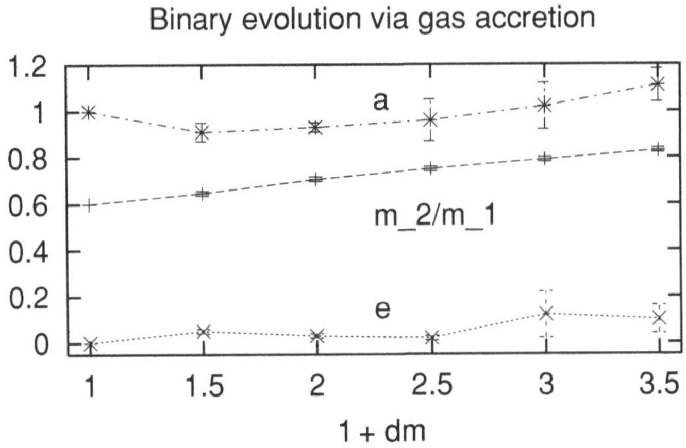

Figure 11.3 The effect of gas accretion by the relative amount dm on the parameters of a binary: a semi-major axis, m_2/m_1 mass ratio and e eccentricity. The error bars describe the variation resulting from different gas dynamical assumptions. The data are based on numerical hydrodynamical calculations by Bate (2000; Fig. 6).

This simple model is consistent with gas dynamical calculations (Fig. 11.3; Bate and Bonnell 1997, Bate 2000). Gas accretion becomes important in the binary black hole evolution if the amount of gas accreted is comparable to the mass of the binary. For example, in the time scale of 10^9 yr, the rate of $1 M_\odot$/yr would significantly change the binary orbital parameters, if the binary mass is about $10^9 M_\odot$. The rate of gas inflow to the centre of M87 has been estimated to be about $1 M_\odot$/yr (Loewenstein and Matthews 1987). However, some of the gas may form clouds and form stars which effectively drop out of the flow (Mathews and Bregman 1978) and it is even possible that the gas is used up completely in star formation before the flow reaches the centre (White and Sarazin 1988). Thus it is presently unclear what role the gas flow plays in the dynamics of binary black holes. The main effect of gas accretion seems to be the evolution towards equal mass components, while the major axis is rather unaffected (Fig. 11.3). The effect of spherically symmetric gas accretion is different: it tends to drive the black holes together towards a merger (Escala *et al.* 2004). It is not clear whether such a situation can arise in galactic nuclei.

We now come to the fourth complication, multiple mergers of galaxies which leads to another application of the three-body problem.

11.2 The problem of three black holes

As described above, it is likely that a merger of two galaxies, each with one central black hole, results in a semi-permanent black hole binary. If the galaxy is rather massive, it is also likely that it focusses the orbits of other nearby galaxies strongly enough to cause yet another merger. Then a third black hole may come to interact with the binary, and we arrive at the problem of three black holes (Valtonen *et al.* 1995, Pietilä *et al.* 1995, Valtonen 1996).

Every now and then it may also happen that two galaxies, each containing a semi-permanent black hole binary, merge and create a strongly interacting four black hole system in the nucleus of the merger. In many ways this situation can be handled as an extension of the three-body problem, and at this level we will also discuss the problem of four black holes.

The problem of three black holes differs from the general three-body problem in several ways. To be exact, the three black hole problem requires solutions in General Relativity theory which are not available at present and probably will not be for some time in the future. The best one can do now is to apply a modified Newtonian theory, the so-called Post-Newtonian formalism. In practice, it is enough to modify Newton's laws only at close encounters between two bodies; even at this approximation the force law between the two close bodies becomes rather complicated (Valtonen *et al.* 1995). Because of the chaotic nature of the three-body orbits, a further simplification is justified: the other deviations from the $1/r^2$ force law, such as terms leading to the relativistic precession of the two-body orbit, may be neglected, except for the term relating to gravitational radiation. The loss of energy and angular momentum from the system via gravitational radiation changes the basic constants of the three-body problem, and therefore must be incorporated.

Gravitational radiation leads to a decay of a binary orbit, and finally to a collapse of the binary black hole system into a single black hole. The final stages of the decay are very rapid, from the point of view of the three-body problem instantaneous. Thus we may separate close two-body encounters in two categories: those leading to a coalescence of the two black holes in less than the crossing time of the three-body system, and those in which the decay time is longer and which may not lead to a decay at all since the third body will modify the orbit. So very roughly we may say that there is a safety limit q for the closeness of the two-body encounter: if the bodies approach within this limit they coalesce without further three-body influences, and if they do not, we may ignore the gravitational radiation altogether for this encounter. This two-step model is very approximate but still quite useful because of the chaotic nature of the three-body problem in general.

The safety limit q may be estimated as follows. We consider the two-body encounter to happen at the pericentre of a highly elliptic orbit; the decay time of such an orbit, with the pericentre distance q and semi-major axis a, is (Peters 1964)

$$T(q) \approx 10T_c \left(\frac{q}{a}\right)^{7/2} \tag{11.20}$$

where T_c is the decay time of a circular orbit with the same value of the semi-major axis (Eq. (11.13)). This initial orbital period of the same orbit is

$$P = 3 \times 10^3 \left(\frac{a}{\text{pc}}\right)^{3/2} \left(\frac{m_1 + m_2}{10^9 M_\odot}\right)^{-1/2} \text{yr}. \tag{11.21}$$

If the decay time is shorter than the initial period, it is unlikely that three-body interactions have time to be effective. Thus we obtain the safety limit q by equating $T(q) = P$, or

$$\frac{q}{a} \approx 10^{-3} \left(\frac{a}{\text{pc}}\right)^{-5/7} \left(\frac{M_L}{M_L^*}\right)^{5/7} \left(\frac{4\mathcal{M}}{m_1 + m_2}\right)^{2/7}. \tag{11.22}$$

Here we have used $m_1 + m_2 = 6 \times 10^8 M_\odot (M_L/M_L^*)$. We see that typically the safety limit $q \approx 10^{-3}$ pc.

However, what is usually more important than q is the ratio q/a. Let us deduce a typical value of a. From the discussion in the previous section it appears that the lifetimes of the black hole binaries in galactic nuclei are likely to exceed the Hubble time, but it is unclear by how much. To get definite numbers, let us say that the binaries involved have exactly $T_c = 1.8 \times 10^{10}$ yr left before coalescence. Then it follows that the semi-major axis

$$a \approx 0.2 \text{ pc} \left(\frac{4\mathcal{M}}{m_1 + m_2}\right)^{1/4} \left(\frac{M_L}{M_L^*}\right)^{3/4}. \tag{11.23}$$

We may use this value of a in the q/a ratio (even though this is clearly only an approximation since the eccentric binary may not even be the same as the original circular binary) and obtain

$$\frac{q}{a} \approx 3 \times 10^{-3} \left(\frac{4\mathcal{M}}{m_1 + m_2}\right)^{3/28} \left(\frac{M_L}{M_L^*}\right)^{5/28}. \tag{11.24}$$

Chapter 8, Eq. (8.39) gives the probability $f(q)$ that the approach is closer than q:

$$f(q) \approx 240 q/a \tag{11.25}$$

at low values of the angular momentum. Substituting q/a from above

$$f(q) \approx 0.72 \left(\frac{4\mathcal{M}}{m_1 + m_2}\right)^{3/28} \left(\frac{M_L}{M_L^*}\right)^{5/28}. \tag{11.26}$$

This suggests that as many as 72% of all three-body systems formed as a result of galactic triple mergers may end up with a two-body collision and merger. Then in effect these galactic nuclei would have a remnant eccentric binary as the end result of the three black hole evolution. Note, however, that the extension of Eq. (8.39) to such large fractions as 0.72 is beyond the range of applicability of the straight line extrapolation. Figure 8.12 suggests that the straight line extrapolation is a factor of 3 too high relative to experimental data at large fractions, and a more correct result would be $\approx 24\%$ loss rate of three-body systems to coalescence. If T_c is reduced to 2×10^9 yr corresponding to the neglect of the loss cone phenomenon, the coalescence fraction becomes 50%. Considering the uncertainties, especially in the typical value of T_c, the range of angular momenta L, the range of black hole masses, the scatter of black hole masses around the mean relation etc., it is difficult to say which number is closer to the truth. Heinämäki (2001) obtained about a 70% loss rate in four black hole simulations which may be indicative also of the loss rate for three black hole systems. This number could be slightly increased by consideration of the Kozai mechanism when the inner and outer binary planes happen to be almost perpendicular to each other (Blaes *et al.* 2002).

What happens to those three black hole systems where no coalescence takes place and all three black holes survive? The end result is, as usual in the three-body problem, an escape of one of the black holes, and the recoil of the binary in the opposite direction. Therefore all three black holes leave the nucleus of the galaxy, unless the mass range is very large and the recoil speed small. In the case of four black hole systems, typically only three out of four black holes are involved in the escape process and the fourth black hole remains in the nucleus of the galaxy.

Let us then consider what happens to the escapers. The key number is the mean orbital speed of the binary before three-body interaction begins: $v_0^2 = G(m_1 + m_2)/a$. Using the above values for $m_1 + m_2$ and a (Eq. (11.23))

$$v_0 \approx 3600\,\mathrm{km/s} \left(\frac{M_L}{M_L^*}\right)^{1/8} \left(\frac{m_1 + m_2}{4\mathcal{M}}\right)^{1/8}. \tag{11.27}$$

For the case of the shorter (no loss cone) binary lifetime, $v_0 \approx 5000$ km/s. Quinlan (1996) estimates that $v_0 \approx \sqrt{\sigma_v c} \approx 9000\,\mathrm{km/s}$. The velocity of escape v_s has a wide distribution which peaks around $v_s = \frac{1}{8} v_0$ (Eq. (7.21), the case of $m_s = m_B$) and extends beyond $v_s = v_0$. It should be compared with the speed of escape from the centre of a galaxy to infinity

$$v_{\mathrm{esc}} \approx 1500\,\mathrm{km/s} \left(\frac{M_L}{M_L^*}\right)^{1/4}. \tag{11.28}$$

Therefore, typically $v_s = v_B \approx 450\,\mathrm{km/s}$, which means that neither the single body nor the binary is able to leave the confines of the galaxy. However, some of the escapers from the tail end of the v_s distribution have high enough speeds to do so. Let us start first by looking at the escapers from the galaxy and ask how frequent they are, how are their escape velocities distributed, and when and how do we expect both the binary and the single black hole to escape from the opposite sides of the galaxy.

The fraction of escape velocities v_s falling in the tail end, $v_s \geq 0.42v_0$, may be obtained from Eq. (7.21) by putting $v = 0.42$ and $m_s = m_B$. It is $F(0.42) = 0.045$. This number is quite sensitive to the type of system: at low angular momentum there can be almost an order of magnitude increase in the tail (see Fig. 7.11) while the opposite happens when the angular momentum becomes very large. The above fraction $F(v_s \geq v_0)$ may be used as a representative value.

For both the binary and the single black hole to escape there is the additional requirement that both masses should be about equal, say $0.7 \lesssim m_s/m_B \lesssim 1.4$, i.e. $0.42 \lesssim m_s/M_B \lesssim 0.58$. The probability for this happening, from Eq. (7.23), is $(M = 1, m_a = m_b = (1 - m_s)/2)$:

$$\begin{aligned} P_s(0.42 \leq m_s \leq 0.58) \\ &= \frac{\int_{0.42}^{0.58} \left(m_s^{-2}/\left(m_s^{-2} + 2[(1-m_s)/2]^{-2}\right)\right) dm_s}{\int_0^1 \left(m_s^{-2}/\left(m_s^{-2} + 2[(1-m_s)/2]^{-2}\right)\right) dm_s} \qquad (11.29) \\ &= 0.16. \end{aligned}$$

Even though the integrals can be evaluated analytically, they are not trivial. We get a simpler result if we use the variable $m_B = 1 - m_s$ instead of m_s. Then the distribution of Eq. (7.23) becomes (for $m_a = m_b$)

$$f(m_B) = \frac{m_B^2}{8m_B^2 + 16m_B + 9} \approx 0.034(m_B - 0.11),$$

where the last form is a good approximation for $m_B \geq 0.11$ (we may put $f(m_B) = 0$ when $m_B < 0.11$). Then the accumulated distribution for m_B becomes

$$F(m_B) = \begin{cases} (m_B - 0.11)^2/(0.89)^2 & \text{if } m_B \geq 0.11 \\ 0 & \text{if } m_B < 0.11. \end{cases} \qquad (11.30)$$

The result on the right hand side of Eq. (11.29) is obtained by using this equation.

Multiplying $F(v_s \geq v_0)$ by $P_s(0.42 \leq m_s \leq 0.58)$ we get the probability for two-sided escapes to be close to 0.7%. This result has been confirmed by numerical simulations (Heinämäki 2001) where the value of about 1% was obtained for four black hole systems. Considering the sensitivity of $F(v_s \geq v_0)$ to the value of total angular momentum, and to a lesser degree the dependence of P_s on the same quantity, the two values are well consistent with each other.

How often would we expect to see black holes escaping from diametrically opposite sides of a triple merger galaxy? Assuming that the black holes are detected one way or another, for example, in the radiation arising from their accretion disks or from jets flowing out of the disk (Valtonen and Heinämäki 2000), we need to know for how long a time the detection is possible. It is not an easy question to answer, but we may estimate the period of visibility of the escaping black holes as follows. After escaping the potential well of the galaxy, the black holes typically have rather small excess speed left, something like ≈ 1000 km/s. Therefore they move ≈ 1000 pc/10^6 yr = 100 kpc/10^8 yr. It could be said that the association of the escaped black holes with the parent galaxy becomes less likely when the black holes are more distant than 100 kpc from the galaxy centre. The 100 kpc scale is still within the range where it should not be difficult to say which galaxy the black holes come from. This suggests that from an observational point of view the period of detection may be of the order of 10^8 yr.

The other limitation is the lifetime of the accretion disk around the escaping black hole. Unlike in the centre of a galaxy where gas inflows replenish the accretion disk, in the intergalactic space very little new replenishment is possible. On the contrary, there is a steady depletion of the accretion disk when the disk gas gradually drifts inwards until it is partly swallowed by the black hole and partly blown out as winds or jets. In order to reach a highly visible state, the accretion rate has to be rather high, say $0.1 M_\odot$/yr. The total mass of the accretion disk should be small compared with the mass of the black hole itself, say $10^7 M_\odot$. At this rate the accretion disk would be completely used up in $(10^7 M_\odot / 0.1 M_\odot)$ yr = 10^8 yr. This is a rough order of magnitude estimate, but it gives some support to the previous visibility age. In reality, the accretion disk mass should decrease exponentially, resulting in an exponential decrease in brightness; the period of observability would represent the time interval when the brightness is above some detection limit.

If the detection period is typically 10^8 yr, the phenomenon of double escape is seen during the fraction 10^8 yr/10^{10} yr = 10^{-2} of the age of the galaxy. We have previously estimated that 76% of triple black hole systems manage to evolve without degeneration into a binary system, and out of them 0.7% end up in the category of the double escape. Therefore the total expected frequency of double escape systems among triple merger galaxies is $10^{-2} \times 0.76 \times 0.007 = 5 \times 10^{-5}$. It is likely that all very bright elliptical galaxies have resulted from at least three galaxies which have merged. Thus we expect that among 20 000 bright elliptical galaxies there would be on average about one which is detectable as a source of double escapers. The result would not change essentially if we were to neglect the loss cones; then the escape from the galaxy becomes easier and the number of double escapes increases. But the increase is compensated by a comparable loss of binaries before the triple black hole systems arise (Problem 11.6). For this reason the results of this present section are valid independent of the details of interactions of binary black holes with their host galaxies.

How could these rare events be possibly seen? The expected frequency of about 1/20 000 among the parent population of bright elliptical galaxies suggests that the escaped black hole pairs may manifest themselves as Fanaroff–Riley type II double radio sources (Saslaw *et al.* 1974, Heinämäki 2001). Without going into further details of double radio sources, it may suffice to say that the radio lobes of double radio sources are typically on opposite sides of the parent galaxy well aligned with the nucleus. Also there are often trails of radio emission connecting the nucleus and the radio lobes, proving that whatever causes the lobe emission has its origin in the nucleus of the parent galaxy. Many times, but not always, the nucleus is also a strong emitter of radio waves. In that case it is reasonable to assume that there is a supermassive black hole also in the centre of the galaxy.

The phenomenology of double radio sources is well described by a model where three or four black hole systems arise in connection with multiple mergers of galaxies and where black holes subsequently escape in a symmetric manner. In the case of three black hole systems the evolution may lead to two escapers, a single body and a binary. The remaining lifetime of the binary is short compared with the travel time out of the galaxy; thus also the initial binary quickly becomes a single black hole. The emission of radio waves from these two receding black holes via mechanisms related to accretion disks and jets is thought to be responsible for the generation of the radio lobes and the trails connecting the lobes to the galactic nucleus. When four black holes are involved, the result is the same except that one black hole now remains in the nucleus and produces the central radio component which is frequently observed.

There are two aspects of double radio sources which follow directly from the three-body problem: the distribution of the double radio source sizes (i.e. maximum linear extent) and the distribution of the symmetry parameter. The latter is the distance of the more distant radio lobe from the galactic nucleus divided by the distance of the nearer component from the nucleus. These distributions could be related to the distributions of escape speeds of the black holes.

Let us assume that a black hole escaped from the nucleus of the galaxy with speed v_s which is greater than the escape speed by a wide margin. The latter statement allows us to ignore the slowing down of the speed of the escaper. We further assume that the black hole is 'visible' for a fixed period of 10^8 yr. Then the maximum distance from the nucleus of galaxy where it can be observed is

$$D_{\max} = 360\ \mathrm{kpc} \left(\frac{v_s}{3600\ \mathrm{km/s}} \right), \tag{11.31}$$

It is actually observed somewhere between the distance $D = 0$ and $D = D_{\max}$.

The distribution of the escape velocities v_s, in units of the binary orbital velocity v_0, is (Eq. (7.21))

$$f(v)\,\mathrm{d}v = \frac{56v\,\mathrm{d}v}{(1 + 8v^2)^{4.5}} \tag{11.32}$$

using $m_s = m_B$ as before. This may be approximated by

$$f(v)\,dv \approx 0.009(v_s/v_0)^{-4.5}\,dv \tag{11.33}$$

when $0.35 \lesssim v \lesssim 0.7$, i.e. in the neighbourhood of the escape velocity $v_{esc} \approx 0.42v_0$. Here we have written the normalisation of v_s in terms of v_0 explicitly, i.e. $v = v_s/v_0$. Equation (11.33) tells us that the number of escapers ending up in the interval $[v, v + dv]$ goes up as $v_0^{4.5}$ when v_0 increases and v_s is fixed (at v_{esc}, say). Increasing v_0 means that we are moving down the v-distribution to the range where the frequency $f(v)$ is higher. The probability of encountering a binary with orbital speed v_0 is proportional to the lifetime of such a system. From Eq. (11.13) the lifetime $T_c \propto a^4 \propto v_0^{-8}$. Therefore the increase in the probability with increasing v_0 is more than compensated by the decrease in lifetime; the total probability $P(v_s)$ of having an escaper in the interval $[v_s, v_s + dv_s]$ is thus proportional to

$$P(v_s) \propto \int_{v_s}^{\infty} v_0^{4.5} \times v_0^{-8}\,dv_0 \propto v_s^{-2.5}. \tag{11.34}$$

The lower limit of integration is based on the fact that the limitation on the binary lifetimes applies only to binaries with orbital speed $v_0 \gtrsim v_s$. Binaries with lower orbital speeds are very long lived but make practically no contribution to the class of double escapers. Because of the one-to-one correspondence between v_s and D_{max}, this is also the distribution for D_{max}; it may also be taken as the first approximation of the distribution of D:

$$P(D) \propto D^{-2.5}. \tag{11.35}$$

Figure 11.4 shows how this distribution follows the observational data on double radio sources. The flattening of the observed distribution below the linear size 400 kpc is expected since the assumptions of the model (neglecting v_{esc}) are not satisfied there; numerical simulations (Valtonen *et al.* 1994) show a similar flattening below $v_s = 2v_{esc} = 3000$ km/s. Note that the average projected (in the sky) distance of each component at this limit is 200 kpc. The typical deprojection factor is $\sqrt{2}$ since there is likely to be an extension of the double source also along the line of sight. Thus the average real distance of the lobes from the nucleus at this limit is just below 300 kpc and the average speed of outward motion is ≈ 300 kpc$/10^8$ yr $= 3000$ km/s, as it should be.

The second quantity of importance is the ratio of distances D_r on the opposite sides of the galaxy. This is equivalent to the ratio of the average outward speeds including the slowing down in the galactic potential. In this case v_0 is a constant since both black hole escapers arise from the same three-body event and we may use Eq. (11.33) for v_s. Because of the difficulty of the double escape, it is reasonable to estimate that for the slower (binary) lobe $v_B \approx v_{esc}$; more definitely, we may put

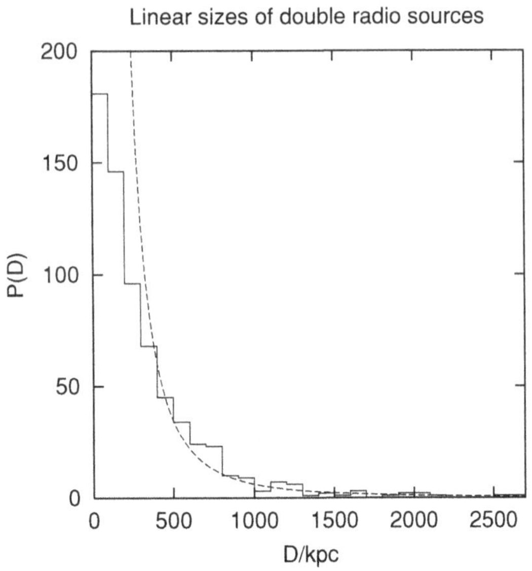

Figure 11.4 The distribution of linear sizes of double radio sources (histogram) compared with with Eq. (11.35), with arbitrary scaling (continuous line).

$v_B = 1.2v_{\rm esc}$ (Valtonen and Heinämäki 2000). Let us estimate the average outward speed to be roughly equal to the asymptotic speed v_∞ after escaping from the galaxy. It is given by $v_\infty^2 = v_s^2 - v_{\rm esc}^2$ for the single body and by a similar expression for the binary. Then the distance ratio D_r is

$$D_r^2 = \frac{v_s^2 - v_{\rm esc}^2}{v_B^2 - v_{\rm esc}^2} \approx \frac{1}{0.44}\left(\frac{v_s^2}{v_{\rm esc}^2} - 1\right).$$

Since we know the distribution of v_s (Eq. 11.33), it is straightforward to calculate the distribution of D_r. It is sometimes convenient to transform to the variable $y = \ln D_r$. Then

$$f(y)\,{\rm d}y \propto \frac{{\rm e}^{2y}\,{\rm d}y}{(1 + 0.44e^{2y})^{2.75}}. \qquad (11.36)$$

This distribution is compared with the observation of double radio sources in Fig. 11.5.

The theoretical peak at $D_r = 1$ ($y = 0$) agrees very well with observations. There is a small but significant difference from the three-body theory in that there are more asymmetric doubles than the theory would predict. It is obvious that the available mass ratios put restrictions on the possible speed ratios. Equal mass systems can only split in such a way that the escape velocities are in a 2:1 ratio, etc. This may explain some of the observed asymmetry. However, numerical simulations produce D_r distributions which are in agreement with the distribution of Eq. (11.36)

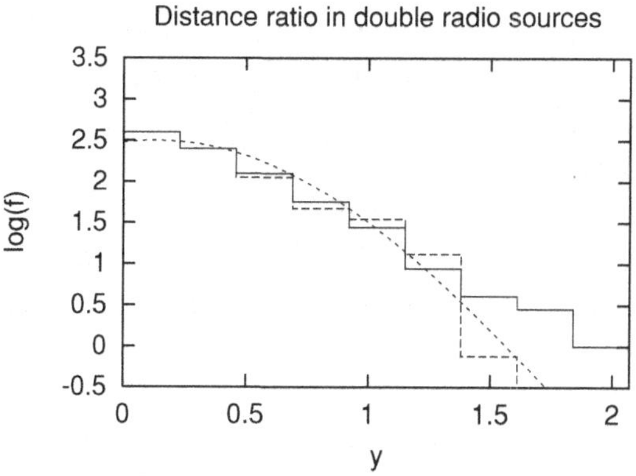

Figure 11.5 The ratio of lobe distances D_r in double radio sources, measured using the parameter $y = \ln D_r$ (solid line histogram). The theoretical distribution of Eq. (11.36) is shown as a continuous line, with arbitrary normalisation. Numerical results from the four-body simulations by Heinämäki (2001) are shown by the dashed-line histogram.

(Valtonen *et al.* 1994, Heinämäki 2001). The results from the four-body experiments of Heinämäki (2001) are also plotted as a histogram in Fig. 11.5. At small D_r they agree with the observational histogram so closely that the two histograms mostly overlap each other. The good agreement between the theory and the double radio source data may be somewhat accidental since it depends on other theories besides the dynamical escape theory.

In spite of many attractive features, the three-body theory for double radio sources (usually called the slingshot theory; Saslaw *et al.* 1974) has not yet gained popularity. The main reason is that no black holes have so far been directly observed inside the radio lobes. However, such an observation is expected to be difficult and may have to wait for a new generation of telescopes (Nilsson *et al.* 1997).

If only one black hole escapes, it follows the velocity distribution given above (Eq. (7.21)). The least massive black hole is the most likely escaper. Then we apply the accumulated distribution for a small escaper, and put $v = 0.42$. We find

$$F(v_s \geq 0.42 v_0) \approx 0.57, \tag{11.37}$$

which is more than ten times the probability for double escapes. Also, it is not necessary to impose the restriction $0.42 \leq m_s \leq 0.58$; this increases the probability of single escapes by another factor of ≈ 6. The dominance of single escapes is a good signature of the black hole escape processes: a single escape is nearly two orders of magnitude more likely than a double escape, as is also found in four-body simulations (Heinämäki 2001).

However, it may be difficult to observe the single escapers. Unless there is a trail connecting the escaper and the nucleus, it may not be clear where the black hole has come from, and it may be confused with distant background objects of similar nature. Also since a single escaper is typically less massive than a double escaper, it is likely to be less luminous and be more easily missed.

Observationally, there are two sided or classical double radio sources (also called D1) as well as one sided (D2) doubles. In the latter sources there is only one outer radio lobe, the second one coincides with the nucleus of the galaxy. Thus D2 doubles could naturally arise from a single escape. The relative numbers D1/D2 vary depending on the type of radio source sample; generally, however, the ratio D1/D2 $\gg 1$, unlike the theoretical number which is D1/D2 $\ll 1$. The difference can be accounted for to some extent by the above mentioned observational selection (Mikkola and Valtonen 1990, Heinämäki 2001).

The black hole which does not escape may still attain considerable speed, well above the velocity dispersion in the nucleus of the galaxy. Then it becomes a radial oscillator in the galaxy. Let us make a simple estimate for the radial extent and the period of oscillation in the galaxy.

We take our model galaxy of mass $4.5 \times 10^{11} M_\odot$; 50% of this mass is inside $r = 10$ kpc radial distance and beyond this distance the mass enclosed by radius r is roughly (Young 1976)

$$M(r) \approx 2.25 \times 10^{11} M_\odot \left(\frac{r}{10\,\text{kpc}} \right)^{0.4}. \tag{11.38}$$

If all this mass were concentrated in the nucleus of the galaxy, the orbital period of a body in a very eccentric orbit around the nucleus, with the apocentre at r, would be

$$t_0 = 2\pi \frac{(r/2)^{3/2}}{\sqrt{GM(r)}} = 3.6 \times 10^6 \left(\frac{r}{\text{kpc}} \right)^{1.3} \text{yr}. \tag{11.39}$$

A radial orbit starting at rest from distance r from the nucleus, which goes through the nucleus and to the other side until it comes to rest again, takes a little longer (7.6% longer to be exact) since the mass is not concentrated in the nucleus and the body feels a reduced amount of central mass when r decreases. The full radial oscillation period is then a little more than $2t_0$.

We next estimate the central velocity v which takes the body out to the maximum distance r from the galactic centre. Let the potentials be U_0 and $U(r)$ at the centre and at the distance r, respectively. Equating the energy per unit mass at the centre and at a very large distance ($r \to \infty$) then gives for the escape velocity v_{esc}

$$\frac{1}{2} v_{\text{esc}}^2 + U_0 = 0 \tag{11.40}$$

since both the velocity and the potential go to zero by definition as $r \to \infty$. The corresponding equation for any other speed v is

$$\frac{1}{2}v^2 + U_0 = U(r) \tag{11.41}$$

since the velocity goes to zero at the distance r. Thus

$$v^2 - v_{\rm esc}^2 = 2U(r). \tag{11.42}$$

The potential is calculated as an integral of the central force from the distance r to $r \to \infty$ using Eq. (11.38):

$$U(r) = -\int_r^\infty \frac{GM(r)}{r^2}\,\mathrm{d}r = -\frac{5}{3}(300\ \mathrm{km/s})^2\left(\frac{r}{10\ \mathrm{kpc}}\right)^{-3/5}. \tag{11.43}$$

Therefore

$$r = \frac{0.35}{\left(1 - v^2/v_{\rm esc}^2\right)^{5/3}}\ \mathrm{kpc}. \tag{11.44}$$

This holds fairly well for $0.8v_{\rm esc} \lesssim v \lesssim 0.99v_{\rm esc}$, i.e. 2 kpc $\lesssim r \lesssim$ 240 kpc.

The oscillation is gradually damped because the black hole interacts with the stars of the galaxy. This interaction may be described as dynamical friction and it is most effective during the crossing of the centre of the galaxy where the star density is high. Therefore we need to model the mass density of stars; for simplicity we take the stars to be one solar mass each, which leads us to the number density

$$n = \begin{cases} 700\ \mathrm{pc}^{-3} & \text{if } r \le 17\ \mathrm{pc} \\ 70\ \mathrm{pc}^{-3}(r/170\ \mathrm{pc})^{-1} & \text{if } 17\ \mathrm{pc} \le r \le 170\ \mathrm{pc} \\ 70\ \mathrm{pc}^{-3}(r/170\ \mathrm{pc})^{-1.75} & \text{if } 170\ \mathrm{pc} \le r \le 1700\ \mathrm{pc}. \end{cases} \tag{11.45}$$

This distribution is partly experimental (Young 1976), partly theoretical (Milosavljevic and Merritt 2001) since observations are difficult especially in the scales below the radial distance $r = 17$ pc.

The expression for dynamical friction from Section 3.14 may be written

$$\frac{\mathrm{d}u}{u} = -4\pi G^2 m_s m_a n \ln\Lambda \frac{\mathrm{d}r}{u^4} \tag{11.46}$$

where u is the speed of the black hole when crossing the galactic centre, m_s is the mass of the star ($m_s = 1M_\odot$), m_a is the mass of the oscillating black hole (here taken to be $m_a = 6 \times 10^8 M_\odot$), n is the local number density of stars, and

$$\Lambda = \frac{b_{\rm max}u^2}{Gm_a}. \tag{11.47}$$

To simplify matters, we put the maximum interaction distance $b_{max} = 170$ pc when $r \geq 170$ pc, and $b_{max} = 17$ pc when $r < 170$ pc. The corresponding values of $\ln \Lambda$ are 5 and 2.7, respectively, assuming that $u = v_{esc} = 1500$ km/s.

Now we may write du/u:

$$\frac{du}{u} = 1.77 \times 10^{-3} \left(\frac{v_{esc}}{u}\right)^4 \times \left[10 \int_0^{0.1} dx + \int_{0.1}^{1} x^{-1}\, dx + \frac{5}{2.7} \int_1^{10} x^{-1.75}\, dx\right] \tag{11.48}$$

where $x = r/170$ pc. We find

$$\frac{du}{u} = 10^{-2} \left(\frac{M_L}{M_L^*}\right)^{-1} \left(\frac{v_{esc}}{u}\right)^4 \tag{11.49}$$

where the dependence on the galaxy mass scale has been added. Note that du/u is directly proportional to the black hole mass m_a which can vary greatly on either side of the mean value.

Even though the speed loss is only about 1% per crossing, the turning point distances decrease rapidly. For $du/u = 10^{-2}$, and starting with $u = 0.99 v_{esc}$, we find the turning point distances in the intervals one cycle (i.e. on the same side of the galaxy) at $r = 240$ kpc, 40 kpc, 17 kpc, 10 kpc, 6.6 kpc, ... For a black hole twice as massive as our basic model we have $r = 240$ kpc, 17 kpc, 6.6 kpc, ... If the turning points of the black hole oscillators are responsible for the creation of radio lobes there would be a succession of lobe pairs in different scales such that the younger pairs are closer to the nucleus and older ones further out (Valtonen 1976b). Such interpretations are complicated by the fact that the gaseous component of the galaxy where the lobes would be stationed may itself be in motion due to galactic winds (Valtonen and Kotilainen 1989). Also the motion of the galaxy through the intergalactic space may drag the radio lobes into the shape of a twin-trail (Valtonen and Byrd 1980).

The lifetime of the oscillator is easy to calculate from Eq. (11.49). Forgetting the mass scaling factor we may write that the total change du of u in dn crossings of the nucleus is (using v_{esc} as a unit speed)

$$\frac{du}{u} = 10^{-2} u^{-4}\, dn \tag{11.50}$$

which is easily integrated from $u = 0$ to the initial value u_0:

$$u_0^4 = 4 \times 10^{-2} n. \tag{11.51}$$

Thus it takes $n \approx 25$ crossings of the nucleus to bring the oscillator with $u_0 \approx v_{esc}$ to a halt. For smaller initial amplitudes fewer crossings are required: e.g. $u_0 = 0.8$ gives $n \approx 10$. Because the oscillation amplitude decreases quickly, so does the oscillation period. The time spent in later oscillations is so short that the total time from the initial ejection to settling back to the nucleus of the galaxy is not much greater than $2t_0$.

When the turning point distance r exceeds $r \approx 10$ kpc, the oscillator lifetime exceeds the typical visibility age of an escaper from the galaxy. Then it is quite likely that the escaper (departing from the other side of the galaxy, opposite to the oscillator) cannot be identified with the three black hole event, and the oscillator is the only reminder of the three-body escape process. The oscillator may also create a double radio source, but of Fanaroff–Riley type I. In type I doubles the brightness fades away from the centre of the galaxy, contrary to the 'classical' type II radio doubles.

In a process with four black holes, the fate of the oscillator is more complicated. Since now one black hole remains in the centre of the galaxy, the oscillator scatters from it on its first crossing of the nucleus. As a result of the two-body scattering, both black holes are ejected and become oscillators of small amplitude. They soon settle back in the nucleus and form a binary there (Valtaoja *et al.* 1989). Thus the lifetime of the oscillator is not much more than t_0 in this case. If the faster escaper, on its way out of the galaxy, is still visible, we may see a double radio source of mixed morphology: a Fanaroff–Riley type II lobe on one side and a type I lobe on the other side of the galaxy. The binary black hole in the nucleus of the galaxy, if it has already formed, may also appear as a strong radio source.

Finally there is the most likely possibility that neither black hole is able to escape from the galaxy, and we have two oscillators along a common line. The black holes necessarily meet before the first oscillation of the larger amplitude is completed, and then they scatter off each other. Since the scattering generally takes place outside the nucleus, the new orbits are not radial. However, the bodies in these non-radial orbits will also be gradually accreted to the nucleus where they form a binary. In addition to the Fanaroff–Riley type I lobe structure we would expect these sources to have a complex central structure, due to non-radial orbits.

One of the signatures of three-body escape is the concentration of escape angles close to the fundamental plane. The orientation of this plane derives from the angular momenta of the orbital motion of the galaxies which have merged and therefore we expect that also the accretion disks around the black holes prefer the same plane. Since the black holes as well as the gas in accretion disks come from at least two separate mergers, each with their own orbital angular momentum, the alignment of the angular momenta of the gaseous disks may not be perfect with the total angular momentum of the black hole system, but anyway strong correlations are expected.

The best indication of the orientation of an accretion disk is the jet or pair of jets. They presumably propagate along the rotation axis of the accretion disk, along the line of the angular momentum vector. On the other hand, escaping black holes propagate in or close to the fundamental plane, nearly perpendicular to the angular momentum vector. Therefore there should be a strong tendency for jets to be perpendicular to the trails leading to radio lobes. This is what is observed: the projections of jets of the central components in the sky are perpendicular to the projections of the trails much more often than is expected from random orientation (Example 7.1; Valtonen 1996).

11.3 Satellite black hole systems

Minor mergers, mergers between galaxies of very different sizes and masses, are presumably much more common than mergers of major galaxies. If minor galaxies also possess supermassive black holes, even though proportionally smaller ones than those discussed so far, then minor mergers produce binaries of very unequal masses. After multiple minor mergers a satellite system develops, dominated by a single massive black hole (say, $10^9 M_\odot$) surrounded by a swarm of smaller black holes (say, $10^5 M_\odot$ category). Even though this cluster may have hundreds of members, it is still much less massive than the central black hole.

The lifetimes of the satellites are in principle much greater than the lifetime of a binary composed of two primaries at the same orbital distance. From the expression for T_c (Eq. (11.13)) we see that lowering the reduced mass $\mathcal{M}$ by a factor 10^4 allows one to lower the orbital radius a by a factor of 10 and still keep the same lifetime. Therefore we may assume that there exist compact black hole clusters also around each component of a massive binary. When the binary orbit shrinks after being involved in three-body processes, and after having suffered energy losses to gravitational radiation, the satellite clusters may become unstable. They should be totally dispersed by the time the binary orbit has shrunk by a factor of ten, i.e. when the binary orbital speed has increased by a factor of $\sqrt{10}$. Then we are discussing an orbital speed around 10 000 km/s and escape speeds of the satellites up to this figure. Thus satellites would mostly leave the parent galaxy.

A small part of the satellite system will be found in ejection orbits instead of escape orbits, and they would approach the shrinking binary again. If they were to escape, their escape speeds are likely to be higher than before since the binary orbital speed is constantly increasing. At the last stable orbit around a non-rotating Schwarzschild black hole the orbital speed is about 120 000 km/s. It is possible that some satellites escape with speeds up to this value if the three-body process takes place just before the binary collapses. Post-Newtonian three-body calculations appear to confirm this (Basu *et al.* 1993, Haque-Copilah *et al.* 1997).

11.4 Three galaxies

Galaxies are 'soft' bodies which can readily stick together when they collide. Dynamical friction brings galaxies together in time scales which are typically much shorter than the Hubble time, and in times of the same order as the period of the initial encounter orbit of the colliding galaxies (see Problem 10.4). Therefore the problem of three galaxies may include collisions in an important way, especially since the distances between galaxies in compact groups are not very much greater than the radii of galaxies. From three-body scattering experiments like the ones shown in Fig. 8.11 and Eq. (8.42) we find that the probability for $q/a_0 \le 0.1$ is about 40%. This is confirmed by numerical experiments with three 'soft' bodies (i.e. galaxies; Zheng *et al.* 1993) which show that indeed about 40% of triples become binary galaxies within the Hubble time. The remaining triple systems become unbound in some cases, but mostly they remain as bound systems, and will perhaps merge at a later time. With enough time, a merger of all three galaxies or the formation of a stable binary system will almost always result. However, the observed triples and also many binary galaxies are still in transitional stages of dynamical evolution because their crossing time is only about 10% of the Hubble time (Wirén *et al.* 1996).

Under special circumstances, especially if a circular binary galaxy exists, passing galaxies may also be accelerated to fairly high speeds. Equation (10.79) tells us that near the stability boundary the relative energy change of the binary can be typically $\Delta E_B/E_B \approx 0.1$. If the orbital speed of the circular binary is v_0, its orbital energy is $|E_B| = \frac{1}{2}\mathcal{M}v_0^2$, $\mathcal{M}$ being the reduced mass. The energy available to the passing third galaxy is then about $0.1\frac{1}{2}\mathcal{M}v_0^2$. If the passing galaxy comes with zero energy (parabolic orbit) in a direct orbit, and leaves with the speed v_s, then

$$\frac{1}{2}mv_s^2 = 0.1\frac{1}{2}\mathcal{M}v_0^2 \tag{11.52}$$

or

$$\frac{v_s}{v_0} \approx \sqrt{0.1\frac{\mathcal{M}}{m}} \approx 0.2 \tag{11.53}$$

for equal mass galaxies.

In typical galaxies with 'flat' rotation curves, the total mass of a galaxy is

$$M = \frac{u^2 R}{G} \tag{11.54}$$

if u is the constant rotation speed and R is the radius of the galaxy. Two such galaxies in contact with each other revolve with orbital speed

$$v_0 = \sqrt{\frac{2GM}{2R}} = u. \tag{11.55}$$

Therefore the asymptotic escape speeds of galaxies reach up to $1/5$ of the internal rotation speeds of galaxies involved, i.e. up to 100 km/s or so. Figure 7.7 tells us that occasionally there may be an escape even with $v_s \approx u$, and such speeds have

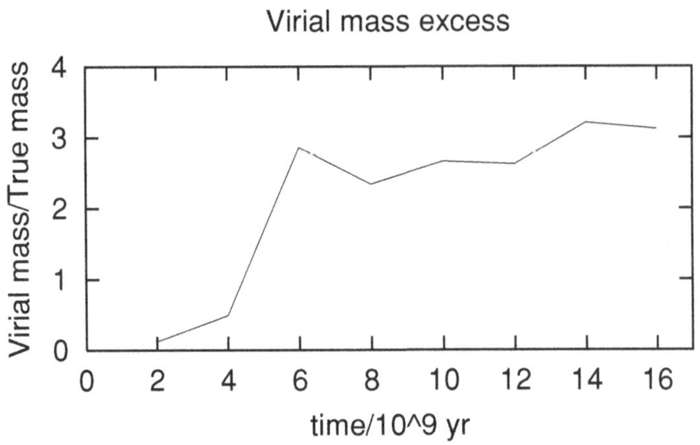

Figure 11.6 The apparent excess mass in a cluster of galaxies like the Coma Cluster, when determined using the virial theorem. Points are from the model calculation by Laine *et al.* (2004).

been confirmed in numerical experiments with 'soft' bodies (Valtonen and Wirén 1994).

This opens up the possibility that in some systems escaper galaxies may play a role (Valtonen *et al.* 1993). This role can be the greater the more massive is the binary which gives the accelerations. For example, in the centre of the Coma Cluster of galaxies there may exist a binary with orbital speed $v_0 \approx 1800$ km/s. Such a binary would accelerate galaxies up to speeds of the same magnitude, and would confuse the analysis of the cluster mass if the escapers are not identified (Valtonen and Byrd 1979). It is very difficult indeed to know which galaxy has passed by the binary and which one has not, and which one has been accelerated to the escape speed by the binary and which one is bound to the cluster as a whole. The net result is that the determination of mass by using observations and the virial theorem (Eq. (2.28)) could be wrong by a factor of two or so.

Figure 11.6 shows the calculated virial mass in a model Coma Cluster divided by true mass (which is of course known in a model). The virial mass determination has been carried out in the same way as observers do, looking at the cluster from a given direction, and using only the 'measurable' information of radial velocities and projected positions in the plane perpendicular to the line of sight. In this simulation which includes soft-body interactions of galaxies in a realistic way, 248 ordinary galaxies of mass-to-light ratio 30–45 fall into the potential well of a circular binary, with a total mass of $3.5 \times 10^{14} M_{\odot}$. Some of the galaxies escape, of the order of a dozen of them, while the rest of the galaxies form a bound cluster around the binary.

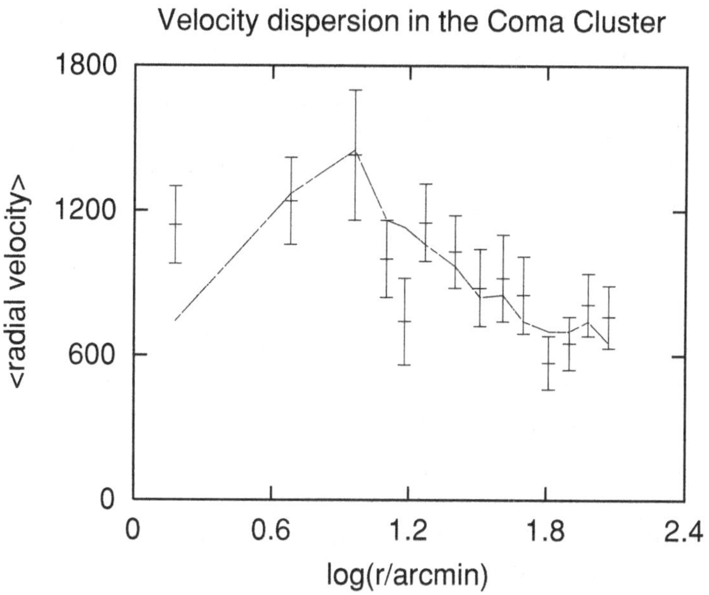

Figure 11.7 The dispersion of radial velocities of galaxies as a function of distance from the centre of the Coma Cluster of galaxies. The points with error bars refer to observational data while the solid line comes from a model calculation by Laine *et al.* (2004).

The radial velocity dispersion of the galaxies as a function of the projected distance from the cluster centre is shown in Fig. 11.7 (solid line). For this model the radial velocity profile matches observations (points with error bars) quite well. The same is also true for the surface density of galaxies as a function of the distance from the centre. Therefore we may say that the model is consistent with the observations. Then we also have to conclude that the mass of the cluster calculated using the virial theorem may be too large by a factor of two to three.

This is a model for a binary + 248 galaxies. But since the dominant interaction is between the binary and one galaxy at a time, we may also use it as representing 248 binary + single galaxy systems. Thus also in such a triple galaxy sample we expect a similar overestimate of the total mass if the virial theorem is used.

11.5 Binary stars in the Galaxy

Numerical simulations of star cluster evolution have shown that three-body interactions take place among cluster stars frequently. In the three-body breakup a binary is often expelled out of the cluster and it becomes a binary in the general field of stars of the Galaxy. There may still be further encounters with other stars later on, but on the whole the 'hard' binaries probably have their properties more or less frozen since their escape from the star cluster of their origin. We will now study what kind of binary star population we expect from this process and how it compares with the observed binaries.

Our statistical theory for three-body break-up in Chapter 7 was derived assuming that all systems have a constant total energy E_0. But in star clusters E_0 may vary greatly from one three-body system to another. According to Eq. (7.22) the available phase space volume σ is inversely proportional to $|E_0|$:

$$\sigma(|E_0|)\mathrm{d}|E_0| \propto |E_0|^{-1}\,\mathrm{d}|E_0|. \tag{11.56}$$

If for any reason the three-body systems are uniformly distributed in the E_0 space then we expect that the binary energies E_B after the three-body breakup also follow Eq. (11.56), i.e. Öpik's law of Eq. (1.2) should be valid. To what extent this is true can be found out by studying young star clusters observationally as well as by simulating star formation processes theoretically.

Hard binaries in star clusters tend to harden further. At the limit of very hard binaries we may put $v = 0$ in Eq. (8.35) and write the average hardening rate

$$\frac{1}{2}\mathcal{M}v_0^2 R_\Delta = \left\langle \frac{\mathrm{d}|E_B|}{\mathrm{d}t} \right\rangle = 3G^2 m_B^3 \frac{\mathcal{M}}{M}\frac{n}{v_3}. \tag{11.57}$$

In a star cluster we may regard the right hand side as a constant in the first approximation, even though in fact the density of stars n and the typical speed of stars do vary during the cluster evolution. But using this assumption, and also putting all stars equal to $1\,M_\odot$, the equation is easily integrated:

$$\frac{E_B}{(E_B)_0} = \frac{16G^2 M_\odot^2 nT}{v_0^2 v_3} \tag{11.58}$$

where $(E_B)_0$ is the initial value of the binary energy, $|(E_B)_0| \ll |E_B|$ and v_0 is the corresponding mean orbital speed. T is the time of escape of the binary from the cluster since the birth of the star cluster.

Because of the evolution in the star cluster as well as the effect of the galactic tides, the cluster is gradually dissolved. The time of dissolution t_d has been estimated at

$$t_d \approx 5.7 \times 10^8 \left(\frac{M_{\text{cluster}}}{250 M_\odot}\right)\left(\frac{1\ \text{pc}}{r_h}\right)^3 \text{yr} \tag{11.59}$$

times a factor depending on the structure of the cluster (Binney and Tremaine 1987). Here M_{cluster} is the mass of the cluster and r_h is its median radius. Since 250 solar mass stars within a sphere of 1 pc in radius makes the average number density $n = 250/(\frac{4}{3}\pi\ \text{pc}^3) \approx 60\ \text{pc}^{-3}$, the equation may be written using this mean number density n:

$$t_d \approx 5.7 \times 10^8 \left(n/60\ \text{pc}^{-3}\right) \text{yr}. \tag{11.60}$$

We may take the typical escape time of the binary to be half of t_d, i.e.

$$T \approx 3 \times 10^8 (n/60 \text{ pc}^{-3}) \text{ yr}. \tag{11.61}$$

From here n may be solved and inserted into Eq. (11.58) above. Then

$$\begin{aligned} \frac{E_B}{(E_B)_0} &\approx \frac{16G^2 M_\odot^2}{v_0^2 v_3} \left(\frac{T}{3 \times 10^8 \text{ yr}}\right) 60 \text{ pc}^{-3} \times T \\ &= 5.3 \left(\frac{T}{3 \times 10^8 \text{ yr}}\right)^2 \left[\frac{v_0^2 v_3}{(\text{km/s})^3}\right]^{-1}. \end{aligned} \tag{11.62}$$

Putting a typical number $v_3 = 0.25$ km/s, and starting from a hard binary with $v_0 = 1$ km/s, we expect to end up with

$$\frac{E_B}{(E_B)_0} \approx 20 \left(\frac{T}{3 \times 10^8 \text{ yr}}\right)^2. \tag{11.63}$$

In a typical hardening period of $T = 10^8$ yr we then expect the average binary binding energy to increase by a factor of two and the corresponding orbital period to shorten by about a factor of three.

Since $|E_B|_0 \propto v_0^2$, the final value of $|E_B|$ does not depend on v_0 (i.e. on the initial orbital period) but only on T. Therefore the distribution of final periods P should depend on the distribution of T.

A numerical simulation of the Pleiades star cluster by Kroupa *et al.* (2001) shows that in its assumed 100 million year lifetime the binary period distribution shifts toward shorter values by about a factor of three at the end of large periods ($P \gtrsim 30$ yr). This agrees with our simple estimate. At the end of short periods no significant shift is detected in the simulation.

Depending primarily on the cluster star density, clusters live different lengths of time, and provide different periods T for the hardening process. We get an idea of the distribution of T from observations of star clusters. The current age τ of a star cluster is a representative time in the history of a cluster, and may well tell us when a typical binary escape happens. The distribution of τ is observed to be (Wielen 1971)

$$f(\tau) \propto \tau^{-1} \tag{11.64}$$

in the interval 2×10^7 yr $\lesssim \tau \lesssim 5 \times 10^8$ yr, and it steepens beyond the upper limit. Let us then suppose that also

$$f(T) \propto T^{-1} \tag{11.65}$$

in this range.

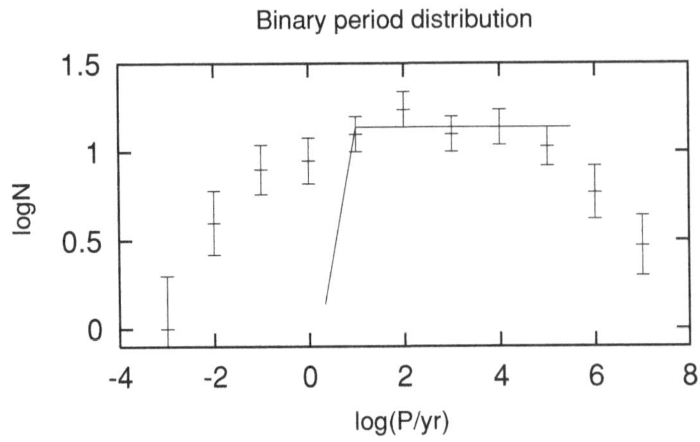

Figure 11.8 The period distribution of a sample of nearby binary stars with a solar type primary (Duquennoy and Mayor 1991, Fig. 7). Lines refer to theoretical expectations.

Since $E_B/(E_B)_0 \propto T^2$, and the corresponding period ratio $P/P_0 \propto T^{-3}$, we find

$$f(P/P_0) = \frac{f(T)\,\mathrm{d}T}{\mathrm{d}(P/P_0)} \propto (P/P_0)^{-1}. \tag{11.66}$$

In a logarithmic scale the distribution of P/P_0 is flat:

$$\frac{f(P/P_0)}{\mathrm{d}\log(P/P_0)} = \text{constant} \tag{11.67}$$

since $\mathrm{d}(P/P_0) = (P/P_0)\,\mathrm{d}(\log(P/P_0))$. This should be valid over one and a half orders of magnitude in T, which corresponds to over four orders of magnitude in P/P_0.

What is the range of validity of this result? At the end of small T, below about $T = 2 \times 10^7$ yr, there is negligible binary hardening. At the other end, $T \geq 5 \times 10^8$ yr, the power-law of Eq. (11.65) steepens and the expected period distribution becomes ($P \lesssim 10$ yr):

$$\frac{f(P/P_0)}{\mathrm{d}\log(P/P_0)} \propto (P/P_0)^{3/2}. \tag{11.68}$$

These distributions are compared with observations (Duquennoy and Mayor 1991) in Fig. 11.8. We notice that the predicted break at the end of low values of P/P_0, below the orbital period of ten years, is not borne out by observations. It appears that these short period binaries come from a binary population which have short periods to start with. Such 'primordial' binaries are observed in star clusters and they make an important contribution to the short period end of the distribution.

The reason for the relative flatness of the short period binary distribution may be in the star formation process. Apparently, Eq. (11.56) applies there at least over a limited range of E_0. The scale free property of the distribution for longer periods may result from binary hardening. The steepening of the period distribution beyond $\log(P/\mathrm{yr}) \approx 5$ is well understood by the disruption of long period binaries in the Galactic field. Relative to the stellar background, these binaries are 'soft' and tend to become even softer until they break up.

The distribution of the eccentricities of binaries leads to the same conclusion: tight binaries, with periods less than 3 yr, have a bell shaped distribution with a peak around $e = 0.3$. Wider binaries, with periods exceeding 3 yr, show a distribution which agrees with $f(e) = 2e$, the distribution expected after three-body evolution (see Fig. 1.7, Duquennoy and Mayor 1991, Kroupa 1995a, b).

The three-body evolution also modifies the binary mass ratios. Binary pairs where both components are massive are more likely to survive than pairs with unequal masses. This makes the mass ratio distribution evolve towards $m_2/m_1 \approx 1$. The mass ratios obtained by picking pairs of stars at random from the initial distribution of stellar masses are therefore subject to later evolution.

Different binaries evolve by different amounts. The most massive binaries tend to settle near cluster centres and they are subject to many strong three-body interactions. As a result, exchanges of binary members take place until the binary is made up of two rather heavy members.

Ordinary binaries are involved in fewer strong three-body interactions. There we may assume that only a single three-body interaction is responsible for the mass ratio distribution. Starting from this assumption, we may pick three mass values at random from the Salpeter (1955) initial mass function $f(m)$ (see Section 1.4). Then we use the probability distribution of Eq. (7.23) to decide which star (m_s) escapes and which are the two others (m_a and m_b) that make up the binary pair. The mass ratio $m = m_b/m_a$ ($m_b < m_a$) is thus obtained. Repeat the process many times and the distribution of mass ratios is built up. The procedure is best carried out by computer in Monte Carlo fashion, i.e. by picking out random numbers from suitable distributions.

The result of this operation is shown in Fig. 11.9 as a dashed line. A comparison of the data points for a sample of binaries with B-type primaries (where the Salpeter mass function is applicable) shows good agreement. It thus appears that these binaries (of typical orbital period 3 yr) have had at least one three-body interaction in the past.

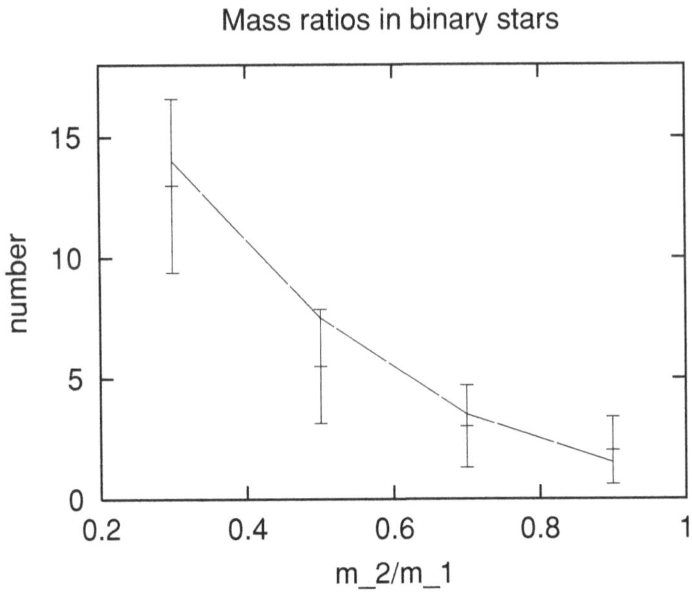

Figure 11.9 The mass ratio of binary star components in an observational sample with a B spectral type primary (points with error bars; Evans 1995). The dashed line is based on a theory where two lower mass companions for the B-type star have been picked at random, and one of the companions has escaped.

In the case of solar type (spectral class G) primaries the Salpeter mass function for single stars is not suitable. However, a flatter power-law, with index $\alpha = 1.25$ may be used (Section 1.4). Then the same process as described above leads to the distribution of Fig. 11.10. The observations by Duquennoy and Mayor (1991) are well described except at the low values of m_2/m_1 where both the observations and the power-law assumption are very uncertain.

For the most massive O-type stars this procedure is not reasonable since numerous three-body encounters have in fact truncated from below the distribution of the possible mass values. Now we may pick three mass values from the power-law distribution of Eq. (1.4) with $\alpha = 3.2$ (applicable to the upper end of the mass range), all of which are above a given lower limit. Then we again ask which one of the three stars escapes, which ones make the binary and what is their mass ratio. The mass ratio distribution built up in this way is shown as a line in Fig. 11.11. It agrees well with the observed O-star primaries sample (Abt 1977).

The rather puzzling situation with the mass ratio distribution varying as a function of the spectral type of the primary is therefore explained as a result of three-body interactions among stars (Valtonen 1997a, b, 1998).

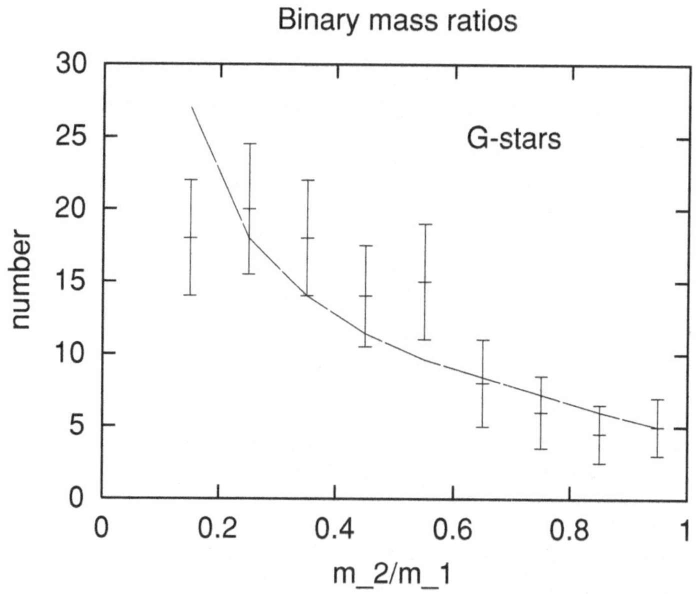

Figure 11.10 The observed distribution of binary star mass ratios when the primary is a solar type star (points with error bars; Duquennoy and Mayor 1991). It is compared with the three-body theory with $\alpha = 1.25$.

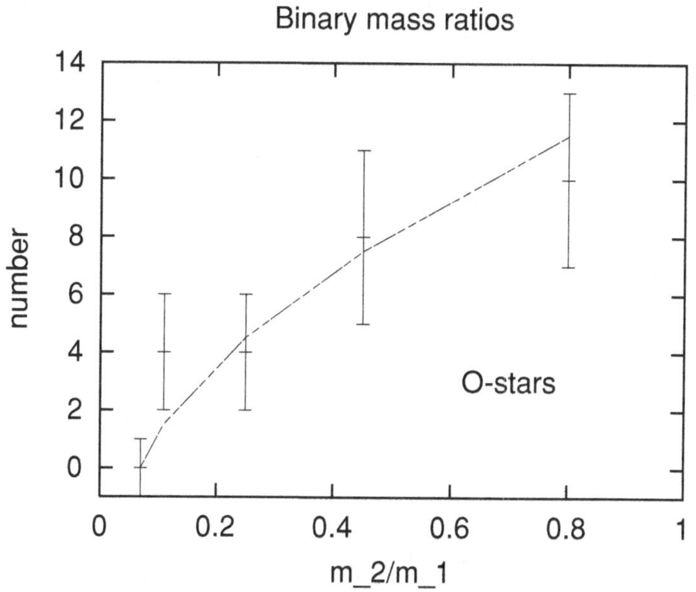

Figure 11.11 The observed distribution of binary star mass ratios when the primary is an O-type star (points with error bars; Abt 1977). It is compared with the three-body theory with $\alpha = 3.2$ and single star mass distribution truncated from below.

11.6 Evolution of comet orbits

We now have the necessary tools to continue the discussion begun in Section 1.3 on the origin of comets. We were left with the question of whether short period comets might evolve from Oort Cloud comets via successive three-body encounters in the Sun–planet–comet system. What are the orbital properties of the comets after several encounters?

We consider encounters with each planet separately and for simplicity suppose that the orbits of the planets are circular. For the current problem this is an acceptable approximation. Then we have a straightforward application of the restricted circular three-body problem which was discussed in Chapter 6.

Any of the planets may be the 'second' body; however, in the comet problem the four largest planets Jupiter, Saturn, Uranus and Neptune dominate. Among them, Jupiter appears to be most important, being the most massive (about 0.1% of the Sun's mass) and having the innermost orbit at about 5.2 AU from the Sun. For this approximation, then, the binary is the Sun–Jupiter system.

Jupiter family comets are usually defined such that their orbital periods are less than 20 years. This translates to orbital semi-major axes being less than $1.4a_B$, where $a_B = 5.2$ AU. A Halley type comet has an orbital period greater than 20 but less than 200 years, and thus its semi-major axis is in the range from $1.4a_B$ to $6.5a_B$. An Oort Cloud comet has a very large semi-major axis, typically $\sim 10^4$ AU.

The minimum energy change U required to bring a comet from the Oort Cloud to the Jupiter family is

$$U_{JF} = 10^3/1.4 = 714, \tag{11.69}$$

while the same quantity for a Halley type comet is

$$U_{HT} = 10^3/6.5 = 154. \tag{11.70}$$

The integrated cross-section for capture into the Jupiter family is (using $F = 4\pi$)

$$\sigma_{JF} = \frac{1}{4} F a_B^2 \frac{1}{U_{JF}^2} \approx 2 \times 10^{-6} \pi a_B^2 \tag{11.71}$$

(Eq. (6.54)). H. A. Newton (1891) earlier obtained essentially the same result. The corresponding cross-section for Halley type comets is

$$\sigma_{HT} \approx 4 \times 10^{-5} \pi a_B^2. \tag{11.72}$$

Therefore the probability of capture directly from a near-parabolic Oort Cloud orbit to a short period orbit is very low, less than 10^{-4}. In this situation we need to investigate other paths which an Oort Cloud comet might take to reach a Halley type orbit. At the other end of the scale we have a diffusion process by which the energy change U_{HT} occurs through many small steps. Every step corresponds to a passage through the planetary region, and U can be either positive or negative.

Let us then consider a random walk in energy space (Fernandez and Gallardo 1994). A typical step size is $U_{\text{step}} = 4$ (Fernandez 1982), and the number of steps required is

$$N_{\text{steps}} \approx \left(U_{HT}/U_{\text{step}}\right)^2 . \tag{11.73}$$

At every step there is a certain probability that a comet escapes from the Solar System along a hyperbolic orbit. The probability that a long period comet survives N_{steps} crossings of the planetary region has been calculated to be (Everhart 1976)

$$p_N \approx 0.5/\sqrt{N_{\text{steps}}}. \tag{11.74}$$

This is also the capture probability of Halley type comets when we substitute N_{steps} from above:

$$p_{HT} \approx 0.5 U_{\text{step}}/U_{HT} \approx 0.013. \tag{11.75}$$

The result has been confirmed by orbit calculation methods by Emel'yanenko and Bailey (1998). For the Jupiter family $U_{JF} = 714$ which gives the capture probability

$$p_{JH} \approx 0.0028 \tag{11.76}$$

which is also found by other methods (Nurmi 2001).

The number of orbital revolutions leading from an Oort Cloud comet to a Halley type comet is $N_{\text{steps}} \approx 1500$. At every crossing of the planetary region there is also the probability $\sigma_{HT}/\pi a_B^2 \approx 5.4 \times 10^{-5}$ that a comet jumps directly into a Halley type orbit; trying to do this 1500 times gives the total probability $5.4 \times 10^{-5} \times 1500 \approx 0.08$ or 8% chance of success. Therefore the diffusion route is more likely to lead to a Halley type orbit than a single jump. The same may also be concluded about the Jupiter family.

The rate of comets coming from the Oort Cloud and passing through the area πa_B^2 of Jupiter's orbit has been estimated to be about one per year if we consider comets brighter than the magnitude limit $H_{10} = 7$ (Bailey and Stagg 1988). This corresponds to bodies which probably have diameters in the range 5–10 km (Weissman 1983, Bailey 1990). Therefore the annual capture rate is about 0.013 Halley type and about 0.0028 Jupiter family comets.

The steady state population may be estimated by calculating how many comets are captured during the visible lifetime of a comet. The lifetime may vary greatly from comet to comet (Wiegert and Tremaine 1999), and also the comets probably fade off gradually rather than disappear from sight suddenly. Therefore the visible lifetime is an average quantity which describes how many revolutions the comet typically takes to become fainter than the limit $H_{10} = 7$. This quantity is very poorly known. We assume that the lifetime is 200 revolutions (Delsemme 1973, Kresak 1987) which translates to 15 000 yr for a typical Halley type comet whose period is about 75 yr. It is less than 10% of the typical dynamical lifetime of a Halley type comet (Bailey and Emel'yanenko 1996). In 15 000 yr about 200 Halley type comets are captured. Therefore this is the expected steady state number of Halley type comets.

The number of known Halley type comets in this absolute brightness category is about 20. It is actually less than 20, but by extrapolation from the brightest magnitudes where the discoveries should be most complete, we arrive at this figure (Hughes 1988). It has been argued that because of incomplete detections the actual number should be more like 200–400 (Fernandez and Gallardo 1994, Shoemaker *et al.* 1994), but there is no unanimity about this (Emel'yanenko and Bailey 1998). In any case one may say that considering the great uncertainties, the captures from the Oort Cloud are numerous enough to explain the current number of Halley type comets. It has also been suggested that part of the Halley type population originates from a relatively low inclination Inner Oort Cloud (Levison *et al.* 2001).

When we repeat the same calculation for a typical Jupiter family comet, with its capture probability about $1/5$ and its period about $1/10$ of the corresponding values for a Halley type comet, we get a steady state population of 4 comets. This is actually the same as the number extrapolated from observations. But if the severe incompleteness in detections is true, then the Oort Cloud captures may explain only about 5% of the current Jupiter family.

The diffusion origin of the Jupiter family would mean that on the way the comets would have spent time as Halley type comets. But there is a good argument against this evolutionary channel: the Tisserand parameter T in the two comet groups is different. Relative to Jupiter, the parameter $T > 2$ for all Jupiter family comets except for seven of them while $T < 2$ for all Halley type comets except for two of them. Remember that the Tisserand parameter is conserved in the capture process as long as the influence of other planets can be ignored. The rather small overlap of the two comet families with respect to the Tisserand parameter suggests that only a small fraction of the Jupiter family ($\approx 5\%$) comes from Halley type orbits.

In reality the comet orbits evolve via many encounters (minor and major), and there are important evolutionary trends also between close encounters. Monte Carlo models have been built for example by Wetherill (1991), Fernandez and Ip (1991), Weissman (1991), Stagg and Bailey (1989), and Nurmi (2001) where the evolution proceeds via a succession of three-body encounters. The conclusions are generally similar to the findings based on the diffusion model. It is especially difficult to get the low inclination Jupiter family when the starting point is the isotropic Oort Cloud (Nurmi *et al.* 2002). Simulations starting from the Kuiper Belt as a source region for the Jupiter family comets have been more successful in this respect (Levison and Duncan 1997).

Other planets besides Jupiter make their own contributions to the comet captures. Since Neptune is the outermost of the giant planets and therefore has the largest orbit cross-section πa_B^2, we will consider its contribution in the following.

According to Fernandez (1982), the total comet flux through Neptune's orbit is about 140 times greater than the flux through Jupiter's orbit, and it could be even another factor of ten greater (Bailey and Stagg 1990). The size of its orbit is 5.8 times greater than Jupiter's while its mass is 18.44 times less than Jupiter's. The energy change required to bring a comet from the Oort Cloud to an orbit with semi-major axis less than 60 AU (twice the orbital radius of Neptune) is

$$U_{TN} \approx \frac{18.44 \times 10^3}{2} \approx 9.2 \times 10^3 \tag{11.77}$$

and the direct capture cross-section

$$\sigma_{TN} \approx 10^{-8} \pi a_B^2. \tag{11.78}$$

The annual direct capture rate is then $140 \times 10^{-8} \approx 10^{-6}$. The lifetime of these Transneptunean comets has been estimated to be 4×10^7 years (Levison and Duncan 1997); in this time span the population of ≈ 40 comets is accumulated; this is also the steady state number of directly captured comets.

The probability of getting a Transneptunean comet by diffusion is

$$p_{TN} \approx 0.0002. \tag{11.79}$$

The annual capture rate becomes 0.03 comets, and the steady state population numbers about one million. This could be up to 10^7 comets depending on the Oort Cloud comet flux at Neptune's orbital distance. The number of steps in the diffusion would be typically five million, and the orbital period would be in millions of years. Therefore there has not been enough time within the age of the Solar System ($\approx 4.5 \times 10^9$ yr) for diffusion to carry the bulk of the comets through. The comets most likely to have been transferred from the Oort Cloud are the low inclination comets (Nurmi 2001), but even among them only a small fraction would have achieved the transition to the Transneptunean region. Still their number could be significant since they represent a flux which starts from the original Oort Cloud at the early Solar System; that flux may have been orders of magnitude greater than the present day comet flux (Chyba *et al.* 1994).

Transneptunean comets are difficult to detect, and cannot be seen at all in the size range (5–10 km diameter) which we have been discussing. There is however, a fairly large number (over 300) of very large, 100 km class, bodies in this region of the Solar System. If they represent a 'tip of an iceberg', the largest of a whole range of comets, then it is possible that around 10^7 comets exist in the size range which we are considering. Thus the captured population could make only a small contribution to the Transneptunean comets; it is usually assumed that most of them are primordial, left over from the formation of the Solar System (so-called Kuiper Belt; Edgeworth 1949, Kuiper 1951). There is some evidence of two distinct populations of Kuiper Belt objects (Levison and Stern 2001); this may have relevance with regard to the captured subpopulation.

It is generally assumed that some of the Transneptunean comets drift inward under the influence of giant planets and finally end up in the Jupiter family (Levison and Duncan 1997). In Monte Carlo simulations (Valtonen *et al.* 1998) about a quarter of Transneptunean comets become visible as Jupiter family comets at the end of a long evolution. Thus Jupiter family comets could have an origin quite distinct from the Halley type comets, and have only a small contribution from direct Oort Cloud or Halley type captures.

Another constraint on the origin of comet families is the distribution of their orbital inclinations. For comets captured from the Oort Cloud to Halley type orbits we may use Eq. (6.59) with $Q = 0.5$ (a typical value according to the simulations by Nurmi *et al.* 2002) and obtain

$$f_{HT}(\iota_0)\,\mathrm{d}\iota_0 = F\sin\iota_0\,\mathrm{d}\iota_0/\pi = 0.5\left[1 + 2.73\,(1+\cos\iota_0)^{1.5}\right]\mathrm{d}\iota_0. \qquad (11.80)$$

As was explained in Section 6.5, the $\sin\iota_0$ factor appearing on the left hand size is due to an isotropic initial distribution of orbital orientations, as in the Oort Cloud. Note that the distribution is given for initial inclinations ι_0; the final inclinations are not very different for Halley type comets (Eq. (6.73)), and therefore we may compare the distribution $f_{HT}(\iota_0)$ with observations. Figure 11.12 shows such a comparison.

When comets are further captured from the Halley type orbits to the Jupiter family, we multiply $f_{HT}(\iota_0)$ by the appropriate capture probability, Eq. (6.59). Now it is suitable to use $Q \approx 1$ (Nurmi *et al.* 2002); we put exactly $Q = 0.938$ in Eq. (6.59) for reasons explained in Section 6.5. Then the inclination distribution becomes

$$f_{JF}(\iota_0)\,\mathrm{d}\iota_0 = f_{HT}(\iota_0)\,\mathrm{d}\iota_0 \times 0.062\pi\left[1 + 127.34\,(1+\cos\iota_0)^{1.72}\right]/\sin\iota_0. \qquad (11.81)$$

The distribution of final inclinations is shifted towards smaller ι and it is also spread considerably (Eqs. (6.73) and (6.74)). Therefore, when we compare Eq. (11.81) with observations (Fig. 11.13) we have to remember that the theoretical line does not include the shift nor the spread; when they are taken into account the agreement between theory and observations is quite good.

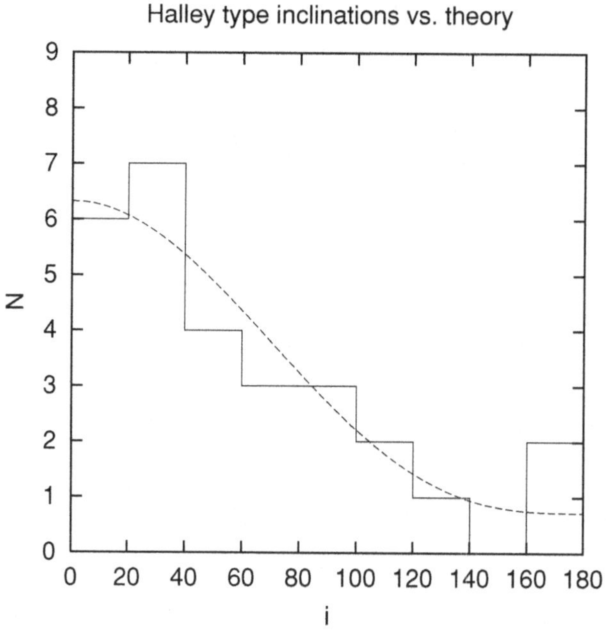

Figure 11.12 A comparison of the inclination distribution of Halley type comets with the theoretical expression (Eq. (11.80)).

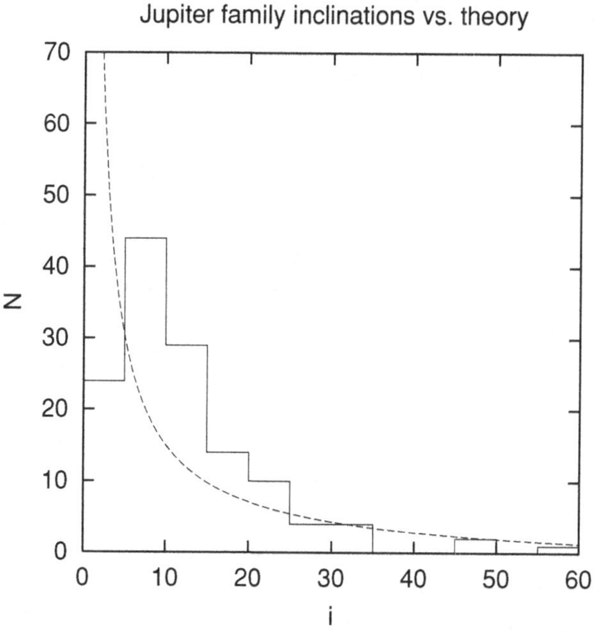

Figure 11.13 The distribution of inclinations in the Jupiter family of comets, and a theoretical line assuming that the Jupiter family is captured from the Halley type comets. The line refers to precapture inclinations, which are more concentrated around $\iota = 0$ than the final inclinations.

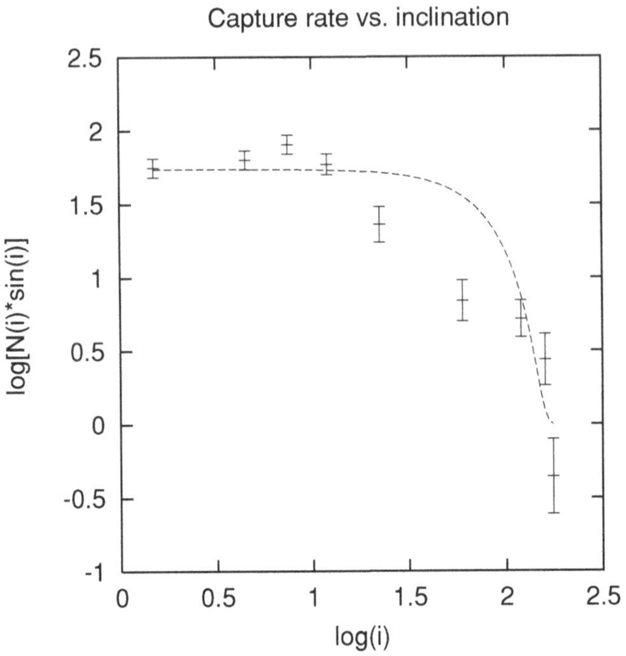

Figure 11.14 Comet capture probability $N(\iota)$ by planet Jupiter from the initial perihelion interval 4 AU $< q <$ 6 AU, per unit interval of initial inclination ι, in units of 10^{-5} (points with error bars, Nurmi 2001). The probability is compared with the single encounter capture cross-section with $Q = 0.85$ (dashed line, Eq. (6.59) scaled by a suitable factor).

Even though our theoretical formulae were derived for a capture via a single encounter, they should also apply to diffusion and to capture after a large number of orbits. In Fig. 11.14 we show calculations by Nurmi (2001) of comet capture by Jupiter. The results show the capture probability after 10 000 orbits when the perihelion of the initial orbit is in the interval $0.77 \leq Q \leq 1.15$. In Eq. (6.59) this interval is effectively covered by using a single value $Q = 0.85$. The comparison of the single encounter theory and the numerical results shows good agreement over most of the range of initial inclinations ι_0, and it justifies to some extent the use of the single encounter theory in the inclination studies.

The Monte Carlo method of calculating small-body orbits has also been used to study the transfer of comets or similar objects between different solar systems (Zheng and Valtonen 1999). It has bearing on the interesting problem of whether simple forms of life (bacteria) can be transferred in life-preserving manner from one planet to another (Mileikowsky *et al.* 2000). The answer appears to be that the transfer of life is indeed possible, but not very common between separate solar systems.

We have now come through a full circle in the study of the three-body problem. In the seventeenth and eighteenth centuries Isaac Newton and his followers wondered whether the orbit of the Earth around the Sun is stable enough for the long-term future of mankind. As we have shown above, the first order three-body perturbations indeed leave the major axis of the Earth's orbit unchanged, and even if we go to the next higher order, the result is the same. Today we ask where life on the Earth came from. Besides the obvious answer that it came from here, we now consider possibilities of elementary life forms travelling through space, carried and protected by small bodies like meteoroids. When they land on a new planet, life may start to flourish and gradually evolve to more complex forms. In these queries the three-body problem plays an important role. However, it is not the perturbation techniques used to solve the question of the stability of the Earth's orbit, but rather the scattering approach that is needed. Using Monte Carlo techniques and orbit calculations we now estimate the probabilities for different kinds of three-body orbits, and in doing so, put our own existence on this planet in perspective.

Problems

Problem 11.1 If the hardening rate πR_a in Eq. (11.7) is lowered by a factor of three, show that the time scale for the coalescence of supermassive binaries is increased by more than a factor of two.

Problem 11.2 Show that the change in the kinetic energy of a gas cloud of mass dm is

$$\mathrm{d}E = -\frac{1}{2}\mathrm{d}m v_0^2 \left[(2m_2/\mathcal{M} - \mathcal{M}/m_2)^2(1-e)/(1+e)\right]$$

when the cloud collides with the binary member m_2 at the apocentre of the binary orbit and at the pericentre of the parabolic cloud orbit (Fig. 11.2). Under what conditions does the term in square brackets equal 1?

Problem 11.3 Show that the change in the potential energy of a binary is

$$\mathrm{d}V = -\frac{\mathrm{d}m}{1+e}\frac{m}{m_2}v_0^2$$

when a mass dm is added to its component at the apocentre of the orbit. Then show using the result of the previous problem that the change in the binary binding energy due to the cloud collision is

$$\frac{\mathrm{d}|E_B|}{|E_B|} = \frac{\mathrm{d}m}{\mathcal{M}}\left[\left(-2m_2/\mathcal{M} + \left(\frac{\mathcal{M}}{m_2}\right)^2(1-e) + 2\frac{m}{m_2}\right) \Big/ (1+e)\right].$$

Under what conditions does the term in square brackets equal 1? Show that for $e = 0$ and $m_2 \ll m_1$, $\mathrm{d}a/a \to 0$, but for all other values $\mathrm{d}a/a > 0$.

Problem 11.4 Show that the angular momentum change in the binary–gas cloud collision is

$$\frac{\mathrm{d}L}{L} = \sqrt{\frac{\mathcal{M}}{m_2}}\left[\sqrt{\frac{2}{1-e}} - 1\right]\frac{\mathrm{d}m}{\mathcal{M}}.$$

Problem 11.5 Use the result of Problem 11.3 to show that

$$\frac{\mathrm{d}a}{a} = \left[\left(2\frac{m_2}{\mathcal{M}} - \left(\frac{\mathcal{M}}{m_2}\right)^2(1-e) - 2\frac{m}{m_2}\right)\Big/(1+e) + \frac{\mathcal{M}}{m_2}\right]\frac{\mathrm{d}m}{\mathcal{M}}.$$

Then calculate using Problem 11.4

$$\mathrm{d}(e^2) = 2(1-e^2)$$
$$\times\left[-\sqrt{\frac{\mathcal{M}}{m_2}}\left(\sqrt{\frac{2}{1-e}} - 1\right) + \left(\frac{m_2}{\mathcal{M}} - \frac{1}{2}\left(\frac{\mathcal{M}}{m_2}\right)^2(1-e) - \frac{m}{m_2}\right)\Big/(1+e)\right.$$
$$\left. + \left(1 + \frac{m}{m_2}\right)\frac{\mathcal{M}}{m_2}\right]\frac{\mathrm{d}m}{\mathcal{M}}.$$

If $m_2/\mathcal{M} = 1$ (large binary mass ratio), what is the largest value of e which gives $d(e^2) \geq 0$? If it is assumed that $\mathrm{d}a/a = 0$, what is the maximum eccentricity that the binary can have due to accretion? Assume (1) that the binary component masses have become equal, (2) that they are very unequal.

Problem 11.6 Calculate the rate of double escapes of black holes per giant elliptical galaxy when the formation of loss cones is neglected.

Problem 11.7 Calculate the probability of direct capture from the Oort Cloud to a Halley type orbit via a close encounter with Saturn. What is the corresponding probability by random walk?

References

Aarseth, S. J. (1971) Direct integration methods of the N-body problem, *Astrophysics and Space Science* **14**, 118–132.

Aarseth, S. J. (1973) Computer simulations of star cluster dynamics, *Vistas in Astronomy* **15**, 13–37.

Aarseth, S. J. (2003a) *Gravitational N-body Simulations: Tools and Algorithms*, Cambridge: Cambridge University Press.

Aarseth, S. J. (2003b) Black hole binary dynamics, *Astrophysics and Space Science* **285**, 367–372.

Aarseth, S. J. and Hills, J. G. (1972) The dynamical evolution of a stellar cluster with initial subclustering, *Astronomy and Astrophysics* **21**, 255–263.

Abt, H. A. (1977) Multiplicity of solar-type stars, *Revista Mexicana de Astronomia Astrofisica* **3**, 47–51.

Abt, H. A. (1983) Normal and abnormal binary frequencies, *Annual Review of Astronomy and Astrophysics* **21**, 343–372.

Agekyan, T. A. and Anosova, J. P. (1967) A study of the dynamics of triple systems by means of statistical sampling, *Astronomicheskii Zhurnal* **44**, 1261 (*Soviet Astronomy* **11**, 1006).

Agekyan, T. A., Anosova, J. P. and Orlov, V. V. (1983) Decay time of triple systems, *Astrophysics* **19**, 66–70.

Ambartsumian, V. (1937) On the statistics of double stars, *Astronomicheskii Zhurnal* **14**, 207.

Anosova, J. P. (1969) Investigation of the dynamics of triple systems by the Monte Carlo method. III. The case of components with different masses, *Astrophysics* **5**, 81–84.

Anosova, J. P. and Orlov, V. V. (1983) The effect of component mass dispersion on the evolution of triple systems, *Trudy Astronomical Observatory of Leningrad* **38**, 142–164.

Anosova, J. P. and Orlov, V. V. (1986) Dynamical evolution of equal-mass triple systems in three dimensions, *Soviet Astronomy* **30**, 380–390.

Anosova, J. P. and Orlov, V. V. (1994) Main features of dynamical escape from three-dimensional triple systems, *Celestial Mechanics and Dynamical Astronomy* **59**, 327–343.

Anosova, J. P. and Polozhentsev, D. D. (1978) Role of mass dispersion in the evolution of triple systems, *Trudy Astronomical Observatory of Leningrad* **34**, 128–144.

Anosova, J. P., Bertov, D. I. and Orlov, V. V. (1984) The influence of rotation on the evolution of triple systems, *Astrofizika* **20**, 327–339.

Anosova, J. P., Orlov, V. V. and Aarseth, S. J. (1994) Initial conditions and dynamics of triple systems, *Celestial Mechanics and Dynamical Astronomy* **60**, 365–372.

Arfken, G. (1970) *Mathematical Methods for Physicists*, San Diego, CA: Academic Press.

Arnold, V. I., Kuzlov, V. V. and Neishtadt, A. I. (1988) Mathematical aspects of classical and celestial mechanics, *Encyclopaedia of Mathematical Sciences*, vol. 3, Berlin: Springer.

Bailey, M. E. (1990) Cometary masses, in *Baryonic Dark Matter*, ed. D. Lynden-Bell and G. Gilmore, Dordrecht: Kluwer, pp. 7–35.

Bailey, M. E. and Emel'yanenko, V. V. (1996) Dynamical evolution of Halley-type comets, *Monthly Notices of the Royal Astronomical Society* **278**, 1087–1110.

Bailey, M. E. and Stagg, C. R. (1988) Cratering constraints on the inner Oort cloud: steady state models, *Monthly Notices of the Royal Astronomical Society* **235**, 1–32.

Bailey, M. E. and Stagg, C. R. (1990) The origin of short-period comets, *Icarus* **86**, 2–8.

Bailyn, C. D. (1984) Numerical computations of coplanar hierarchical three-body systems, private communication.

Bailyn, C. D. (1987) Compact X-ray binaries in hierarchical triples. II. Separation of the inner binary, *Astrophysical Journal* **317**, 737–745.

Bailyn, C. D. (1989) The formation of hierarchical triple systems by single-star–binary encounters, *Astrophysical Journal* **341**, 175–185.

Bailyn, C. D. and Grindlay, J. E. (1987) Compact X-ray binaries in hierarchical triples. I. Tidal angular momentum loss and GX17+2, *Astrophysical Journal* **312**, 748–754.

Basu, D., Valtonen, M. J., Valtonen, H. and Mikkola, S. (1993) Jets from mergers of binary black holes, *Astronomy and Astrophysics* **272**, 417–420.

Bate, M. R. (2000) Predicting the properties of binary stellar systems: the evolution of accreting protobinary systems, *Monthly Notices of the Royal Astronomical Society* **314**, 33–53.

Bate, M. R. and Bonnell, I. A. (1997) Accretion during binary star formation II. Gaseous accretion and disc formation, *Monthly Notices of the Royal Astronomical Society* **285**, 33–48.

Becker, L. (1920) On capture orbits, *Monthly Notices of the Royal Astronomical Society* **80**, 590–603.

Begelman, M. C., Blandford, R. D. and Rees, M. J. (1980) Massive black hole binaries in active galactic nuclei, *Nature* **287**, 307–309.

Binney, J. and Tremaine, S. (1987) *Galactic Dynamics*, Princeton, NJ: Princeton University Press.

Blaes, O., Lee, M. H. and Socrates, A. (2002) The Kozai mechanism and the evolution of binary supermassive black holes, *Astrophysical Journal* **578**, 775–786.

Blandford, R. D. and Rees, M. J. (1974) A 'twin-exhaust' model for double radio sources, *Monthly Notices of the Royal Astronomical Society* **169**, 395–415.

Brasser, R., Innanen, K. A., Connors, M., Veillet, C., Wiegert, P., Mikkola, S. and Chodas, P. W. (2004) Transient co-orbital asteroids, *Icarus* **171**, 102–109.

Burdet, C. A. (1967) Regularization of the two-body problem, *Zeitschrift für Angewandte Mathematik und Physik* **18**, 434.

Burrau, C. (1913) Numerische Berechnung eines Spezialfalles des Dreikörperproblems, *Astronomische Nachrichten* **195**, 113–118.

Campbell, W. W. (1910) Nine stars having variable radial velocities, *Lick Observatory Bulletin* **6**, 57–58.

Chandrasekhar, S. (1942) *Principles of Stellar Dynamics*, New York: Dover.

Chernin, A. D. and Valtonen, M. J. (1998) Intermittent chaos in three-body dynamics, *New Astronomy Reviews* **42**, 41–66.

Chyba, C. F., Owen, T. C. and Ip, W.-H. (1994) Impact delivery of volatiles and organic molecules to Earth, in *Hazards Due to Comets and Asteroids*, ed. T. Gehrels *et al.*, Tucson, AZ: University of Arizona Press, pp. 9–58.

Danby, J. M. A. (1962) *Fundamentals of Celestial Mechanics*, Richmond, VA: Willman-Bell, 2nd revised edition 1988.

Danby, J. M. A. (1964) Stability of the triangular points in the elliptic restricted problem of three bodies, *Astronomical Journal* **69**, 165–172.

Delaunay, C. (1860) *Théorie du Mouvement de la Lune*, Paris: T. T. Malet-Bachelier.

Delsemme, A. H. (1973) Origin of the short-period comets, *Astronomy and Astrophysics* **29**, 377–381.

Duncan, M., Quinn, T. and Tremaine, S. (1987) The formation and extent of the solar system comet cloud, *Astronomical Journal* **94**, 1330–1338.

Duquennoy, A. and Mayor, M. (1991) Multiplicity among solar-type stars in the solar neighborhood. II. Distribution of the orbital elements in an unbiased sample, *Astronomy and Astrophysics* **248**, 485–524.

Edgeworth, K. E. (1949) The origin and evolution of the solar system, *Monthly Notices of the Royal Astronomical Society* **109**, 600–609.

Eggleton, P. and Kiseleva, L. (1995) An empirical condition for stability of hierarchical triple systems, *Astrophysical Journal* **455**, 640–645.

Emel'yanenko, V. V. and Bailey, M. E. (1998) Capture of Halley-type comets from the near parabolic flux, *Monthly Notices of the Royal Astronomical Society* **298**, 212–222.

Escala, A., Larson, R. B., Coppi, P. S. and Mardones, D. (2004) The role of gas in the merging of massive black holes in galactic nuclei. I. Black hole merging in a spherical gas cloud, *Astrophysical Journal* **607**, 765–777.

Euler, L. (1772) Theoria Motuum Lunae, Typis Academiae Imperialis Scientiarum Petropoli (reprinted in Opera Omnia, Series 2, ed. L. Courvoisier, Vol. 22, Lausanne: Orell Fussli Turici, 1958).

Evans, N. R. (1995) The mass ratios of Cepheid binaries, *Astrophysical Journal* **445**, 393–405.

Everhart, E. (1968) Change in total energy of comets passing through the Solar System, *Astronomical Journal* **73**, 1039–1052.

Everhart, E. (1969) Close encounters of comets and planets, *Astronomical Journal* **74**, 735–750.

Everhart, E. (1976) The evolution of comet orbits, in *The Study of Comets*, ed. B. Donn *et al.*, Proc. IAU Coll. 25, NASA SP-393, Washington, DC: NASA, pp. 445–464.

Fernández, J. A. (1982) Dynamical aspects of the origin of comets, *Astronomical Journal* **87**, 1318–1332.

Fernández, J. A. and Gallardo, T. (1994) The transfer of comets from parabolic orbits to short-period orbits: numerical studies, *Astronomy and Astrophysics* **281**, 911–922.

Fernández, J. A. and Ip, W.-H. (1991) Statistical and evolutionary aspects of cometary orbits, in *Comets in the Post-Halley Era*, Vol. 1, ed. R. L. Newburn Jr. *et al.*, Dordrecht: Kluwer, pp. 487–535.

Frenk, C. S., White, S. D. M., Davis, M. and Efstathiou, G. (1988) The formation of dark halos in a universe dominated by cold dark matter, *Astrophysical Journal* **327**, 507–525.

Goldstein (1950) *Classical Mechanics*, New York: Addison-Wesley.

Gould, A. (1991) Binaries in a medium of fast low-mass objects, *Astrophysical Journal* **379**, 280–284.

Gurevich, L. E. and Levin, B. Y. (1950) On the origin of double stars, *Astronomicheskii Zhurnal* **27**, 273 (in Russian).
Hagihara, Y. (1976) *Celestial Mechanics, V: Topology of the Three-Body Problem*, Tokyo: Japan Society for the Promotion of Science.
Hamilton, W. R. (1834) On a general method in dynamics, and Second essay on a general method in dynamics, *Philosophical Transactions of the Royal Society* (1834) 247–308; (1835) 95–144.
Haque-Copilah, S., Basu, D. and Valtonen, M. J. (1997) Ejections of population III objects seen as blueshifted QSOs?, *Journal of Astrophysics and Astronomy* **18**, 73.
Harrington, R. S. (1972) Stability criteria for triple stars, *Celestial Mechanics* **6**, 322–327.
Harrington, R. S. (1975) Production of triple stars by the dynamical decay of small stellar systems, *Astronomical Journal* **80**, 1081–1086.
Heggie, D. C. (1973) Regularization using a time-transformation only, in *Recent Advances in Dynamical Astronomy*, ed. B. D. Tapley and V. Szebehely, Dordrecht: Reidel, p. 34.
Heggie, D. C. (1975) Binary evolution in stellar dynamics, *Monthly Notices of the Royal Astronomical Society* **173**, 729–787.
Heggie, D. C. and Hut, P. (1993) Binary–single star scattering. IV. Analytic approximations and fitting formulae for cross sections and reaction rates, *Astrophysical Journal Supplement Series* **85**, 347–409.
Heggie, D. and Hut, P. (2003) *The Gravitational Million-Body Problem: a multidisciplinary approach to star cluster dynamics*, Cambridge: Cambridge University Press.
Heggie, D. C. and Sweatman, W. L. (1991) Three-body scattering near triple collision or expansion, *Monthly Notices of the Royal Astronomical Society* **250**, 555–575.
Heggie, D. C., Hut, P. and McMillan, S. L. W. (1996) Binary-single star scattering. VII. Hard binary exchange cross sections for arbitrary mass ratios: numerical results and semianalytic fits, *Astrophysical Journal* **467**, 359–369.
Heinämäki, P. (2001) Symmetry of black hole ejections in mergers of galaxies, *Astronomy and Astrophysics* **371**, 795–805.
Heinämäki, P., Lehto, H. J., Valtonen, M. J. and Chernin, A. D. (1998) Three-body dynamics: intermittent chaos with strange attractor, *Monthly Notices of the Royal Astronomical Society* **298**, 790–796.
Heinämäki, P., Lehto, H. J., Valtonen, M. J. and Chernin, A. D. (1999) Chaos in three-body dynamics: Kolmogorov–Sinai entropy, *Monthly Notices of the Royal Astronomical Society* **310**, 811–822.
Hénon, M. (1970) Numerical exploration of the restricted problem. VI. Hill's case: non-periodic orbits, *Astronomy and Astrophysics* **9**, 24–36.
Hénon, M. (1976a) A two-dimensional mapping with a strange attractor, *Communications in Mathematical Physics* **50**, 69–78.
Hénon, M. (1976b) A family of periodic solutions of the planar three-body problem, and their stability, *Celestial Mechanics* **13**, 267–285.
Hénon, M. (1977), Stability of interplay motions, *Celestial Mechanics* **15**, 243–261.
Hénon, M. (1982) On the numerical computation of Poincaré maps, *Physica D: Nonlinear Phenomena* **5**, 412–414.
Hénon, M. (1997) *Generating Families in the Restricted Three-Body Problem*, Lecture Notes in Physics **m 52**, Berlin: Springer.
Hénon, M. (2001) *Generating Families in the Restricted Three-Body Problem. II. Qualitative Study of Bifurcations*, Lecture Notes in Physics **m 65**, Berlin: Springer.

Hietarinta, J. and Mikkola, S. (1993): Chaos in the one-dimensional gravitational three-body problem, *Chaos*, **3**, 183–203.

Hill, G. W. (1878). Researches in the lunar theory, *Annals of the Journal of Mathematics* **1**, 5–26, 129–147, 245–261.

Hills, J. G. (1975) Encounters between binary and single stars and their effect on the dynamical evolution of stellar systems, *Astronomical Journal* **80**, 809–825.

Hills, J. G. (1981) Comet showers and the steady-state infall of comets from the Oort cloud, *Astronomical Journal* **86**, 1730–1740.

Hills, J. G. (1984) Close encounters between a star–planet system and a stellar intruder, *Astronomical Journal* **89**, 1559–1564.

Hills, J. G. (1989) Effect of intruder mass on collisions with hard binaries. I. Zero impact parameter, *Astronomical Journal* **97**, 222–235.

Hills, J. G. (1990) Encounters between single and binary stars: the effect of intruder mass on the maximum impact velocity for which the mean change in binding energy is positive, *Astronomical Journal* **99**, 979–982.

Hills, J. G. (1991) Computer simulations of encounters between massive black holes and binaries, *Astronomical Journal* **102**, 704–715.

Hills, J. G. (1992) Effect of intruder mass on collisions with hard binaries. II. Dependence on impact parameter and computations of the interaction cross sections, *Astronomical Journal* **103**, 1955–1969.

Huang, T. Y. and Innanen, K. A. (1983) Stability and integrability in the planar general three body problem, *Astronomical Journal* **88**, 1064–1073.

Huang, T.-Y. and Valtonen, M. J. (1987) An approximate solution to the energy change of a circular binary in a parabolic three-body encounter, *Monthly Notices of the Royal Astronomical Society* **229**, 333–344.

Hughes, D. W. (1988) Cometary magnitude distribution and the ratio between the numbers of long- and short-period comets, *Icarus* **73**, 149–162.

Hut, P. (1983) Binary–single star scattering. II. Analytic approximations for high velocity, *Astrophysical Journal* **268**, 342–355.

Hut, P. (1984) Hard binary–single star scattering cross sections for equal masses, *Astrophysical Journal Supplement Series* **55**, 301–317.

Hut, P. (1993) Binary–single star scattering. III. Numerical experiments for equal-mass hard binaries, *Astrophysical Journal* **403**, 256–270.

Hut, P. and Bahcall, J. N. (1983) Binary–single star scattering. I. Numerical experiments for equal masses, *Astrophysical Journal* **268**, 319–341.

Hut, P. and Inagaki, S. (1985) Globular cluster evolution with finite-size stars: cross sections and reaction rates, *Astrophysical Journal* **298**, 502–520.

Hut, P., Shara, M. M., Aarseth, S. J., Klessen, R. S., Lombardi, J. C. Jr., Makino, J., McMillan, S., Pols, O. R., Teuben, P. J. and Webbink, R. F. (2003) MODEST-1: Integrating stellar evolution and stellar dynamics, *New Astronomy* **8**, 337–370.

Innanen, K. A. (1979) The limiting radii of direct and retrograde satellite orbits with applications to the solar system and stellar systems, *Astronomical Journal* **84**, 960–963.

Innanen, K. A. (1980) The Coriolis asymmetry in the classical restricted 3-body problem and the Jacobian integral, *Astronomical Journal* **85**, 81–85.

Innanen, K. A., Zheng, J. Q., Mikkola, S. and Valtonen, M. J. (1997) The Kozai mechanism and the stability of planetary orbits in binary star systems, *Astronomical Journal* **113**, 1915–1919.

Ivanov, A. V. and Chernin, A. D. (1991) Signs of stochasticity in the dynamics of model triplets of galaxies, *Pisma v Astronomicheskii Zhurnal* **17**, 569–574.

Ivanov, A. V., Filistov, E. A. and Chernin, A. D. (1995) Evolution of triple systems: stochastic behaviour and dynamic instability, *Astronomy Reports* **39**, 368–381.

Jacobi, C. G. J. (1836) Sur le mouvement d'un point et sur un cas particulier du probleme des trois corps, *Comptes Rendus* **3**, 59.

Jeans, J. H. (1919) The origin of binary systems, *Monthly Notices of the Royal Astronomical Society* **79**, 408–416.

Jeans, J. H. (1928) *Astronomy and Cosmogony*, Cambridge: Cambridge University Press.

Karttunen, H., Kröger, P., Oja, H., Poutanen, M. and Donner, K. J. (eds.) (2003) *Fundamental Astronomy*, Berlin: Springer, 4th edition.

Kinoshita, H. and Nakai, H. (1999) Analytical solution of the Kozai resonance and its application, *Celestial Mechanics* **75**, 125–147.

Kolmogorov, A. N. (1954) On conservation of conditionally periodic motions under small perturbations of the Hamiltonian, *Doklady Akademii Nauk SSSR* **98**(4), 527–530.

Komberg, B. V. (1967) A quasar model as a double system, *Astronomicheskii Zhurnal* **44**, 906–907.

Kovalevsky, J. (1967) *Introduction to Celestial Mechanics*, Dordrecht: Reidel.

Kozai, Y. (1962) Secular perturbations of asteroids with high inclination and eccentricity, *Astronomical Journal* **67**, 591–598.

Kozai, Y. (2004) Extension of secular perturbation theory for asteroids, *Proceedings of the Japan Academy, Series B* **80**, 157–165.

Kresak, L. (1987) Dormant phases in the aging of periodic comets, *Astronomy and Astrophysics* **187**, 906–908.

Kroupa, P. (1995a) Inverse dynamical population synthesis and star formation, *Monthly Notices of the Royal Astronomical Society* **277**, 1491–1506.

Kroupa, P. (1995b) The dynamical properties of stellar systems in the Galactic disk, *Monthly Notices of the Royal Astronomical Society* **277**, 1507–1521.

Kroupa, P., Aarseth, S. and Hurley, J. (2001) The formation of a bound star cluster: from the Orion nebula to the Pleiades, *Monthly Notices of the Royal Astronomical Society* **321**, 699–712.

Kuiper, G. P. (1935a) Problems of double-star astronomy. I, *Publications of the Astronomical Society of the Pacific* **47**, 15–42.

Kuiper, G. P. (1935b) Problems of double-star astronomy. II, *Publications of the Astronomical Society of the Pacific* **47**, 121–150.

Kuiper, G. P. (1951) On the origin of the solar system, in *Astrophysics*, ed. J. A. Hynek, New York: McGraw-Hill, pp. 357–424.

Kustaanheimo, P. (1964) Spinor regularization of the Kepler motion, *Annales Universitatis Turkuensis, Series A1* **73**.

Kustaanheimo, P. and Stiefel, E. (1965) Perturbation theory of Keplerian motion based on spinor regularization, *Journal of Mathematics* **218**, 204.

Lagrange, J. L. (1778) *Oeuvres*, ed. M. J.-A. Serret, Paris: Gauthier-Villars, 1867–1892, Vol. 6, 1873.

Lagrange, J. L. (1811) *Mécanique Analytique*, Paris.

Laine, S., Zheng, J.-Q. and Valtonen, M. J. (2004) Improved models for the evolution of the Coma cluster of galaxies, *Astronomical Journal* **127**, 765–770.

Laplace, P. S. (1799–1825) *Traité de Mechanique Céleste*, Vols. 1, 2, 3, Paris: J. B. M. Duprat, Vol. 4, Paris: V. Courrier, Vol. 5, Paris: Bachelier.

Laplace, P. S. (1805) Theorie des Comètes, *Méchanique Céleste*, **4**, 193.

Lehto, H. J. and Valtonen, M. J. (1996) OJ287 outburst structure and a binary black hole model, *Astrophysical Journal* **460**, 207–213.

Lehto, H. J., Heinämäki, P., Chernin, A. and Valtonen, M. J. (2000) Maasai warrior shield: the life time of a three body system, in *Small Galaxy Groups*, IAU Coll. 174, ASP Conf. ser. 209, ed. M. J. Valtonen and C. Flynn, San Francisco, CA: PASP, pp. 314–317.

Levison, H. F. and Duncan, M. J. (1997) From the Kuiper Belt to Jupiter-family comets: the spatial distribution of ecliptic comets, *Icarus* **127**, 13–32.

Levison, H. F. and Stern, S. A. (2001) On the size dependence of the inclination distribution of the main Kuiper belt, *Astronomical Journal* **121**, 1730–1735.

Levison, H. F., Dones, L. and Duncan, M. J. (2001) The origin of Halley-type comets: probing the inner Oort cloud, *Astronomical Journal* **121**, 2253–2267.

Lexell, A. J. (1778) *Réflexions sur le Temps Périodique des Comètes en Général*, St. Petersburg: Impr. de l'Académie Impériale des Sciences.

Lexell, A. J. (1779) Disquisitio de tempore periodico comatae anno 1770 observati, *Philosophical Transactions of the Royal Society of London* **69**, 68–85.

Li, T.-Y. and Yorke, J. A. (1975) Period three implies chaos, *American Mathematics Monthly* **82**, 985–992.

Loewenstein, M. and Mathews, W. G. (1987) Evolution of hot galactic flows, *Astrophysical Journal* **319**, 614–631.

Lyttleton, R. A. and Yabushita, S. (1965) The effect of stellar encounters on planetary motions, *Monthly Notices of the Royal Astronomical Society* **129**, 105–125.

Mandelbrot, B. B. (1982) *The Fractal Geometry of Nature*, San Francisco, CA: Freeman.

Marchal, C. (1990) *The Three-body Problem*, Studies in Astronautics 4, Amsterdam: Elsevier.

Marchal, C., Yoshida, J. and Sun, Y. S. (1984) Three-body problem, *Celestial Mechanics* **34**, 65–93.

Mardling, R. and Aarseth, S. (1999) Dynamics and stability of three-body systems, in *The Dynamics of Small Bodies in the Solar System*, ed. B. A. Steves and A. E. Roy, Dordrecht: Kluwer, pp. 385–392.

Marsden, B. G. and Williams, G. V. (1999) *Catalogue of Cometary Orbits*, IAU Central Bureau of Astronomical Telegrams, Minor Planet Center, Smithsonian, 13th edition.

Mathews, W. G. and Bregman, J. N. (1978) Radiative accretion flow onto giant galaxies in clusters, *Astrophysical Journal* **224**, 308–319.

McCarthy, P. J., Spinrad, H. and van Breugel, W. (1995) Emission-line imaging of 3CR radio galaxies. I. Imaging data, *Astrophysical Journal Supplement Series* **99**, 27–66.

Merritt, D. and Ferrarese, L. (2001) Black hole demographics from the $M_\bullet \sigma$ relation, *Monthly Notices of the Royal Astronomical Society* **320**, L30–L34.

Mihalas, D. and Binney, J. (1981) *Galactic Astronomy: Structure and Kinematics*, San Francisco, CA: Freeman, p. 231.

Mikkola, S. (1983) Encounters of binaries. I. Equal energies, *Monthly Notices of the Royal Astronomical Society* **203**, 1107–1121.

Mikkola, S. (1984a) Encounters of binaries. II. Unequal energies, *Monthly Notices of the Royal Astronomical Society* **207**, 115–126.

Mikkola, S. (1984b) Encounters of binaries. III. Fly-bys, *Monthly Notices of the Royal Astronomical Society* **208**, 75–82.

Mikkola, S. (1986) A comparison of initially bound and unbound three-body systems, *Monthly Notices of the Royal Astronomical Society* **223**, 757–762.

Mikkola, S. (1994) A numerical exploration of the phase-space structure of chaotic three-body scattering, *Monthly Notices of the Royal Astronomical Society* **269**, 127–136.

Mikkola, S. and Aarseth, S. J. (1990) A chain regularization method for the few-body problem, *Celestial Mechanics* **47**, 375–390.

Mikkola, S. and Aarseth, S. J. (1993) An implementation of N-body chain regularization, *Celestial Mechanics* **57**, 439–459.

Mikkola, S. and Aarseth, S. J. (1996) A slow-down treatment for close binaries, *Celestial Mechanics and Dynamical Astronomy* **64**, 197–208.

Mikkola, S. and Hietarinta, J. (1989) A numerical investigation of the one-dimensional Newtonian three-body problem, *Celestial Mechanics* **46**, 1–18.

Mikkola, S. and Hietarinta, J. (1990) A numerical investigation of the one-dimensional three-body problem. II: positive energies, *Celestial Mechanics and Dynamical Astronomy* **47**, 321–331.

Mikkola, S. and Hietarinta, J. (1991) A numerical investigation of the one-dimensional three-body problem. III: mass dependence in the stability of motion, *Celestial Mechanics* **51**, 379–394.

Mikkola, S. and Innanen, K. (1997) Orbital stability of planetary quasi-satellites, in *The Dynamical Behaviour of our Planetary System*, ed. R. Dvorak and J. Henrard, Dordrecht: Kluwer, pp. 345–355.

Mikkola, S. and Valtonen, M. J. (1986) The effect of total angular momentum on the break-up of a three-body system, *Monthly Notices of the Royal Astronomical Society* **223**, 269–278.

Mikkola, S. and Valtonen, M. J. (1990) The slingshot ejections in merging galaxies, *Astrophysical Journal* **348**, 412–420.

Mikkola, S. and Valtonen, M. J. (1992) Evolution of binaries in the field of light particles and the problem of two black holes, *Monthly Notices of the Royal Astronomical Society* **259**, 115–120.

Mikkola, S., Innanen, K. A., Muinonen, K. and Bowell, E. (1994). A preliminary analysis of the orbit of the Mars Trojan asteroid (5261) Eureka, *Celestial Mechanics and Dynamical Astronomy* **58**, 53–64.

Mikkola, S., Brasser, R., Wiegert, P. and Innanen, K. (2004) Asteroid 2002 VE68, a quasi-satellite of Venus, *Monthly Notices of the Royal Astronomical Society* **351** (3), L63–L65.

Mileikowsky, C., Cucinotta, F. A., Wilson, J. W., Gladman, B., Horneck, G., Lindegren, L., Melosh, J., Rickman, H., Valtonen, M. J. and Zheng, J. Q. (2000) Natural transfer of viable microbes in space. 1. From Mars to Earth and Earth to Mars, *Icarus* **145**, 391–427.

Miller, M. C. and Hamilton, D. P. (2002) Four-body effects in globular cluster black hole coalescence, *Astrophysical Journal* **576**, 894–898.

Milosavljevic, M. and Merritt, D. (2001) Formation of galactic nuclei, *Astrophysical Journal* **563**, 34–62.

Monaghan, J. J. (1976a) A statistical theory of the disruption of three-body systems I, *Monthly Notices of the Royal Astronomical Society* **176**, 63–72.

Monaghan, J. J. (1976b) A statistical theory of the disruption of three-body systems II, *Monthly Notices of the Royal Astronomical Society* **177**, 583–594.

Monaghan, J. J. (1977) The mass distribution of disrupting three-body systems, *Monthly Notices of the Royal Astronomical Society* **179**, 31–32.

Murray, C. D. and Dermott, S. F. (1999) *Solar System Dynamics*, Cambridge: Cambridge University Press.

Nash, P. E. and Monaghan, J. J. (1978) A statistical theory of the disruption of three-body systems III. Three-dimensional motion, *Monthly Notices of the Royal Astronomical Society* **184**, 119–125.

Newton, H. A. (1891) Capture of comets by planets, *Astronomical Journal* **11**, 73–75.
Newton, I. (1687) *Philosophiae Naturalis Principia Mathematica*, London: Royal Society (reprinted in *The Mathematical Principles of Natural Philosophy*, New York: Philosophical Library, 1964).
Nilsson, K., Valtonen, M. J., Jones, L. R., Saslaw, W. C. and Lehto, H. J. (1997) Optical emission in the radio lobes of Cygnus A, *Astronomy and Astrophysics* **324**, 888–898.
Nurmi, P. (2001) Long-term evolution of Oort cloud comets: methods and comparisons, *Monthly Notices of the Royal Astronomical Society* **323**, 911–922.
Nurmi, P., Valtonen, M. J., Zheng, J.-Q. and Rickman, H. (2002) Long-term evolution of Oort cloud comets: capture of comets, *Monthly Notices of the Royal Astronomical Society* **333**, 835–846.
Oort, J. H. (1950) The structure of the cloud of comets surrounding the solar system and a hypothesis concerning its origin, *Bulletin of the Astronomical Institutes of the Netherlands* **11**, 91–110.
Öpik, E. J. (1924) Statistical studies of double stars, *Tartu Observatory Publications* **25**, No. 6, 1–167.
Öpik, E. J. (1951) Collision probabilities with the planets and the distribution of interplanetary matter, *Proceedings of the Royal Irish Academy, Section A* **54A**, 165–199.
Peters, P. C. (1964) Gravitational radiation and the motion of two point masses, *Physical Revue* **136**, B1224–1232.
Petit Bois, G. (1961) *Tables of Indefinite Integrals*, New York: Dover.
Pietilä, H. (1999) The role of binary black holes in galactic nuclei, Ph.D. Thesis, *Annales Universitatis Turkuensis, Series A1* **239**.
Pietilä, H., Heinämäki, P., Mikkola, S. and Valtonen, M. J. (1995) Anisotropic gravitational radiation in the problems of three and four black holes, *Celestial Mechanics and Dynamical Astronomy* **62**, 377–394.
Poincaré, H. (1890) Sur le problème des trois corps et les équations de la dynamique, *Acta Mathematica* **13**, 259.
Poincaré, H. (1892–1899) *Les Methodes Nouvelles de la Mecanique Celeste*, in 3 volumes, Paris: Gauthier-Villars, reprinted by Dover, New York.
Quinlan, G. D. (1996) The dynamical evolution of massive black hole binaries. I. Hardening in a fixed stellar background, *New Astronomy* **1**, 35–36.
Roy, A. E. (2005) *Orbital Motion*, Bristol: Inst. Physics Publ., 4th edition.
Roy, A. and Haddow, M. (2003) Energy change in a hard binary due to distant encounters, *Celestial Mechanics and Dynamical Astronomy* **87**, 411–435.
Ruelle, D. and Takens, F. (1971) On the nature of turbulence, *Communications in Mathematical Physics* **20**, 167–192, **23**, 343–344.
Saari, D. G. (1974) The angle of escape in the three body problem, *Celestial Mechanics* **9**, 175–181.
Salpeter, E. (1955) The luminosity function and stellar evolution, *Astrophysical Journal* **121**, 161–167.
Saslaw, W. C., Valtonen, M. J. and Aarseth, S. J. (1974) The gravitational slingshot and the structure of extragalactic radio sources, *Astrophysical Journal* **190**, 253–270.
Schubart, J. (1956) Numerische Aufsuchung periodischer Lösungen in Dreikörperproblem, *Astronomische Nachrichten* **283**, 17–22.
Shoemaker, E. M., Weissman, P. R. and Shoemaker, C. S. (1994) The flux of periodic comets near Earth, in *Hazards Due to Comets and Asteroids*, ed. T. Gehrels, Tucson, AZ: University of Arizona Press, p. 313.

Sidlichovsky, M. (1983) On the double averaged three-body problem, *Celestial Mechanics* **29**, 295–305.

Sigurdsson, S. and Phinney, E. S. (1993) Binary–single star interactions in globular clusters, *Astrophysical Journal* **415**, 631–651.

Sillanpää, A., Haarala, S., Valtonen, M. J., Sundelius, B. and Byrd, G. G. (1988) OJ287: Binary pair of supermassive black holes, *Astrophysical Journal* **325**, 628–634.

Stagg, C. P. and Bailey, M. E. (1989) Stochastic capture of short-period comets, *Monthly Notices of the Royal Astronomical Society* **241**, 507–541.

Standish, E. M. (1972) The dynamical evolution of triple star systems: a numerical study, *Astronomy and Astrophysics* **21**, 185–191.

Sundman, K. F. (1912) Memoire sur le probleme des trois corps, *Acta Mathematica* **36**, 105–179.

Szebehely, V. (1967) *The Theory of Orbits*, New York: Academic Press.

Szebehely, V. (1972) Mass effects in the problem of three bodies, *Celestial Mechanics* **6**, 84–107.

Szebehely, V. and Peters, C. F. (1967) Complete solution of a general problem of three bodies, *Astronomical Journal* **72**, 876–883.

Tanikawa, K. and Umehara, H. (1998) Oscillatory orbits in the planar three-body problem with equal masses, *Celestial Mechanics and Dynamical Astronomy* **70**, 167–180.

Tisserand, F. (1889) Sur la theorie de la capture des cometes periodiques; Note sur l'integrale de Jacobi, et sur son application a la theorie des cometes, *Bulletin Astronomique* **6**, 289.

Tisserand, F. (1896) *Traité de Mécanique Céleste*, Vol. 4, Paris: Gauthier-Villars.

Trimble, V. (1990) The distributions of binary system mass ratios: a less biased sample, *Monthly Notices of the Royal Astronomical Society* **242**, 79–87.

Valtaoja, L., Valtonen, M. J. and Byrd, G. G. (1989) Binary pairs of supermassive black holes: formation in merging galaxies, *Astrophysical Journal* **343**, 47–53.

Valtonen, M. J. (1974) Statistics of three-body experiments, in *Stability of the Solar System and of Small Stellar Systems*, Proc. IAU Symp. 62, ed. Y. Kozai, Dordrecht: Reidel, pp. 211–223.

Valtonen, M. J. (1975a) Statistics of three-body experiments: probability of escape and capture, *Memoirs of the Royal Astronomical Society* **80**, 77–91.

Valtonen, M. J. (1975b) A gravitational three-body scattering experiment, *Memoirs of the Royal Astronomical Society* **80**, 61–75.

Valtonen, M. J. (1976a) Statistics of three-body experiments: escape energy, velocity and final binary eccentricity, *Astrophysics and Space Science* **42**, 331–347.

Valtonen, M. J. (1976b) Radio trails in the slingshot theory, *Astrophysical Journal* **209**, 35–45.

Valtonen, M. J. (1988) The general three-body problem in astrophysics, *Vistas in Astronomy* **32**, 23–48.

Valtonen, M. J. (1996) Triple black hole systems formed in mergers of galaxies, *Monthly Notices of the Royal Astronomical Society* **278**, 186–190.

Valtonen, M. J. (1997a) Mass ratios in wide binary stars, *Astrophysical Journal* **485**, 785–788.

Valtonen, M. J. (1997b) Wide binaries from few-body interactions, *Celestial Mechanics and Dynamical Astronomy* **68**, 27–41.

Valtonen, M. J. (1998) Mass ratios in multiple star systems, *Astronomy and Astrophysics* **334**, 169–172.

Valtonen, M. J. and Aarseth, S. J. (1977) Numerical experiments on the decay of three-body systems, *Revista Mexicana de Astronomia y Astrofisica* **3**, 163–166.

Valtonen, M. J. and Byrd, G. G. (1979) A binary model for the Coma cluster of galaxies, *Astrophysical Journal* **230**, 655–666.

Valtonen, M. J. and Byrd, G. G. (1980) Orbital dynamics of the radio galaxy 3C129. II. Internal tail structure, *Astrophysical Journal* **240**, 442–446.

Valtonen, M. J. and Heggie, D. C. (1979) Three-body gravitational scattering: comparison between theory and experiment, *Celestial Mechanics* **19**, 53–58.

Valtonen, M. J. and Heinämäki, P. (2000) Double radio sources: two approaches, *Astrophysical Journal* **530**, 107–123.

Valtonen, M. J. and Innanen, K. A. (1982) The capture of interstellar comets, *Astrophysical Journal* **255**, 307–315.

Valtonen, M. J. and Kotilainen, J. (1989) Wide-angle-tailed radio galaxies in the slingshot model, *Astronomical Journal* **98**, 117–123.

Valtonen, M. J. and Mikkola, S. (1991) The few-body problem in astrophysics, *Annual Review of Astronomy and Astrophysics* **29**, 9–29.

Valtonen, M. J. and Wirén, S. (1994) Two-body escape speeds in merged four-galaxy groups, *Monthly Notices of the Royal Astronomical Society* **266**, 353–356.

Valtonen, M. J., Zheng, J.-Q. and Mikkola, S. (1992) Origin of Oort cloud comets in the interstellar space, *Celestial Mechanics and Dynamical Astronomy* **54**, 37–48.

Valtonen, M. J., Byrd, G. G., McCall, M. L. and Innanen, K. A. (1993) A revised history of the Local Group and a generalized method of timing, *Astronomical Journal* **105**, 886–893.

Valtonen, M. J., Mikkola, S., Heinämäki, P. and Valtonen, H. (1994) Slingshot ejections from clusters of three and four black holes, *Astrophysical Journal Supplement Series* **95**, 69–86.

Valtonen, M. J., Mikkola, S. and Pietilä, H. (1995) Burrau's three-body problem in the post-Newtonian approximation, *Monthly Notices of the Royal Astronomical Society* **273**, 751–754.

Valtonen, M. J., Zheng, J.-Q., Mikkola, S., Nurmi, P. and Rickman, H. (1998), Monte Carlo simulations of comet capture from the Oort cloud, *Celestial Mechanics and Dynamical Astronomy* **69**, 89–102.

Valtonen, M. J., Mylläri, A. A., Orlov, V. V. and Rubinov, A. V. (2003) Dynamics of rotating triple systems, *Astronomy Letters* **29**, 41.

Valtonen, M. J., Mylläri, A. A., Orlov, V. V. and Rubinov, A. V. (2004) Dynamics of rotating triple systems: statistical escape theory versus numerical simulations, *Monthly Notices of the Royal Astronomical Society*, in press.

Valtonen, M. J., Mylläri, A. A., Orlov, V. V. and Rubinov, A. V. (2006) Long term stability of hierarchical triple systems, *Monthly Notices of the Royal Astronomical Society*, submitted.

Walters, M. H. H. (1932a) The variation of eccentricity and semi-major axis for the orbit of a spectroscopic binary, *Monthly Notices of the Royal Astronomical Society* **92**, 786–805.

Walters, M. H. H. (1932b) Variations in the eccentricity and semi-axis major of the orbit of a spectroscopic binary, *Monthly Notices of the Royal Astronomical Society* **93**, 28–77.

Weissman, P. R. (1983) The mass of the Oort cloud, *Astronomy and Astrophysics* **118**, 90–94.

Weissman, P. R. (1991) Dynamical history of the Oort cloud, in *Comets in the Post-Halley Era*, Vol. 1, ed. R. L. Newburn Jr. *et al.*, Dordrecht: Kluwer, pp. 463–486.

Wen, L. (2003) On the eccentricity distribution of coalescing black hole binaries driven by the Kozai mechanism in globular clusters, *Astrophysical Journal* **598**, 419–430.

Wetherill, G. W. (1991) End products of cometary evolution: cometary origin of Earth-crossing bodies of asteroidal appearance, in *Comets in the Post-Halley Era*, Vol. 1, ed. R. L. Newburn Jr. *et al.*, Dordrecht: Kluwer, pp. 537–556.

White, R. E. and Sarazin, C. L. (1988) Star formation in the cooling flows of M87/Virgo and NGC1275/Perseus, *Astrophysical Journal* **335**, 688–702.

Wiegert, P. and Tremaine, S. (1999) The evolution of long-period comets, *Icarus* **137**, 84–121.

Wiegert, P. A., Innanen, K. A. and Mikkola, S. (1997) An asteroidal companion to the Earth, *Nature* **387**, 685–686.

Wielen, R. (1971) The age distribution and total lifetimes of galactic clusters, *Astronomy and Astrophysics* **13**, 309–322.

Wirén, S., Zheng, J.-Q., Valtonen, M. J. and Chernin, A. D. (1996) Computer simulations of interacting galaxies in compact groups and the observed properties of close binary galaxies, *Astronomical Journal* **111**, 160–167.

Yabushita, S. (1966) Lifetime of binary stars, *Monthly Notices of the Royal Astronomical Society* **133**, 133–143.

Young, P. J. (1976) Tables of functions for a spherical galaxy obeying the $R^{1/4}$ law in projection, *Astronomical Journal* **81**, 807–816.

Young, P. J., Westphal, J. A., Kristian, J., Wilson, C. P. and Landauer, F. P. (1978) Evidence for a supermassive object in the nucleus of the galaxy M87 from SIT and CCD area photometry, *Astrophysical Journal* **221**, 721–730.

Yu, Q. (2002) Evolution of massive binary black holes, *Monthly Notices of the Royal Astronomical Society* **331**, 935–958.

Zheng, J.-Q. (1994) Orbital changes of Oort cloud comets due to strong planetary influences, *Astronomy and Astrophysics Supplement Series* **108**, 253–270.

Zheng, J. Q. and Valtonen, M. J. (1999) On the probability that a comet that has escaped from another solar system will collide with the Earth, *Monthly Notices of the Royal Astronomical Society* **304**, 579–582.

Zheng, J.-Q., Valtonen, M. J. and Chernin, A. D. (1993) Computer simulations of interacting galaxies in compact groups and the observed properties of triple galaxies, *Astronomical Journal* **105**, 2047–2053.

Zier, C. and Biermann, P. L. (2001) Binary black holes and tori in AGN. I. Ejection of stars and merging of the binary, *Astronomy and Astrophysics* **377**, 23–43.

Author index

Aarseth, 4, 6, 18, 34, 41, 210, 215, 272, 273, 274, 290, 317
Abt, 13, 14, 320, 321
Agekyan, 40, 210
Ambartsumian, 13
Anosova, 40, 188, 190, 192, 207, 210, 212
Arfken, 72, 76
Arnold, 139

Bahcall, 205, 280, 283, 285
Bailey, 323, 324, 325
Bailyn, 275
Basu, 312
Bate, 295, 298
Becker, 3
Begelman, 290
Biermann, 295
Binney, 15, 68, 316
Blaes, 299
Blandford, 18, 290
Bonnell, 295, 298
Brasser, 128
Bregman, 298
Burdet, 37
Burrau, 3
Byrd, 310, 314

Campbell, 13
Chandrasekhar, 68
Chernin, 43, 313
Chyba, 325

Delaunay, 2
Delsemme, 324
Dermott, 6
Dones, 324
Duncan, 11, 323, 324, 325, 326
Duquennoy, 13, 14, 318, 320, 321

Edgeworth, 11, 326
Eggleton, 272, 273
Emelyanenko, 323, 324

Escala, 298
Euler, 1, 133
Evans, 15, 320
Everhart, 8, 154, 157, 158, 159, 160, 323

Fernandez, 323, 324, 325
Frenk, 290

Gallardo, 323, 324
Goldstein, 80
Gould, 141, 215
Grindlay, 275
Gurevich, 147

Haddow, 249, 257
Hagihara, 125
Hamilton, 81, 275
Haque-Copilah, 312
Harrington, 273
Heggie, 6, 37, 170, 178, 200, 201, 202, 203, 207, 212, 213, 214, 215, 233, 249, 257, 281, 282, 283, 284
Heinämäki, 42, 43, 172, 299, 300, 301, 304, 305, 306
Hernquist, 295
Herschel, W., 12
Hietarinta, 136
Hill, 2, 121
Hills, 11, 154, 203, 206, 215, 218, 277, 279, 281, 286
Huang, 265, 266, 271, 272, 273, 274, 275
Hughes, 324
Hurley, 317
Hut, 6, 201, 202, 203, 205, 206, 211, 212, 215, 218, 233, 249, 280, 281, 282, 283, 285
Hénon, 5, 42, 133, 136

Inagaki, 218
Innanen, 131, 133, 157, 233, 234, 238, 271, 272, 273, 274, 275
Ip, 324, 325
Ivanov, 43

Jacobi, 2
Jeans, 13

Kepler, 64
Kinoshita, 233, 235, 237
Kiseleva, 272, 273
Kolmogorov, 139
Kotilainen, 310
Kovalevsky, 70
Kozai, 233, 235
Kresak, 324
Kroupa, 317
Kuiper, 11, 14, 326
Kustaanheimo, 37

Lagrange, 2, 81, 133
Laine, 314, 315
Laplace, 8
Lehto, 41, 290
Leverrier, 8
Levin, 147
Levison, 323, 324
Lexell, 8
Li, 41
Loewenstein, 298
Lyttleton, 249

Mandelbrot, 41
Marchal, 5, 139, 233
Mardling, 272, 273, 274
Matthews, 298
Mayor, 13, 14, 318, 320, 321
McCarthy, 194
Merritt, 290, 291, 292, 293, 295, 309
Mihalas, 15
Mikkola, 41, 128, 133, 136, 150, 152, 158, 166, 179, 185, 186, 188, 200, 201, 308
Mileikowsky, 328
Miller, 275
Milosavljevic, 290, 291, 292, 293, 295, 309
Monaghan, 173, 178, 188
Murray, 6

Nakai, 233, 235, 237
Newton, H.A., 8
Newton, Isaac, 1, 20, 329
Nilsson, 307
Nurmi, 323, 324, 325, 326, 328

Öpik, 10
Orlov, 188, 190, 192, 207, 210, 212, 275
Oscar II, 2
Owen, 325

Peters, 3, 216, 294, 299
Petit Bois, 234
Phinney, 218, 219
Pietilä, 152, 299
Poincaré, 2, 41
Polozhentsev, 210, 212

Quinlan, 152, 293, 295, 301
Quinn, 11

Rees, 18, 290
Roy, 249, 257
Ruelle, 41
Rutherford, 68

Saari, 191
Salpeter, 14, 319
Sarazin, 298
Saslaw, 18, 37, 182, 188, 192, 206, 210, 249, 271, 290, 303, 307
Schubart, 136, 138
Shoemaker, 324
Sidlichovsky, 233
Sigurdsson, 218, 219
Sillanpää, 290
Spinrad, 194
Stagg, 323, 324, 325
Standish, 188
Stiefel, 38
Sundman, 2, 37
Sweatman, 200
Szebehely, 3, 125, 209, 210, 216

Takens, 41
Tanikawa, 40
Tisserand, 2, 8
Tremaine, 11, 68, 316, 323
Trimble, 14

Umehara, 40

Valtaoja, 311
Valtonen, 18, 37, 43, 150, 152, 157, 158, 165, 166, 183, 186, 188, 192, 193, 208, 210, 212, 265, 266, 282, 284, 290, 299, 302, 305, 306, 307, 308, 310, 312, 313, 314, 320, 326, 328
van Breugel, 194

Walters, 248
Weissman, 323, 324
Wen, 275
Wetherill, 324
White, 298
Wiegert, 128, 323
Wirén, 312

Yabushita, 249
Yorke, 41
Young, 291, 309
Yu, 295

Zheng, 158, 166, 313, 328
Zier, 295

Subject index

AA-map, *see* Agekian–Anosova map
accretion, 298
accretion disk, 295
action-angle variables, 109
Agekian–Anosova map, 40
angular momentum, 26, 31, 38, 49, 55, 184, 188
angular momentum distribution, 190
argument of perihelion, 55
attractor, 40

barycentre, 26
beam theory, 18
Bessel function, 72
binary
  evolution of, 245
  hard, 147
  hierarchical, 245
  soft, 147
binary stars, 12
  eccentricity, 13, 14
  period, 13
binding energy
  rate of change, 214
binding energy distribution, 204
black holes, 17

canonical transformations, 92
capture, 152, 203
capture cross-section, 153, 205, 206
centre of mass, 26, 48
centrifugal force, 45
chaos, 41, 171
collision, 216
collision plane, 156
Coma cluster, 16
comets, 8
  inclination, 11
configuration space, 81
conic section, 53
conjugate variables, 89
constraint, 80

coordinates
  canonical, 89
  cyclic, 87
  generalised, 81
  inertial, 21, 26, 48, 116
  Jacobian, 30, 56, 57
  Lagrangian, 29
  polar, 52, 53, 86
  rectangular, 22
  rotating, 43, 116
  sidereal, 116
  synodic, 116
Coriolis force, 45
cross-section of interaction, 198
Cruithne, 128
cyclic coordinate, 87
Cygnus A, 18

Delaunay's elements, 108, 222
detailed balance, 207
distance, as a function of time, 75
dynamical friction, 68

e-folding time, 43
eccentric anomaly, 58, 59
  series, 70
eccentricity, 53, 55, 149, 264
eccentricity distribution, 179, 180, 181, 187
ecliptic, 54
ejection, 207
elimination of mean anomalies, 113
elimination of nodes, 111
energy change, 156
energy distribution, 178
energy integral, 32, 51
equation of motion, 22
equations of motion
  Hamiltonian, 89
    three-body, 109
    two-body, 97
  Lagrangian, 85
  three-body, 116

equations of motion (*Cont.*)
Hamiltonian, 116
Newtonian, 118
two-body, 47
escape, 171
escape angle, 188, 193
escape velocity, 180, 182
escaper mass, 183
Euler's theorem, 90
Euler–Lagrange equation, *see* Lagrange equation
Eureka, 128
event horizon, 17
exchange, 197, 211
exchange cross-section, 211, 212, 213

fast encounter, 274
flyby, 197, 211
flyby cross-section, 212
flyby orbits, 41
focussing factor, 148, 149
fractal dimension, 41
fractal geometry, 41

galaxy groups, 15
generating function, 93
gravitational constant, 39
gravitational potential, 24

Halley's comet, 9
Hamilton–Jacobi equation, 95
Hamiltonian, 88
properties, 89
Hamiltonian mechanics, 80
Hamiltonian principle, 81
hard binary, 215
Heggie's law, 147
hierarchical systems, 221, 226
Hill's equations, 131
Hill's sphere, 131, 132
holonomic constraint, 80
horseshoe orbit, 128, 129
hyperbolic orbit, 66
Hénon map, 42

impact distance, 66
impact parameter, 66
impulsive approximation, 274
inclination, 55, 157, 166
integrals of equations of motion, 25
integration of orbits, 34
invariable plane, 32
ionisation, 197

Jacobian integral, 120, 121
Janus, 128

K-S regularisation, 37
Kepler's equation, 61
Kepler's first law, 53
Kepler's second law, 53
Kepler's third law, 64
kinetic energy, 27
Kolmogorov–Arnold–Moser theorem, 139
Kolmogorov–Sinai entropy, 43
Kozai cycle, 235
Kozai resonance, 273
Kuiper Belt, 8, 326

Lagrange equation, 83
Lagrange–Jacobi identity, 33
Lagrangian equilateral triangle, 2, 133, 134
Lagrangian function, 81
Lagrangian planetary equations, 225, 232
Lagrangian point, 123, 124
stability, 125
law of gravity, 21
Legendre polynomials, 76
Legendre transformation, 87
lifetime, 41, 207
lifetime distribution, 210
line of nodes, 55
longitude of perihelion, 55
longitude of the ascending node, 55
loss cone, 174, 175
Lyapunov exponent, 43

M87, 289, 296
map, 40
mass function, 319
mean anomaly, 60
mean motion, 60, 64
moment of inertia, 32
momentum
canonical, 89
generalised, 87

Newton's laws, 20
node, 55

Oort Cloud, 8, 159, 322
Öpik's law, 13
orbital elements, 54
orbital energy, 167
orbital velocity, 57
orbits of satellites, 130
osculating elements, 56, 221, 222

parameter, 55
of a conic section, 53
pericentre distance, 157
perihelion, 55
perihelion time, 52
perturbations, 221
of *a* and *e*, 240
perturbing function, 222
averaged, 227, 231
point transformation, 95
position of a planet, 64
potential, 76
effective, 119

potential energy, 23, 24
Principia, 1, 20
Pythagorean problem, 3, 4

radio source, 193, 195
reduced mass, 32
regularisation, 36
relative energy change, 197
resonance, 197
Rutherford differential cross section, 68

scaling, 34, 38
scattering, 141, 197
scattering angle, 67
scattering cross-section, 198
Schwarzschild radius, 17
semi-latus rectum, 53, 55
semi-major axis, 55, 148
separation of variables, 98
slingshot theory, 18
slow encounter, 248
Solar System evolution, 238
speed of a planet, 65
stability, of triple systems, 270
star clusters, 214
stars
  mass distribution function, 14
strange attractor, 41
Sundman series, 39
surface velocity, 53, 54, 63

tadpole orbit, 128, 129
three-body problem
  one-dimensional, 136, 138
  restricted, 115
Tisserand parameter, 123, 165, 168, 322
Tisserand's criterion, 139
total energy, 92
triple stars, 275
Trojan asteroid
  of Jupiter, 128
  of Mars, 128
true anomaly, 53, 58, 59

units, 39, 56
  in three-body problem, 115

variation, 83
variational calculus, 82
Venus, 128
virial theorem, 27

weak chaos, 41

Printed in the United States
By Bookmasters